Representing Electrons

# REPRESENTING ELECTRONS

## A Biographical Approach to Theoretical Entities

**Theodore Arabatzis**

The University of Chicago Press
Chicago and London

Theodore Arabatzis is assistant professor in the Department of Philosophy and History of Science at the University of Athens.

The University of Chicago Press, Chicago 60637
The University of Chicago Press, Ltd., London
© 2006 by The University of Chicago
All rights reserved. Published 2006
Printed in the United States of America

15  14  13  12  11  10  09  08  07  06      1  2  3  4  5

ISBN: 0-226-02420-2 (cloth)
ISBN: 0-226-02421-0 (paper)

Library of Congress Cataloging-in-Publication Data

Arabatzis, Theodore, 1965 –
    Representing electrons : a biographical approach to theoretical entities /
    Theodore Arabatzis.
        p. cm.
    Includes bibliographical references and index.
    ISBN 0-226-02420-2 (alk. paper) — ISBN 0-226-02421-0 (pbk. : alk. paper)
    1. Electrons—History.  2. Science—Philosophy.  3. Realism.  I. Title.
QC793.5.E62A73 2006
539.7′2112—dc22

                                                        2005004218

⊚ The paper used in this publication meets the minimum requirements of the American National Standard for Information Sciences—Permanence of Paper for Printed Library Materials, ANSI Z39.48 –1992.

*To the memory of my teacher and friend, Yorgos Goudaroulis*

# Contents

# Acknowledgments

This book was long in the making. It began as a doctoral dissertation in the Program in History of Science at Princeton University, where I was given the opportunity to jointly pursue my historical and philosophical interests. I am grateful for the interdisciplinary education I obtained at Princeton, which made it possible to write a book of this kind. Over the years I worked on it, I had the privilege of getting acquainted with many scholars, colleagues, and friends, who enriched beyond measure my intellectual life. First, I would like to thank those who read in its entirety an early draft of this book.

Words will inevitably fail me if I try to express my debt to Kostas Gavroglu. Time and again, he went out of his way to help me and made me feel that my work was his top priority. Without his persistent encouragement, unfaltering support, and constructive advice, this book would probably never have reached the publication stage.

I am equally indebted to Nancy Nersessian. Her involvement with this project began in her capacity as my dissertation adviser. Well beyond that stage, her interest in my work never abated. Her constant encouragement and insightful guidance helped me substantially to transform my dissertation into a book. Among innumerable other things, I owe to her the title of this book.

The philosophical point of view that permeates the following pages was inspired by my interaction with Bas van Fraassen. His powerful defense of a subtly agnostic stance in the philosophy of science cast a spell on me and motivated me to look at familiar historical episodes from a new perspective. More important, his extensive feedback on my work enabled me to develop and clarify my philosophical views.

The reaction of Norton Wise to my—sometimes half-baked—ideas was always a source of encouragement and inspiration. His enthusiasm for my biographical approach strengthened my confidence in it and prompted me to develop it further. Furthermore, he helped me to avoid various pitfalls, especially in my attempt to navigate the murky waters of the old quantum theory.

I am grateful to the original referee appointed by the University of Chicago Press for his penetrating criticism and very helpful suggestions on how to turn my dissertation into a book. I am also indebted to two additional referees who read carefully a penultimate version of the manuscript and offered detailed suggestions for its improvement. I profited considerably from their comments, even if I haven't always followed their advice.

I would also like to thank the following historians and philosophers of science: Aristides Baltas, for our many stimulating discussions on the philosophical issues that occupy this book and for his friendly interest in how my work was coming along; Jemima Ben-Menahem, for her encouraging and helpful response to my ideas on scientific discovery; Jed Buchwald, for his sharp and constructive comments on my work on the early history of the electron; Isobel Falconer, for her expert assessment of my analysis of the discovery of the electron, a topic she has thoroughly explored; Charles Gillispie, for his supportive and kind words about my work, which I tried to remember whenever my confidence was waning; Ernan McMullin, for his extensive and perceptive comments on my work on scientific discovery and scientific realism; and Stathis Psillos, for our numerous arguments over scientific realism, which have restrained my antirealist predisposition.

Part of the work for this book was carried out while I was a postdoctoral fellow at the Dibner Institute for the History of Science and Technology. I am grateful for the opportunity to pursue my research in such a hospitable institution and to take advantage of its unparalleled resources for the history of science. I am also grateful to my home institution, the Department of Philosophy and History of Science at the University of Athens, for its generous sabbatical policy, and to my colleagues there, for providing a warm and friendly working environment.

Many of the ideas in this book were first presented at various conferences and workshops in Europe and the United States. I am particularly grateful to Jed Buchwald and Andrew Warwick, for inviting me to participate in two conferences celebrating the discovery of the electron, which were held in London under the auspices of the Royal Society and the Science Museum, and in Cambridge, Massachusetts, under the auspices of the Dibner Institute, respectively; to Friedrich Steinle and Jutta Schickore for inviting me to talk on scientific discovery at the Max Planck Institute for the History of Science in Berlin; to David Rowe, for his invitation to lecture on the discovery of the electron in the Department of Mathematics at Johannes Gutenberg University in Mainz; and to Stella Vosniadou, for inviting me to present my work on conceptual change and scientific realism at a European Symposium on Conceptual Change in Delphi. I would like to thank the audiences on all of these occasions for stimulating discussion and criticism.

I want to express my appreciation for the late Susan Abrams, who warmly welcomed my original manuscript and book proposal. I will always be grateful for her interest in my work, and for strongly encouraging me to turn my dissertation into a book. I am also indebted to her successor, Catherine Rice, for her considerable interest in my project and for the efficiency and care with which she handled the review process. Finally, I would like to thank the wonderful staff of the University of Chicago Press, and especially Erik Carlson for his meticulous editing of the manuscript.

And now a few acknowledgments of a more personal nature:

This book is devoted to the memory of my beloved mentor and friend, Yorgos Goudaroulis. Yorgos was a gifted and inspiring teacher, who incited my passion for history and philosophy of science. If I had not had the good fortune to meet him, this book would never have been written.

All those years that I was following the messy life of the representation of the electron, my family and friends were always there for me. Their bewildered faces, whenever I tried to explain what my book was all about, helped me to retain a sense of perspective. I couldn't possibly detail here how much I owe to each of them and, especially, to those who had to put up with my frustration when I was trying to finish the book: my sister Julia Arabatzis and my friends Nick Albanis and Mitsos Triantopoulos. I thank them all.

Last, but not least, I want to thank my wife, Sylvie Papamantellou, for her love and support, which made the end of this project easier than it sometimes appeared.

I would like to thank the Cambridge University Library, the Royal Society of London, and the University College London Library for facilitating my research in their archival collections.

Early versions of parts of this book have appeared elsewhere:

Part of section 4 in chapter 1 draws on my "Rethinking the 'Discovery' of the Electron," *Studies in History and Philosophy of Modern Physics*, 27 (1996): 405–435.

Chapters 3 and 4 draw on my "Rethinking the 'Discovery' of the Electron,"

*Studies in History and Philosophy of Modern Physics,* 27 (1996): 405–435; and my "The Zeeman Effect and the Discovery of the Electron," in J. Z. Buchwald and A. Warwick (eds), *Histories of the Electron: The Birth of Microphysics* (Cambridge, Mass.: MIT Press, 2001), pp. 171–194.

Chapter 7 draws on my joint paper with Kostas Gavroglu, "The Chemists' Electron," *European Journal of Physics,* 18 (1997): 150–163; and on my "How the Electrons Spend Their Leisure Time: Philosophical Reflections on a Controversy between Chemists and Physicists," in P. Janich and N. Psarros (eds), *The Autonomy of Chemistry: 3rd Erlenmeyer-Colloquy for the Philosophy of Chemistry* (Würzburg: Königshausen and Neumann, 1998), pp. 149–159.

Part of section 4 in chapter 9 draws on my "Can a Historian of Science Be a Scientific Realist?" *Philosophy of Science,* 68, suppl. (2001): S531–S541.

# Introduction

This book reconstructs some aspects of the historical development of the representation of the electron from the late nineteenth century till the mid-1920s, addressing explicitly the historiographical and philosophical issues involved in such a project. The aim of the book indicates the need for an interaction between the history and philosophy of science. The very idea of structuring a historical narrative around the representation of the electron spawns pressing philosophical questions. In the philosophical literature the electron is usually portrayed as the paradigm of an unobservable entity. For those who disbelieve in the existence of unobservable entities, and ipso facto of the electron, a historical project devoted to its representation might seem vacuous, an attempt to write the history of a theoretical entity which has no real counterpart in nature.[1] Furthermore, in telling the story of the electron's representation one has to tackle the issue of the electron's identity over time. The representation in question was an evolving entity, and it could be argued that the scientists who used the term "electron" in, say, 1900 and those who used it in, say, 1925 were not talking about the same thing. If that were the case, the representation of the electron could not be the subject of a coherent historical narrative. Finally, the evolution of that representation has implications for the issue of scientific realism, since it might be interpreted as throwing doubt on the existence of the electron.

1. The possibility of this antirealist reaction was pointed out to me by Ernan McMullin.

1

In engaging with this historical project, besides its "intrinsic" interest the possibility of a fruitful exchange between the history and the philosophy of science provided my main motivation. The significance of the philosophy of science for understanding historically scientific practice has been underrated. Even though the relationship between the history of science and the philosophy of science has been discussed extensively, the focus of the discussion has been on the importance of history of science for the philosophical understanding of science.[2] To the best of my knowledge, there has been very little discussion of the ways in which the philosophy of science can enrich historiographical practice.[3] Some authors have even denied that the philosophy of science has anything to offer to the history of science. Thomas Kuhn's words are characteristic: "I do not think current philosophy of science has much relevance for the historian of science."[4] Kuhn made that statement in the 1970s, but it captures the attitude of many historians ever since. The historians' skepticism toward the value of the philosophy of science may have been justified, in view of some crude attempts to "apply" philosophical theories of scientific change to historical case studies. As I will try to show in this book, however, the philosophy of science has two historiographically significant functions: first, to provide a metahistorical analysis of conceptual issues in the history of science, and second, to examine the philosophical presuppositions of historiographical categories (e.g., of the notion of scientific discovery) and choices (e.g., of the subject of a historical narrative).

While my research was well under way, I came across some features of the electron's representation that had not been previously apparent to me. I had tacitly assumed that the representation in question was a plastic resource that physicists and chemists could manipulate at will, in order to solve the conceptual and empirical problems they faced. While attempting to understand how

2. See, e.g., A. Donovan, L. Laudan, and R. Laudan (eds.), *Scrutinizing Science: Empirical Studies of Scientific Change* (Baltimore: Johns Hopkins Univ. Press, 1992). This collection "addresses diverse and frequently conflicting claims about how science changes. *It seeks to test these claims against well-researched historical cases*" (p. xii; my emphasis).

3. This is confirmed by recent literature on the relationship between the history and philosophy of science. See T. Nickles, "Philosophy of Science and History of Science," in A. Thackray (ed.), *Constructing Knowledge in the History of Science, Osiris,* 10 (1995): 139–163; C. Pinnick and G. Gale, "Philosophy of Science and History of Science: A Troubling Interaction," *Journal for General Philosophy of Science,* 31 (2000): 109–125; and H. Radder, "Philosophy and History of Science: Beyond the Kuhnian Paradigm," *Studies in History and Philosophy of Science,* 28 (1997): 633–655. The historiography of experimentation provides, perhaps, the only notable case where philosophical questions and issues have motivated and guided historical work. See, e.g., F. Steinle, "Experiments in History and Philosophy of Science," *Perspectives on Science,* 10 (2002): 408–432.

4. T. S. Kuhn, *The Essential Tension* (Chicago: Univ. of Chicago Press, 1977), p. 12.

a new property, spin, was attributed to the electron, however, I realized that the electron's representation did not tolerate the newly suggested property. In particular, the attempt to incorporate spin within a classical representation of the electron's structure resulted in a violation of the special theory of relativity—a theory that was also supposed to govern the electron's behavior. Thus, the physicists' attempt to portray the electron as a tiny spinning ball resulted in an incoherent representation of its behavior. The physicists were forced, in turn, to restore the coherence in the electron's representation, which had not conformed to their expectations and desires.[5] In other words, the representation in question turned out to have a life of its own.

The autonomy of the electron's representation suggested to me that theoretical entities are active agents that participate in the development of scientific knowledge.[6] Since I was familiar with Karl Popper's suggestion that the world of representations has some independence from (and interacts with) the world of humans, my "discovery" fell on prepared ground. From that point on, I started to develop the historiographical implications of this idea and attempted to read my historical materials in that light. It occurred to me that certain episodes in the history of science, which involved the electron qua theoretical entity as an active participant, could be told from its point of view. The emphasis in a narrative of that kind would be on the heuristic resources embodied in the electron's representation and on the resistance that it exhibited to manipulation. Such a narrative would have the structure of a biography, with the electron's representation as the main actor.

This approach will be fully spelled out in chapter 2. Here a clarification is necessary, to avoid a confusion that the term "theoretical entity" may give rise to. In the philosophical literature the expressions "theoretical entity" and "unobservable entity" are used interchangeably to denote entities that are postulated within a certain theoretical context and are not accessible to observation. Bas van Fraassen has argued that "[s]uch expressions as 'theoretical entity' and 'observable-theoretical dichotomy' are, on the face of it, examples of category mistakes. Terms or concepts are theoretical (introduced or adapted for the purposes of theory construction); entities are observable or unobserv-

---

5. This episode is reconstructed in detail in chapter 8.

6. Here I assume that "a life of its own" implies "autonomy." This assumption is not idiosyncratic. For example, Ian Hacking's famous slogan that experimentation has a life of its own was meant to emphasize its independence from theory. Furthermore, the term "agent" should not be interpreted as implying intentional action. In fact, in current English usage there is no necessary link between "agency" and "intentionality." According to my *Pocket Oxford Dictionary,* an "agent" is just "one who or *that which* exerts power or produces effect" (my emphasis). See also chapter 2, p. 46.

able."[7] Even though his observation is correct, strictly speaking, one might still retain the term "theoretical entity" as a shorthand expression of more cumbersome phrases like "the representation associated with an unobservable entity." In what follows, I use that term in the sense specified. Thus, by "writing the history of theoretical entities" I mean "writing the history of the corresponding representations." The unobservable entities that the representations stand for, on the other hand, are, in most cases, supposed to be invariant and, therefore, do not have a history.

The ambiguity between the electron and its representation is quite common in the historical and philosophical literature. For instance, a recent collection of essays is titled *Histories of the Electron,* whereas it clearly concerns representations of the electron. Furthermore, one sees repeatedly expressions like "Lorentz's electron" or "Bohr's electron." As I will argue in chapters 2 and 9, these expressions do not stand for different entities (different kinds of electrons), but for different representations of the same entity. Thus, "Bohr's electron" should be interpreted as a shorthand expression of "Bohr's representation of the electron."

Historical investigation can be daunting. As Fernand Braudel has aptly remarked, "There is a whole past to be reconstructed. Endless tasks rear up and demand our attention, if we are to deal with even the simplest realities."[8] The more my research progressed the more aware I became of the immense complexity of the life of the electron's representation. I would not exaggerate if I said that its story from the late nineteenth century to the mid-1920s is the story of physics and chemistry during that period. It is hard to think of developments that did not somehow implicate the electron. A historical narrative adequate to the task of telling the life of its representation would require a much longer and significantly more detailed treatment than the one I have been able to provide. I came around this problem by focusing on some historical developments that were directly relevant to my historiographical and philosophical concerns. Thus, the story that I tell is highly selective and told with an eye toward the methodological issues raised in chapters 1 and 2. Furthermore, in choosing the topics of the historical chapters I attempted, first, to focus on historical developments that altered significantly the representation of the electron and, second, to complement existing historical scholarship on the electron.

So this book does not aim at comprehensive coverage of the history of the electron's representation. Rather, its primary objective is metahistorical—

7. B. C. van Fraassen, *The Scientific Image* (New York: Oxford Univ. Press, 1980), p. 14.·

8. F. Braudel, *On History* (Chicago: Univ. of Chicago Press, 1980), p. 13.

namely, to understand the process of scientific discovery in microphysics, how theoretical entities are constructed, and how they function in scientific practice. Each of the historical chapters addresses a particular metahistorical issue: chapter 4 the "discovery" of unobservable entities, chapters 5, 6, and 8 the agency and recalcitrance of theoretical entities, and chapter 7 the identity of theoretical entities across disciplines and over time. However, the case studies I present do not have the character of "illustrations" of pre-established philosophical positions. Rather, my philosophical approach to the above issues has evolved in an attempt to come to grips with the historical material. The following chapters can be seen as an extended argument that these issues are not artificially imposed on the early history of the electron's representation, but are raised by the attempt to come to terms with that history.

The episodes I have chosen to reconstruct took place in the period from 1891, when the term "electron" was introduced, to 1925, when the notion of spin was put forward. That was a turning point in the history of the electron's representation. The proposal of spin was the last step in the development of the old quantum theory before the creation of quantum mechanics by Werner Heisenberg, Max Born, Pascual Jordan, and Paul Dirac. The new mechanics constituted a radical break with the older theory and was accompanied by severe interpretative difficulties, to a large extent absent from pre-1925 electron physics, whose repercussions are still debated by physicists and philosophers.

Within that period (1891–1925), the development of the representation of the electron can be construed as the outcome of two distinct (even though partially interacting) research programs. The first aimed at an increasingly precise measurement of certain electronic parameters (initially the charge-to-mass ratio and subsequently the charge of the electron). The other research program concerned the nature and behavior of electrons: the origin of their mass (mechanical or electromagnetic?) and its dependence on velocity, their size (do they occupy a finite space or are they point particles?), their self-energy, the laws that they obey, their degrees of freedom, their distribution within the atom, and their wave properties. With the exception of Pieter Zeeman's and J. J. Thomson's experiments, which initiated the former research program and which I discuss, I have focused on some aspects of the latter research program. Most of those episodes are familiar to specialists. The originality of the book, as I see it, consists in presenting a synthesis of primary sources and secondary scholarship on some important aspects of the development of the electron's representation, in light of the pertinent historiographical and philosophical issues.

## The Structure of the Book

Chapter 1 provides an overview of these issues. First, it starts with an explication of the notion of the problem situation, a notion that pervades the historical chapters to follow. Second, it gives a critical presentation of the issue of scientific discovery, with an emphasis on the "discovery" of unobservable entities. This part aims, among other things, to show the interdependence of the historiographical category of scientific discovery and the philosophical issue of scientific realism. The final section sketches the debate on the implications of meaning change for scientific realism and gives a historical twist to that debate, a twist that is fully developed in chapter 9.

Chapter 2 develops my biographical approach to writing the history of theoretical entities, an approach that, I hope, dispels the antirealist's skepticism toward the significance of such historical projects. I suggest a view of theoretical entities as constructions from experimental data. Viewing theoretical entities in this way enables us to understand their agency and to tackle successfully the problem of their identity over time. Furthermore, I point out the advantages of biography as a means of tracing the historical development of theoretical entities. Finally, I contrast this biographical perspective with recent approaches that also employ the notion of biography to reconstruct the history of "scientific objects."

Chapter 3 is an attempt to reconceptualize the "discovery of the electron." I provide a critical appraisal of the received view of that discovery, with an eye to the methodological issues about discovery that were raised in chapter 1. I attempt to show that the question Who discovered the electron? is not merely factual, but requires conceptual analysis and is entangled with the problem of scientific realism. I discuss some realist accounts of the discovery of the electron and find them wanting. Furthermore, I present evidence from early-twentieth-century sources which supports the view that the establishment of the electron as a new and fundamental constituent of matter was not an event, but a gradual process that was intertwined with the parallel debate over the existence of atoms.

Chapter 4 attempts to situate the principal historical actors within the wider process that led to the acceptance of the electron as an element of the ontology of physics. This chapter could very well expand into a book. I have not aimed at completeness; rather, I take up only those theories and experiments that were crucial for convincing the scientific community of the existence of the electron. Zeeman's and Thomson's measurements of the charge-to-mass ratio of the electron ($e/m$) are presented as the first steps in the construction of the quantitative aspects of the electron's representation. Unlike traditional narratives of the "discovery of the electron," mine puts special emphasis on the discovery of the Zeeman effect, the magnetic splitting of spectral lines. That discovery not only led to a determination of $e/m$ before Thomson's classic experiments on cathode rays, but also played a very important role in establishing the reality of the electron. Furthermore, it turned out to be crucial for the subsequent development of the representation of the electron. Key aspects of that development (e.g., the exclusion principle and spin) were the direct outcome of successive attempts to account for the "anomalous" Zeeman effect. These attempts are examined in chapter 8.

Chapters 5, 6, and 8 provide a reading of the development of the old quantum theory of the atom, which aimed at understanding the behavior and distribution of electrons bound within the atom, from the perspective of the electron. A defining characteristic of biographical studies as well as their main historiographical asset is that they offer a unique perspective on the historical developments in which their subject participated. Read in this way, the evolution of the old quantum theory can be construed as an extended episode from the life of the electron qua theoretical entity.

In particular, chapter 5 examines the birth of a quantum representation of the electron. It reads Niels Bohr's papers, the locus of that birth, from the point of view of the electron and highlights the active role of its representation in both guiding and constraining Bohr's thought. Bohr's revolutionary proposal transformed the electron's representation into that of a nonclassical particle, that is, a particle that did not fully obey the laws of classical mechanics and electromagnetic theory. Chapters 6 and 8 explore the further metamorphosis of the representation of the electron after that initial "quantum leap." They focus on some aspects of the history of quantum numbers, selection rules, the correspondence principle, transition probabilities, the exclusion principle, and spin. Each of these developments has a history that I have not attempted to follow. The emphasis of my narrative is on the introduction of these important innovations rather than on their subsequent elaboration and application. Again, I portray the electron qua theoretical entity as an active agent and highlight its role in the development of the old quantum theory of the atom.

I should point out that chapters 5, 6, and 8 are not put forward as a novel interpretation of the historical development of quantum theory. They are too selective and schematic to fulfill that purpose. Moreover, with few exceptions, my interpretation of the developments that I reconstruct is in agreement with the results of the considerable and solid historical scholarship of the past three decades. The aim of these chapters is, rather, to read some aspects of that history from the perspective of the electron, as opposed to the atom and, more important, to introduce the electron's representation as an autonomous actor who participated in the construction of that theory.

Chapter 7 discusses the chemists' representation of the electron and contrasts it with its physical counterpart. Here we see that the notion of the problem situation is crucial for explaining the diverging outlooks of chemists and physicists. I sketch the first indications of the nature of the reconciliation that was eventually achieved with the advent of quantum mechanics. Furthermore, I discuss the transdisciplinary character of the electron qua theoretical entity and its significance for bringing chemistry and physics closer.

Chapter 9 provides an analysis of the philosophical debate concerning the meaning variance of scientific terms and its implications for scientific realism. I argue against the widespread view that meaning change is incompatible with scientific realism and, thus, provide a way out for the aspiring realist, without, however, committing myself to a realist position. Furthermore, I indicate the importance of a historicist approach to the ontological status of unobservable entities. I suggest that the historical reconstruction of the concept associated with an unobservable entity and, more important, of its putative referent are indispensable for an adequate realist construal of that entity's ontological status. Thus, realism can make sense only locally (i.e., with respect to particular entities). In concluding the book I discuss the compatibility of the evolution of the concept of the electron with a realist interpretation of its ontological status and argue that certain aspects of that evolution allow, but not require, a realist attitude toward the electron.

# Chapter 1 | Methodological Preliminaries

## 1. Introduction

This chapter provides a survey of the historiographical and philosophical issues that form the backbone of the book. Section 2 is devoted to an analysis of the concept of the problem situation, a concept that permeates the subsequent historical chapters. Sections 3 and 4 discuss the issue of scientific discovery, with an emphasis on the "discovery" of unobservable entities. The fifth section outlines the debate on the implications of meaning change for scientific realism and gives a historical twist to that debate.

## 2. Karl Popper and the Notion of the "Problem Situation"

One of the most significant and neglected elements of Karl Popper's philosophy of science is his emphasis on scientific problems and problem situations. In his view, "[T]he history of science should be treated not as a history of theories, but as a history of problem situations and their modifications (sometimes imperceptible, sometimes revolutionary) through the intervention of attempts to solve the problems."[1] For Popper, scientific problems can be classified as two different kinds, empirical and theoretical. A scientist faces an empirical problem when "he wants to find a new theory capable of explaining certain experimental facts; facts which the earlier theories successfully explained; others which they could not explain; and some by which they

---

1. K. R. Popper, *Objective Knowledge: An Evolutionary Approach,* rev. ed. (Oxford: Oxford Univ. Press, 1979; first publ. 1972), p. 177.

were actually falsified."[2] Theoretical problems, on the other hand, consist of "theoretical difficulties (such as how to dispense with certain ad hoc hypotheses, or how to unify two theories)."[3]

Problems, according to Popper, always arise against a background which "consists of at least a *language,* which always incorporates many theories in the very structure of its usages . . . and of many other theoretical assumptions, unchallenged at least for the time being."[4] This background of theoretical beliefs and assumptions provides a framework that enables the emergence and formulation of a problem. "A problem together with its background . . . constitutes what I call a *problem situation.*"[5] The background supplies the constraints that a solution to the problem should satisfy, as well as the conceptual resources that are employed to obtain that solution. The former, it should be emphasized, do not always originate from scientific knowledge, but may have an extrascientific origin.[6] For instance, Galileo's attempt to provide an explanation for the tides was constrained, according to Popper, by his opposition to astrology: "Galileo rejected the lunar influence [as an explanation of the tides] because he was an opponent of astrology which essentially identified the planets with the gods."[7] The latter are not necessarily provided by the discipline which gave rise to the problem, but may come from a different source. Even though problems belong to certain disciplines in the sense that they arise "out of a discussion characteristic of the tradition of the discipline in question," "the solution of problems may cut through the boundary of many sciences."[8] For example, "the [physical] problem of explaining certain spec-

2. K. R. Popper, *Conjectures and Refutations: The Growth of Scientific Knowledge* (1962; repr., New York: Harper and Row, 1968), p. 241.

3. Ibid.

4. Popper, *Objective Knowledge,* p. 165.

5. Ibid. Popper had already talked about problem situations in 1934, but at that time he conceived of them as mere "contradictions" or "falsifications." See his *Logik der Forschung* (Vienna: Springer, 1935), p. 222.

6. I am aware that the attempt to draw a clear line of demarcation between science and pseudoscience or metaphysics is beset with difficulties. However, I disregard them at this point, since Popper thought that his falsifiability requirement provided an adequate demarcation criterion.

7. Popper, *Objective Knowledge,* p. 173. Popper's suggestion that a problem situation may involve extrascientific considerations was further developed by J. Agassi and W. Berkson, who emphasized the importance of metaphysical problems for understanding scientific practice. See J. Agassi, "On the Nature of Scientific Problems and Their Roots in Metaphysics," in his *Science in Flux,* Boston Studies in the Philosophy of Science 28 (Dordrecht: Reidel, 1975), pp. 208–239; and W. Berkson, *Fields of Force: The Development of a World View from Faraday to Einstein* (New York: John Wiley and Sons, 1974).

8. Popper, *Conjectures and Refutations,* pp. 67, 74.

tral terms (with the help of a hypothesis concerning the structure of atoms) may turn out to be soluble by purely mathematical calculations."[9]

From a Popperian perspective, the aim of the historiographical enterprise would be to describe and explain past scientific beliefs, decisions, and actions as reasonable solutions to the specific problem situation faced by the scientist under consideration. In Popper's view, "The historian's task is . . . so to reconstruct the problem situation *as it appeared to the agent* [emphasis added], that the actions of the agent become *adequate* to the situation."[10] This is an instance of what he elsewhere calls *situational analysis,* that is, "a certain kind of *tentative or conjectural* explanation of some human action which appeals to the situation in which the agent finds himself" (emphasis added).[11] His approach can be implemented with sensitivity to the categories and values of historical actors, a sensitivity that has for some time now been highly valued in the historical profession.[12]

Thus, Popper's historiographical thesis is "that the main aim of all historical understanding is the hypothetical reconstruction of a historical *problem-situation.*"[13] The reconstruction in question is hypothetical because it cannot be just read off the historical record. There is often room for different interpretations of how a problem situation "appeared to the agent." Although I agree with Popper's thesis, I think that his notion of a problem situation needs further development to become an effective historiographical guide. A step in this direction has been taken by Larry Laudan, who has extended Popper's insight to a comprehensive problem-oriented approach to scientific development.[14] Laudan's major contribution is that he emphasized the importance of conceptual problems for scientific practice and provided a detailed analysis of their character.[15] According to Laudan, there are two kinds of conceptual

9. Ibid., p. 74.

10. Popper, *Objective Knowledge,* p. 189.

11. Ibid., p. 179. See also Popper's "Situational Logic in History," in his *The Poverty of Historicism* (London: Routledge and Kegan Paul, 1957). For Popper "historicism" amounted to the belief in historical laws and, therefore, should not be confused with what we nowadays take a historicist approach to be.

12. My endorsement of Popper's historiographical methodology does not mean I applaud his—rather unimpressive—attempts to practice history.

13. Popper, *Objective Knowledge,* p. 170.

14. See L. Laudan, *Progress and Its Problems: Towards a Theory of Scientific Growth* (Berkeley: Univ. of California Press, 1977); and Laudan, "A Problem-Solving Approach to Scientific Progress," in I. Hacking (ed.), *Scientific Revolutions* (Oxford: Oxford Univ. Press), pp. 144–155.

15. While it is true, as Paul Feyerabend pointed out, that Laudan tends to overemphasize the novelty of his approach, his analysis of conceptual problems is far more detailed and systematic than Popper's. Even though Popper acknowledged the existence of theoretical problems, his analysis was focused on empirical problems. This is hardly surprising, given his view

problems, "internal" and "external." Internal or intratheoretic problems arise when a theory is "internally inconsistent or the theoretical mechanisms it postulates are ambiguous" or circular.[16] External problems are created when a theory "makes assumptions about the world that run counter to other theories or to prevailing metaphysical assumptions, or when . . . [it] makes claims about the world which cannot be warranted by prevailing epistemic and methodological doctrines . . . [or when it] fails to utilize concepts from other, more general theories to which it should be logically subordinate."[17] Some examples that will be more fully discussed in later chapters will illustrate the nature of certain kinds of conceptual problems that played a significant role in the development of the representation of the electron. The old quantum theory of black body radiation provides an example of an internally inconsistent theory.[18] The derivation of Max Planck's distribution law, a formula that represented the energy distribution over the entire frequency spectrum of black body radiation, was accomplished by combining the quantum hypothesis with results deduced from classical electrodynamics. The inconsistency arose because classical electrodynamics implied that the energy of electric resonators within a black body was a continuous magnitude, whereas, according to the quantum hypothesis, the resonator energy could take only discrete values. The conceptual problem generated by that inconsistency was rather decisive in the development of the old quantum theory.[19]

The atomic theories of matter and electricity in the beginning of the twentieth century illustrate the kinds of conceptual problems that are created when some of the explanatory mechanisms proposed by a theory are circu-

---

that problems "*as a rule* arise from the clash between, on the one side, expectations inherent in our background knowledge and, on the other side, some new findings, such as our observations or some hypotheses suggested by them" (Popper, *Objective Knowledge*, p. 71). The rest of Feyerabend's critical remarks concern the adequacy of Laudan's model as a normative theory of scientific rationality and, even if valid, do not undermine its historiographical potential. See P. Feyerabend, "More Clothes from the Emperor's Bargain Basement: A Review of Laudan's *Progress and Its Problems*," *British Journal for the Philosophy of Science*, 32 (1981): 57–71; repr. in his *Philosophical Papers*, vol. 2, *Problems of Empiricism* (Cambridge: Cambridge Univ. Press, 1981), pp. 231–246.

16. Laudan, "A Problem-Solving Approach," p. 146.

17. Ibid. For further discussion and concrete examples of conceptual problems see Laudan, *Progress and Its Problems*, pp. 48–64.

18. This example is abstracted from J. Norton, "The Logical Inconsistency of the Old Quantum Theory of Black Body Radiation," *Philosophy of Science*, 54 (1987): 327–350. Norton's paper was brought to my attention by Aristides Baltas.

19. Norton reads Einstein's probabilistic derivation of Planck's formula (1916) as an attempt to deal with the "consistency problem." See ibid., p. 330.

lar. Both of those theories explained certain macroscopic properties of bodies (e.g., charge distribution) by endowing their microscopic constituents with the same properties. Thus the same properties figured both in the thing to be explained (the *explanandum*) and the explanatory mechanism (the *explanans*). As G. N. Lewis pointed out in his assessment of the atomic theory,

> in giving up the continuous theory of matter, and replacing it by the theory of discrete centers which we call atoms (or electrons and nuclei), we have some-how failed in consistency. A race with more limited sense perceptions than our own might study the properties of sand and conclude these properties to be due to the existence of grains, but would they then be justified in regarding the grains as composed of sand? Yet this is the kind of inference that modern science has sanctioned. The properties of electricity have been explained by assuming it to be composed of electrons, after which we naively consider the electrons as made up of electricity, and speculate concerning the distribution of electricity about the electron center. We also have regarded the atoms as possessing prop-erties similar to those of the larger bodies which they compose.[20]

This conceptual problem took care of itself, so to speak, in the sense that the circularity inherent in those theories became a "kind of inference that mod-ern science has sanctioned." It should be noted, however, that, usually, "[w]hat requires explanation cannot itself figure in the explanation; . . . [to paraphrase Molière] we would not be satisfied were the sleep-inducing qualities of opium explained by reference to its soporific properties."[21]

Turning to external conceptual problems, we find in the incompatibility between Bohr's theory of the atom and classical electrodynamics an example of the kind of conceptual problems that arise when a novel theory contradicts an already established theory. In the initial formulation of the old quantum theory of the atom in 1913 by Niels Bohr it was assumed that the electron, in any of its "stationary" orbits around the nucleus, does not radiate. This assumption was inconsistent with the then accepted electromagnetic the-ory, which implied that any charge in accelerated motion, and therefore the electron within the atom, emits radiation.[22] The subsequent development of the quantum theory of the atom was partly an attempt to come to terms

20. G. N. Lewis, *Valence and the Structure of Atoms and Molecules* (New York: Chemical Catalog Co., 1923), p. 42.

21. N. R. Hanson, *The Concept of the Positron: A Philosophical Analysis* (Cambridge: Cam-bridge Univ. Press, 1963), p. 42. Hanson gives several examples of scientists' refusal to sanction circular patterns of explanation.

22. This is, of course, an oversimplified description. For details see chapter 5.

with this problem, that is to provide a theoretical justification of Bohr's bold conjecture.

Bohr's theory was also in conflict with a then prevalent methodological tenet, namely, that an acceptable physical theory should provide causal explanations of phenomena. When Bohr put forward his theory, Ernest Rutherford criticized it for failing to give a causal account of an electron's transition between two different stationary states. Rutherford's penetrating criticism revealed a major conceptual problem that was eventually dissolved with the formulation of quantum mechanics, which abandoned the demand for causal explanation.[23]

Finally, another kind of external conceptual problem is illustrated by a controversy between chemists and physicists which took place during the second and third decades of the twentieth century. The controversy concerned the behavior of electrons within the atom. The chemists advocated a static electron, whereas the physicists favored a dynamic electron that was in constant motion within the atom (see chapter 7). The problem arose because the chemists chose to "adopt the concepts of atomic physics—electrons, nuclei, and orbits—and try to explain the chemical facts in terms of these," without "accept[ing] the physical conclusions in full, and . . . assign[ed] to these entities properties which the physicists have found them not to possess."[24] Thus, they were "open to the reproach of an eminent physicist, that 'when chemists talk about electrons they use a different language from the physicists.'"[25] This problem was resolved with the advent of quantum mechanics, which incorporated aspects of both the chemists' and the physicists' representations into the quantum-mechanical conception of the electron.

Laudan's treatment of conceptual problems enriched significantly the Popperian conception of a problem situation. However, further development of that conception is necessary to make it an adequate explanatory tool. In particular, the cultural context within which a problem arises should also be taken as part of the problem situation. For philosophers of science like Popper and Laudan problems and problem situations belong to the Popperian third world, that is, to the world "of *ideas in the objective*

---

23. Again, I am oversimplifying. There are interpretations of quantum mechanics which do not dispense with causality. However, it would be fair to say that the original (and dominant) Copenhagen interpretation was at odds with the requirement for causal accounts of phenomena.

24. N. V. Sidgwick, *The Electronic Theory of Valency* (London: Oxford Univ. Press, 1927); the quote is from the preface. I would like to thank Kostas Gavroglu for bringing Sidgwick's book to my attention.

25. Ibid.

*sense."*[26] Given their concern with scientific progress and rationality, their idealized treatment of problems is understandable. If one's aim is to appraise the scientific rationality of a certain scientific episode, then one can legitimately restrict her analysis to the epistemic factors that played a role in the given event. If, however, our aim as historians of science is to explain past scientific episodes, then consideration of further nonepistemic parameters *may,* in some cases, turn out to be indispensable. For explanatory purposes the notion of a problem situation should be extended to include aspects of the scientists' micro- and macrosocial situation. Macrosocial parameters (e.g., class membership) and microsocial factors (e.g., the professional interests of the participants in the episode under study) should also be considered *potential* aspects of a scientist's problem situation, in the sense that they *might* act as constraints on the range of admissible solutions.[27]

Such a proposal has been recently made by Nancy Nersessian. The cultural context in which scientific activity takes place functions in two ways: as a source of problems and as a repository of resources upon which scientists draw to resolve their problems. An example of the former function is provided by Michael Faraday's Sandemanian religion, which motivated his belief in the unity of forces and thus generated the problem of developing scientific representations of electromagnetic phenomena compatible with that belief.[28] For an example of the latter function consider the development of quantum mechanics. Paul Forman's well-known study suggests that the pervasiveness of indeterminism in the intellectual atmosphere of Weimar culture was a significant heuristic factor in the construction of that theory.[29]

The formation of scientific concepts is another area where the notion of

26. Popper, *Objective Knowledge,* p. 154.

27. While I think that sociological accounts of scientific practice reveal important aspects of the scientific enterprise, I do not share the currently popular view that sociological analysis provides the only proper way to understand science. See my "Rational versus Sociological Reductionism: Imre Lakatos and the Edinburgh School," in K. Gavroglu, J. Christianidis, and E. Nicolaidis (eds.), *Trends in the Historiography of Science,* Boston Studies in the Philosophy of Science 151 (Dordrecht: Kluwer, 1994), pp. 177–192.

28. N. J. Nersessian, "How Do Scientists Think? Capturing the Dynamics of Conceptual Change in Science," in R. N. Giere (ed.), *Cognitive Models of Science,* Minnesota Studies in the Philosophy of Science 15 (Minneapolis: Univ. of Minnesota Press, 1992), pp. 3–44; on p. 38.

29. Here I draw on Norton Wise's modification of Forman's thesis. See M. N. Wise, "Forman Reformed," unpublished manuscript. Cf. also N. J. Nersessian, *Opening the Black Box: Cognitive Science and History of Science,* Cognitive Science Laboratory Report 53 (Princeton: Princeton Univ., Jan. 1993), esp. pp. 29–30; this article has been published in an abbreviated form in A. Thackray (ed.), *Constructing Knowledge in the History of Science, Osiris,* 10 (1995): 194–214. In its original version, Forman's thesis was very controversial, and it has generated a substantial body of critical literature.

a problem situation is applicable. In particular, one could employ that notion to understand conceptual change.[30] For some time now many philosophers have favored a theory of meaning in which the meaning of scientific terms is determined by their place within the overall theoretical structure in which they are embedded (see chapter 9). Strictly speaking, this conception of meaning could be an applicable historiographical and philosophical tool only if the theoretical framework in question were coherent and fixed. If the framework evolves, one has the difficulty of specifying which of the changes that the framework undergoes transform the meaning of the terms embedded in it. Furthermore, if there is no coherent framework to begin with, it is not clear how, according to this conception, the terms in question get their meaning. For instance, consider the case of the term "electron" in the context of the old quantum theory of the atom. It would be difficult to specify how the meaning of that term was determined by the theory in question, since neither of the above conditions is fulfilled. On the one hand, the old quantum theory evolved considerably from its initial formulation in 1913 to its eventual demise in 1925. On the other hand, it did not constitute a proper theoretical framework, that is, a theoretical structure based on consistent foundations that provides theoretical justifications for all of its explanatory mechanisms (see chapters 5, 6, and 8).

For these reasons it is preferable to focus on scientific terms themselves and the theoretical and experimental practices out of which their meaning is determined. The meaning of such a term amounts to the characteristics that are ascribed, by the scientists who are using it, to the corresponding entity or process (see chapter 9). Those characteristics are proposed in response to specific problem situations and, therefore, are the outcome of the actual scientific practices that give rise to those situations and lead to their solution.[31] The evolution of a term's meaning reflects the evolution of the problem situations that implicate the term in question and should be reconstructed by explicating those situations. Thus, the "problem situation framework" provides the means through which to examine how meanings are constructed.

---

30. It is worth pointing out that conceptual change or meaning change was a nonproblem for Popper.

31. The idea that any adequate theory of meaning should be grounded on scientific practice and seen in the context of evolving problem situations has been developed by Nersessian. See her *Faraday to Einstein: Constructing Meaning in Scientific Theories* (Dordrecht: Martinus Nijhoff, 1984); and "How Do Scientists Think?" esp. p. 12. Kostas Gavroglu and Yorgos Goudaroulis have followed a similar approach in their attempt to reconstruct the development of low-temperature physics. See their *Methodological Aspects of the Development of Low Temperature Physics: Concepts out of Contexts* (Dordrecht: Kluwer, 1989).

So far I have tacitly assumed that the notion of a "problem" is unproblematic. It is not, and the difficulties associated with it have been extensively discussed in the philosophical literature.[32] Since I will not be able to address this issue in detail, let me discuss briefly the position that I find most satisfactory, namely, that of Thomas Nickles. From his analysis we can derive the proposal that a problem can be conceived as a question together with a set of constraints on any admissible solution.[33] He suggested that problems should be represented in a "matrix form," that is, "as ordered or structured sets, in which the constraints are classified according to function and weighted as to importance." This is necessary because "not all constraints are equally important; nor is simply weighing them enough, since . . . the various constraints in the set do not serve the same function."[34] His suggestion provides a very helpful analytical tool for the construction of historical explanations. It captures the central challenge faced by the historian when trying to explain a past scientific event, namely, to judge the relative importance of various factors (intellectual, social, material, etc.) in determining the outcome of the episode under scrutiny. Its main asset as a historiographical tool is that it aids the historian in meeting that challenge in an explicit and systematic way. For historiographical purposes, I will restrict the notion of a "problem" to *known* constraints, since, otherwise, problems obtain a Popperian "third world" quality that detracts from their utility for historical explanation. To put it another way, only constraints that are recognized by the historical actors themselves can be incorporated in a historical narrative, since only if they are known can they guide the problem-solving process.[35]

Subject to these qualifications, I follow Nickles in regarding a problem as a question together with the constraints that the answer should satisfy. Some of these constraints (the demand for empirical adequacy, inter- and intratheoretic consistency, etc.) come from the scientific culture itself. Other constraints (e.g., compatibility with a political, religious, or metaphysical doctrine) origi-

32. See, e.g., W. Berkson, "Research Problems and the Understanding of Science," in N. J. Nersessian (ed.), *The Process of Science* (Dordrecht: Martinus Nijhoff, 1987), pp. 83–93.

33. See T. Nickles, "Scientific Problems and Constraints," in P. D. Asquith and I. Hacking (eds.), *PSA 1978: Proceedings of the 1978 Biennial Meeting of the Philosophy of Science Association* (East Lansing, Mich.: Philosophy of Science Association, 1978), vol. 1, pp. 134–148. This conception of problems is not explicitly formulated in Nickles's article, but it follows from his remark that "[a]lthough constraints do much to determine a scientific problem, . . . there are at least some cases in which they do not determine the question asked completely" (p. 143).

34. Ibid., p. 141.

35. Of course, scientists may not be aware that some factors (e.g., some of their views) function as constraints on their theorizing. My point is, rather, that they must be familiar with those factors, without necessarily realizing their constraining role.

nate in the wider cultural environment. Furthermore, the social situation of the researcher who is engaged with the problem (e.g., his professional interests) might also constrain the range of acceptable solutions. The function and importance of each type of constraint vary from case to case (i.e., for different actors) and cannot be determined a priori. Finally, the micro- and macrocultural contexts provide the resources that are employed to tackle the problem, and they put a premium on the quest for a solution.

In the chapters that follow I have employed the methodology discussed in this section for reconstructing some aspects of the formation of the representation of the electron. Since key aspects of that representation emerged in the attempt to resolve specific problem situations, an understanding of its development is contingent on the historical reconstruction of the emergence and resolution of those problem situations.

My analysis will, in turn, suggest a further enrichment of the notion of a problem situation. As I will argue in the next chapter, theoretical entities give rise to a distinct kind of problem. When scientists employ the representation of an unobservable entity for explanatory or predictive purposes, they often attribute novel properties to that entity. It may turn out, however, that these novel properties are in conflict with the already established properties of the entity. In these cases the attempted theoretical move renders the representation incoherent; that is, it creates a new problem situation. Furthermore, theoretical entities sometimes provide heuristic guidance for resolving the problems in which they are implicated. A plausible problem-solving strategy is to eliminate the idealizations or simplifications in the theoretical entities in question (see chapter 2, pp. 36 – 41).

### 3. Scientific Discovery as a Philosophical and Historiographical Category

A problem-oriented approach to scientific practice can also elucidate the issue of scientific discovery. To show that, it is imperative to examine how the problem of scientific discovery has been posed in the history and philosophy of science. Before the historicist turn in the philosophy of science, it was generally regarded that scientific activity takes place within two distinct contexts, the context of discovery and the context of justification. The former consists in the processes of generation of scientific hypotheses and theories; the latter in their testing and validation. According to Hans Reichenbach, who codified the distinction, the context of discovery was the province of historians, psychologists, and sociologists and was not susceptible to logical analysis. On the other hand, the context of justification was an area which could be rigorously explored and formalized and thus fell within the province of logic and philosophy.[36]

This distinction historically derived from several premises. First, it was based on a conception of the philosophy of science as a normative enterprise, that is, an enterprise whose aim was to lay out rules that should govern any activity that deserves to be called science. Second, it was grounded on a conflation of scientific discovery with the generation of novel ideas. Thus, the study of discovery had to be the study of scientific creativity. Third, it rested on the widespread view that there are no rules whose application can enhance one's creativity. The latter two assumptions precluded the possibility of a normative theory of discovery and, along with the first one, rendered impossible the philosophical exploration of discovery. Finally, the distinction required justification to be a rule-governed process so as to be the subject of a normative project.

All of these assumptions have for some time now been under attack, and, consequently, the distinction has been undermined. To begin with, there has been a gradual shift toward a more "naturalistic" conception of the philosophy of science, namely, a conception that stresses the descriptive and hermeneutic aspects of the philosophical study of science as opposed to its normative ones.[37] Furthermore, the conflation of discovery with generation has been exposed and criticized. This point is crucial for my purposes and will be extensively discussed below. Even if this conflation and the concomitant identification of the study of discovery with the study of creativity were valid, one could still deny that creativity is an unanalyzable, totally mysterious phenomenon that precludes the possibility of a normative theory of discovery. Indeed, there has been overwhelming evidence that hypothesis generation and theory construction are reasoned processes whose explication can (and should) be carried out by philosophers of science.[38] Some have even argued

36. See H. Reichenbach, *The Rise of Scientific Philosophy* (Berkeley: Univ. of California Press, 1951), p. 231. Reichenbach's *Experience and Prediction* (Chicago: Univ. of Chicago Press, 1938) is usually cited as the primary site of that distinction. However, as Nickles has pointed out, the distinction found there "is merely one between scientific activity itself and that activity as logically reconstructed." See his "Introductory Essay: Scientific Discovery and the Future of Philosophy of Science," in T. Nickles (ed.), *Scientific Discovery, Logic, and Rationality,* Boston Studies in the Philosophy of Science 56 (Dordrecht: Reidel, 1980), pp. 1–59, on p. 12.

37. See P. Kitcher, "The Naturalists Return," *Philosophical Review,* 101, no. 1 (1992): 53–114.

38. See, e.g., Nersessian, "How Do Scientists Think?"; and T. Nickles, "Can Scientific Constraints Be Violated Rationally?" in Nickles (ed.), *Scientific Discovery, Logic, and Rationality,* pp. 285–315.

that it is possible to devise a normative theory of scientific discovery that would specify heuristic procedures which would improve the efficiency of scientific inquiry and, thus, facilitate the discovery process.[39] Finally, the notion of justification as a rule-governed process has been challenged. Justification itself requires many discovery tasks. For example, to justify a hypothesis one needs to "discover" an appropriate test as well as the auxiliary statements to render the hypothesis testable.[40]

The distinction has also been undermined on different grounds. It has been argued that the kind of reasoning that is involved in generating a hypothesis is not fundamentally different from the kind of reasoning employed in justification.[41] Moreover, hypothesis generation and theory construction are extended problem-solving processes with many stages, each of which involves partial justification. At each particular stage one's aim is to satisfy some of the constraints posed by the problem. The satisfaction of those constraints amounts to partial justification of the evolving solution.[42] Furthermore, justification in science often takes the form of heuristic appraisal, that is, of evaluating the future problem-solving potential of a theory.[43] For someone who views discovery as an instance of problem-solving, this form of theory appraisal amounts to judging the capacity of a theory to generate discoveries and, therefore, it is closely linked with discovery itself. Finally, Nickles has stressed the importance of "generative justification" or "discoverability," namely, a form of appraisal that justifies a claim by deriving it from already established knowledge. Justification in this case amounts to specifying a rationally reconstructed (not necessarily the actual) discovery path.[44] Thus, it is

39. See, e.g., P. Langley et al., *Scientific Discovery: Computational Explorations of the Creative Process* (Cambridge, Mass.: MIT Press, 1987).

40. See T. Nickles, "Introductory Essay," esp. p. 13; Nickles, "Beyond Divorce: Current Status of the Discovery Debate," *Philosophy of Science,* 52 (1985): 177–206, esp. p. 193; and Nickles, "Discovery," in R. C. Olby et al. (eds.), *Companion to the History of Modern Science* (London: Routledge, 1990), pp. 148–165, esp. p. 162. Cf. also H. Putnam, "The 'Corroboration' of Theories," in R. Boyd et al. (eds.), *The Philosophy of Science* (Cambridge, Mass.: MIT Press, 1991), pp. 121–137.

41. P. Achinstein, "Discovery and Rule-Books," in Nickles (ed.), *Scientific Discovery, Logic, and Rationality,* pp. 117–132.

42. See Langley et al., *Scientific Discovery;* and Nickles, "Introductory Essay."

43. See Nickles, "Beyond Divorce," p. 194; "'Twixt Method and Madness," in Nersessian (ed.), *The Process of Science,* pp. 41–67, esp. pp. 47–48; and "Heuristic Appraisal: A Proposal," *Social Epistemology,* 3, no. 3 (1989): 175–188.

44. See T. Nickles, "Positive Science and Discoverability," in P. D. Asquith and P. Kitcher (eds.), *PSA 1984: Proceedings of the 1984 Biennial Meeting of the Philosophy of Science Association* (East Lansing, Mich.: Philosophy of Science Association, 1984), vol. 1, pp. 13–27; Nickles, "Beyond Divorce," pp. 194–195; and Nickles, "Truth or Consequences? Generative versus Conse-

reasonably established that justification and discovery are much more closely related than formerly thought.

The distinction has also been attacked from a historical perspective. Thomas Kuhn, for instance, has argued that "[c]onsiderations relevant to the context of discovery are . . . relevant to justification as well; scientists who share the concerns and sensibilities of the individual who discovers a new theory are ipso facto likely to appear disproportionately frequently among that theory's first supporters."[45] However, many of the critics of the distinction continue to share with its proponents the same conception of scientific discovery. The context of scientific discovery, on this view, consists in the processes that lead to the formulation of new hypotheses or theories. In other words, both sides of the debate equate discovery either with the "generation"[46] of hypotheses or with the construction of scientific theories.[47] With some exceptions, justification is still not seen to be part of the discovery process.[48]

This view of scientific discovery is, I think, considerably misleading, and without it the debate on the validity of the distinction between the two contexts could not even start. To begin with, the term discovery is used to designate many different kinds of processes: the discovery of phenomena through controlled experiment (e.g., the discovery of the Zeeman effect—the magnetic splitting of spectral lines), the discovery of entities which are accessible

---

quential Justification in Science," in A. Fine and J. Leplin (eds.), *PSA 1988: Proceedings of the 1988 Biennial Meeting of the Philosophy of Science Association,* 2 vols. (East Lansing, Mich.: Philosophy of Science Association, 1989), vol. 2, pp. 393–405, esp. p. 394.

45. T. S. Kuhn, *The Essential Tension* (Chicago: Univ. of Chicago Press, 1977), p. 328. For an analysis of Kuhn's criticism of the discovery-justification distinction see P. Hoyningen-Huene, "Context of Discovery and Context of Justification," *Studies in History and Philosophy of Science,* 18 (1987): 501–515.

46. I borrow the term from Nickles, "Introductory Essay."

47. See, e.g., R. Burian, "Why Philosophers Should Not Despair of Understanding Scientific Discovery," in Nickles (ed.), *Scientific Discovery, Logic, and Rationality,* pp. 317–336, esp. pp. 322–323; M. V. Curd, "The Logic of Discovery: An Analysis of Three Approaches," in Nickles (ed.), *Scientific Discovery, Logic, and Rationality,* pp. 201–219, esp. pp. 201–202; and L. Laudan, "Why Was the Logic of Discovery Abandoned?" in Nickles (ed.), *Scientific Discovery, Logic, and Rationality,* pp. 173–183, esp. pp. 174–175.

48. The exceptions are important. See, e.g., G. Gutting, "Science as Discovery," *Revue Internationale de Philosophie,* 131–132 (1980): 26–48; Hoyningen-Huene, "Context of Discovery and Context of Justification"; N. Koertge, "Explaining Scientific Discovery," in P. D. Asquith and T. Nickles (eds.), *PSA 1982: Proceedings of the 1982 Biennial Meeting of the Philosophy of Science Association,* 2 vols. (East Lansing, Mich.: Philosophy of Science Association, 1983), vol. 1, pp. 14–28; C. R. Kordig, "Discovery and Justification," *Philosophy of Science,* 45 (1978): 110–117; and Ernan McMullin's contribution to "(Panel Discussion) The Rational Explanation of Historical Discoveries," in T. Nickles (ed.), *Scientific Discovery: Case Studies,* Boston Studies in the Philosophy of Science 60 (Dordrecht: Reidel, 1980), pp. 28–33.

to immediate inspection (e.g., the discovery of a previously unknown species), the discovery of objects which are not accessible to unaided observation (e.g., the discovery of the planet Neptune), the discovery of entities which are unobservable in principle (e.g., the discovery of the electron), the discovery of new properties of well established entities (e.g., the discovery of electron spin), the discovery of new principles (e.g., the discovery of energy conservation), and the discovery of new theories (e.g., the discovery of the special theory of relativity).

In all of these cases the two contexts are inextricably linked. Consider, for example, the discovery of unobservable entities. Individuals or groups can acquire the status of "the discoverer" only after they have convinced the rest of the scientific community of the existence of the entity in question. A mere hypothesis that a new entity exists would not qualify as a discovery of that entity. The justification of that hypothesis would be a constitutive characteristic of that discovery. The context of discovery is "laden" with the context of justification because "discovery" is a term that implies success and has realist presuppositions: if one succeeds in discovering something, then, no doubt, this something exists.[49] That this is the case is witnessed by the fact that in the historical literature the historiographical issue of scientific discovery has been discussed only in relation to entities that remain part of the accepted scientific ontology (e.g., oxygen). No historian or philosopher, to the best of my knowledge, has ever used the term "discovery" to characterize the proposal and acceptance of an entity (e.g., phlogiston) that we now believe was a fictitious one.[50] The example of phlogiston is instructive. Contemporary historians and philosophers do not think that phlogiston was discovered, despite the fact that some eighteenth-century chemists referred to phlogiston as one of the most significant discoveries in the history of chemistry. In Joseph Priestley's words, phlogiston "was at one time thought to have been the greatest discovery that had ever been made in the science."[51] This suggests that there is a retrospec-

49. After formulating this idea, I discovered that Nickles ("Introductory Essay," p. 9) had also put it forward. Nickles, in turn, credits G. Ryle (*The Concept of Mind* [Chicago: Univ. of Chicago Press, 1949], pp. 303–304). Cf. also the references in n. 48.

50. A possible exception is Langley et al., *Scientific Discovery*.

51. Quoted in J. B. Conant, "The Overthrow of the Phlogiston Theory," in J. B. Conant and L. K. Nash (eds.), *Case Histories in Experimental Science* (Cambridge, Mass.: Harvard Univ. Press, 1957), p. 13. Cf. also Priestley's remark that "[i]t was the *great discovery* of STAHL, that . . . [phlogiston] is transferrable from one substance to another, how different soever in their other properties, such as sulphur, wood, and all the metals, and therefore is the same thing in them all" (emphasis added). J. Priestley, "Experiments Relating to Phlogiston, and the Seeming Conversion of Water into Air," *Philosophical Transactions of the Royal Society*, 73 (1783): 398–434, on p. 399.

tive, evaluative dimension to discovery. Only beliefs that have remained immune to revision can be designated with that term.

Despite the critical remarks that have been raised against the distinction,
one can still distinguish between the original historical mode of hypothesis
generation and the "final" form of justification. These two aspects of the discovery process need not coincide. The actual path that led to the hypothesis
(theory) in question might be "edited" out of the presentation of the hypothesis before the community.[52] Furthermore, justification itself is a constantly
evolving process: it is rarely the case that the justification of a hypothesis re-
tains its original form. As science develops, the justification of scientific beliefs
undergoes continuous reconstruction.[53] Thus, the distinction becomes a tem-
poral one, as opposed to a logical one, between two aspects of the discovery
process.

This brings me to the application of the term "discovery." "Discovery"
should not be confused with either "generation" or "construction." Even
though the terms "generation" and "construction" (or "extended generation")
do not preclude that the outcome of the corresponding processes is a true
statement about nature, they do not imply it either. Furthermore, they carry
the connotations of "creation"; with construction something comes into being as a result of human action. "Discovery," on the other hand, implies truth.
Moreover, it carries the connotations of "revelation"; some truth about nature is disclosed to a passive intellect.[54] It should be noted that by using the
term "construction" I do not thereby commit myself to the currently fash-
ionable view that scientific facts are socially constructed. Much of the work
carried out under this approach is strongly relativist and antirealist. While this
is not the place to take up the challenge posed by the contemporary sociology
of scientific practice, I should point out that viewing science as a constructive
activity does not necessarily carry relativist or antirealist implications. The
neutrality of such a view as regards the issue of relativism is shown by the fact
that one might be able to specify canons of sound construction that would
transcend the local practices of particular scientific groups. Moreover, it is

---

52. It is interesting that this was also Reichenbach's original notion of justification. See
his *Experience and Prediction*, p. 6.

53. The function and importance of reconstruction in science have been emphasized
by T. Nickles in his "Justification and Experiment," in D. Gooding, T. Pinch, and S. Schaffer
(eds.), *The Uses of Experiment: Studies in the Natural Sciences* (Cambridge: Cambridge Univ.
Press, 1989), pp. 299–333.

54. Cf. J. Stachel, "Scientific Discoveries as Historical Artifacts," in Gavroglu et al. (eds.),
*Trends in the Historiography of Science*, pp. 139–148; and K. L. Caneva, *The Form and Function of
Scientific Discoveries*, Dibner Library Lecture Series (Washington, D.C.: Smithsonian Institu-
tion Libraries, 2001).

conceivable that one could come along and show that sound constructions result in genuine facts about nature. This possibility shows that constructionism, properly understood, is neutral with respect to the realism debate.[55]

In view of the distinction between discovery and construction, I would like to revise and extend the tentative classification of scientific discoveries that I offered above. Some of those discoveries (D) will be reclassified as constructions (C) or inventions (I). The utility of this revision will become apparent below. In the domain of application of the term "discovery" I will include individual observable entities (e.g., Neptune), observable natural kinds (e.g., tigers), and phenomena (e.g., the Zeeman effect). The term "construction" will apply to problems, solutions to problems, theories, theoretical entities (e.g., the representation of the electron), principles (e.g., energy conservation), and representations of unobservable properties (e.g., electron spin). Finally, I will use the term "invention" to characterize the development of novel theoretical and experimental techniques, and the creation of new instruments. All these different kinds of discovery and construction are interrelated. For instance, the construction of a problem (e.g., the incompatibility of two established scientific theories) might result in the construction of a new theory that will resolve the problem in question.

It is worth exploring the similarities and differences between these kinds of discoveries and constructions. The aim would be to determine whether the generative and justificatory procedures are similar in all cases and whether different kinds of discoveries are valued differently by the scientific community, that is, whether they are assigned a different social status. It seems to me that the items in the above classification differ from each other in important respects. For instance, discovery cannot be identified with problem solving. The proposal of, say, phlogiston solved several eighteenth-century chemical problems, but it does not count as a discovery. In what follows I will focus on problems, solutions, phenomena, and theoretical entities, simply because these are the historiographical categories that are significant for the historical episodes that will be reconstructed in the chapters to follow.

In regard to problems, what exactly is involved in the claim that problems are constructed? By "constructing a problem" I mean articulating some difficulties (hitherto unrecognized) in the established body of scientific knowledge. The nature of these difficulties has been already discussed in the above section on "problem situations." The construction of a solution, on the other hand,

---

55. The connections between the discovery issue and the realism debate will be explicitly drawn below.

amounts to devising an outcome (e.g., a representation) that satisfies the constraints imposed by the problem. This outcome need not be a true proposition or set of true propositions and, thus, should not be conflated with the outcome of a discovery process. When we proceed to phenomena, their discovery involves the following circumstances: the observation of a novel situation and the construction of an argument that the observations obtained are not artifacts of the apparatus employed and that all perturbing factors ("noise") have been eliminated. Furthermore, the validity of the argument in question must not be affected by subsequent theoretical and experimental developments.[56] Finally, the "discovery" of unobservable entities can be seen as the first stage of the construction of their representation. During that stage scientists construct a representation of a novel entity, attempting to resolve particular (empirical or conceptual) problems. If the emerging representation provides an adequate solution to those problems, then this is taken as an indication that the corresponding entity exists. Thus, the "discovery" of an unobservable entity and the early phase of the construction of its representation are two aspects of a single process and cannot be sharply distinguished. The "discovery" ends when the "discoverers" persuade the rest of the community that the entity in question is real. Only in this qualified sense can one claim that an unobservable entity was discovered. I defer a detailed discussion of this case to the following section.

Further, the widespread view that a discovery is an isolated event that can be credited to a single individual is misconceived, at least in those cases that concern me here. Both the discovery of phenomena and the discovery of unobservable entities involve many complex tasks and, thus, cannot take place at a single moment. Furthermore, the discovery of unobservable entities is rarely the accomplishment of a single individual. These are Kuhnian insights and they are reinforced by the realization that the context of discovery comprises both the context of generation and the context of justification.[57] I think, however, that Kuhn's claim that discoveries of phenomena, which could not be predicted from accepted theory, cannot be attributed to particular individuals is not, in general, true. The criteria that enable us to claim that such a discovery has been accomplished are the criteria that are involved in judg-

56. For further discussion I refer the reader to my "On the Inextricability of the Context of Discovery and the Context of Justification," in J. Schickore and F. Steinle (eds.), *Revisiting Discovery and Justification,* preprint 211 (Berlin: Max Planck Institute for the History of Science, 2002).

57. See T. S. Kuhn, *The Structure of Scientific Revolutions,* 2nd ed. (Chicago: Univ. of Chicago Press, 1970), pp. 52–65; and *The Essential Tension,* pp. 165–177.

ing the reliability of the experiment that exhibits the new phenomenon. Regardless of whether the phenomenon can be given a theoretical explanation, the experimental result, along with the demonstration of its validity (usually based on experimental background knowledge) constitute the occurrence of a discovery. Both of these achievements might be the product of a single scientist.

One might not want to use the term "discovery" to characterize the products of scientific activity, but undoubtedly discovery occupies a central place in the scientists' own image of their enterprise. Discoveries are seen as the units of scientific progress and are accordingly valued. Usually they are constructed in the light of knowledge that was not available to the historical actors at the time when the presumed discovery took place and are intimately tied to the reward structure of science. They tend to be post hoc reconstructions of specific episodes, whose aim is to propagate and reward certain practices and beliefs that are deemed significant for contemporary scientific activity.[58] The "discovery" of the electron provides a good example of what I have in mind. It is a discovery that supposedly took place in 1897 and was the exclusive achievement of J. J. Thomson. Neither of these claims can stand historical scrutiny (see chapters 3 and 4). This poses an interesting historiographical problem regarding the aims and function of the retrospective construction of that discovery. This problem, in turn, suggests that the construction and continuous reconstruction of scientific discoveries can be fruitfully studied from a sociological perspective.[59] The study of discovery transcends both the psychological exploration of scientific creativity and the philosophical analysis of scientific justification, since in many cases discoveries serve specific purposes within the scientific community. The psychological, philosophical, and sociological approaches to the study of discovery are complementary and should not be undertaken at the expense of one another.[60]

In what follows I concentrate on the "discovery" of unobservable entities. The historiographical and philosophical issues associated with such discoveries are discussed in a relatively abstract fashion. Concrete illustrations are provided in chapters 3 and 4, where I examine in some historical detail the discovery of the electron.

58. Cf. S. Schaffer, "Scientific Discoveries and the End of Natural Philosophy," *Social Studies of Science,* 16 (1986): 387–420.

59. Cf. A. Brannigan, *The Social Basis of Scientific Discoveries* (New York: Cambridge Univ. Press, 1981); Schaffer, "Scientific Discoveries"; and Caneva, *The Form and Function of Scientific Discoveries.*

60. Cf. Nersessian, *Opening the Black Box.*

## 4. Several Approaches to the Discovery of Unobservable Entities: A Taxonomy and Critique

In order to identify an event or a process as the discovery of an unobservable entity, it is necessary to specify some criteria for what constitutes a discovery of this kind.[61] Several possible stances on the problem of what constitutes a discovery can be adopted. The first two possibilities depend on the position that one favors in the debate on scientific realism, a salient aspect of which concerns the grounds that we have for believing in the reality of the unobservable entities postulated by science (electrons, protons, fields, etc.). First, one might favor an antirealist perspective, that is, maintain that one has to be at least agnostic with respect to the existence of unobservable entities. From such a point of view discoveries of unobservables never take place. To quote from an eminent contemporary representative of this approach, "[S]cientific activity is one of construction rather than discovery: construction of models that must be adequate to the phenomena, and not discovery of truth concerning the unobservable."[62] According to this stance, "discovery" has nothing to do with truth. Rather, it is a process of constructing empirically adequate models. The unobservable entity is a convenient fiction. To put it in terms of the discovery-justification distinction, existence claims concerning the unobservable can never be sufficiently justified. This view of scientific activity is compatible with (but not necessary for) the approach to the issue of scientific discovery that I favor; but more on this below.

One might adopt a second (realist) stance, that is, propose certain epistemological criteria whose satisfaction would provide adequate grounds for believing in the existence of an unobservable entity. From this point of view a discovery takes place when an individual or a group has managed to meet the required criteria. As an example consider Ian Hacking's proposal that a belief in the reality of an, in principle, unobservable entity is justified to the extent

---

61. Even though I do not believe that unobservable entities are discovered, in the traditional sense of the term "discovery," I will continue to use this term in this section for two reasons. First, it is used by the proponents of views that I will be arguing against. Only after having argued against those views might I be justified in dropping that term. Second, as I have already indicated, the term might still be used to capture a distinction between two stages. The first stage is usually characterized by ontological debates, where the existence of an entity is contended. After that stage is over, a realist might claim that the entity has been discovered (i.e., that we know that it exists).

62. B. C. van Fraassen, *The Scientific Image* (New York: Oxford Univ. Press, 1980), p. 5.

that the entity in question can be manipulated.[63] It follows then that an unobservable entity has been discovered only if a scientist has found a way to manipulate this entity. Justification is considered an essential aspect of the discovery process and is identified with manipulability.

It is evident that the adequacy of the proposed way for deciding when something qualifies as a genuine discovery depends on the adequacy of the epistemological criteria for what constitutes unobservable reality. Any difficulties that might plague the latter would cast doubt on the adequacy of the former. Although this approach can be, in principle, realized, no adequate proposal of the kind outlined has been made so far. That is, no epistemological criteria have been formulated whose satisfaction would amount to an existence proof of an unobservable entity. In particular, Hacking's proposal that manipulability provides such a proof leaves much to be desired (see chapter 9). The merits and limitations of Hacking's view with respect to the discovery of the electron will be discussed in chapter 3.

If these two possibilities capture the philosopher's main stances toward the issue of scientific discovery, a third possibility captures the scientist's perspective.[64] The recipe in this case involves two steps. First, one identifies the most central aspects of the modern concept associated with the particular entity in question. Second, one looks at the historical record and tries to identify the scientist who first articulated those salient aspects and who, furthermore, provided an experimental demonstration of the validity of his conception. This person (or persons) would then qualify as the discoverer of the given entity. However, there is a serious difficulty—what I will call the *problem of knowledge*—which undermines this approach. Kuhn formulated this problem very succinctly, in relation to the discovery of oxygen: "Apparently to discover something one must also be aware of the discovery and know as well what it is that one has discovered. But, that being the case, how much must one know?"[65] Any entity that forms part of the accepted ontology of contemporary science is endowed with several properties. The electron, for instance, has a given mass, a certain charge, an intrinsic magnetic disposition (spin), a dual nature (particle versus wave), and many other features. The question then arises, How many properties must one have discovered in order to be granted the status of the discoverer of the entity in question?

---

63. See I. Hacking, *Representing and Intervening* (Cambridge: Cambridge Univ. Press, 1983), esp. pp. 262–266.

64. This perspective usually characterizes scientists who write retrospective accounts of scientific discoveries.

65. See Kuhn, *The Essential Tension*, p. 170.

Another aspect of the problem of knowledge concerns mistaken beliefs. If knowing what one has discovered is a prerequisite for being credited with the discovery, then can one be considered the discoverer of, for example, the electron even though he entertained wrong beliefs about it? For instance, in 1897 J. J. Thomson thought of his corpuscle, an entity later identified with the electron, as a structure in the ether. Leading physicists at the time (e.g., Joseph Larmor and H. A. Lorentz) entertained similar "wrong" conceptions of the electron. To put the problem in terms of the discovery-justification distinction, how many beliefs about an entity should be justified for the entity in question to be discovered? Until the proponents of this approach manage to tackle these two aspects of the problem of knowledge, this approach will remain indefensible. In chapter 3 I will discuss the concrete manifestations of this approach as revealed in some of the historiography of the discovery of the electron.

A related problem, faced by the "friends of discovery," is the *problem of identification*.[66] If most, or even some, of the beliefs that the putative "discoverer" had about the "discovered" entity are not true, it is not at all evident that the entity in question is the same as its contemporary counterpart. It has to be shown, for instance, that J. J. Thomson's corpuscles, which were conceived as classical particles and structures in the ether, can be identified with contemporary electrons, which are endowed with quantum numbers, wave-particle duality, indeterminate position-momentum, and so on. The "friends of discovery" should propose some criteria that enable us to identify the original entity with its present counterpart. In the next section and in chapter 9 I will attempt to come to terms with the problem of identification, for reasons which are not directly related to scientific discovery. Nevertheless, I will sketch below a more neutral approach to scientific discovery, which has the advantage of avoiding that problem altogether.

Because of these problems, I would be extremely reluctant to base a historical narrative about an unobservable entity on the traditional, realist notion of scientific discovery. Rather than trying to resolve these problems, there is another way to approach discovery episodes that avoids philosophical pitfalls. One should simply try to historicize the notion of scientific discovery, by adopting the perspective of the relevant historical actors, without worrying whether that perspective can be justified philosophically.[67] According to this final approach—and the one I favor—the discovery of an entity amounts

---

66. I owe the expression "friends of discovery" to Tom Nickles.

67. Cf. Arthur Fine's "Natural Ontological Attitude" in his *The Shaky Game: Einstein, Realism, and the Quantum Theory* (Chicago: Univ. of Chicago Press, 1986).

to the formation of consensus within the scientific community about its existence.[68] Given the realist connotations of the term "discovery," one might even avoid using it when writing the history of a concept denoting an unobservable entity. In undertaking such a task, one would show how the given concept was introduced into the scientific literature and would reconstruct the experimental and theoretical arguments that were given in favor of the existence of the entity it denoted. The next step would be to trace the developmental process that followed this initial stage and gradually transformed the concept in question. The evolution of any such concept resembles a process of gradual construction that takes place in several stages and, thus, can be periodized.[69] A realist might want to label the first stage of that process "the stage of discovery," but this would make no difference whatsoever with respect to the adequacy of the historical reconstruction. Only in this weak sense can the term "discovery" be used with respect to unobservable entities. In its stronger form (i.e., as implying existence) two further conditions are required. First, the consensus with respect to the reality of the entity in question should be maintained to this very day. If the consensus has changed and the entity has been eliminated from the ontology of science, the proper conclusion would be that it had never been discovered.[70] Second, one should propose some criteria that enable us to identify the original entity with its present counterpart.

This historicist approach is by no means novel, and many historians of science would subscribe to it. As a result of considerable discussion in science studies about "scientific discovery" as a historiographical category, it is widely accepted that scientific discoveries are not straightforward revelations of a preexisting natural order, but reflect "the local practices of contemporary research communities."[71] This shift in attitudes toward scientific discovery can be seen in recent historical scholarship. Consider, for example, the histo-

---

68. Note that here I am not referring to retrospective community judgments regarding the identity of the putative discoverer and the nature of his or her achievement. These judgments are usually at odds with historical reality.

69. Cf. I. Hacking, *The Social Construction of What?* (Cambridge, Mass.: Harvard Univ. Press, 1999), p. 50.

70. This condition, which derives from the retrospective character of scientific discovery, takes care of the worry that on my "account one might say, for example, that the ether was discovered in the early nineteenth century, only to have been undiscovered sometime around 1900." J. Z. Buchwald and A. Warwick, "Introduction," in Buchwald and Warwick (eds.), *Histories of the Electron: The Birth of Microphysics* (Cambridge, Mass.: MIT Press, 2001), pp. 1–17, on p. 6.

71. Schaffer, "Scientific Discoveries." The quote is from the abstract.

riography of energy conservation. Several parallel developments, from the study of steam engines to theoretical mechanics to physiology, contributed to the formulation of that principle. Until recently, historians portrayed those developments as "simultaneous discoveries." This view, however, has been plausibly challenged, because all those putative discoverers of energy conservation were concerned with different problems and came up with different theoretical hypotheses. It was only in the 1850s that those different approaches were retrospectively interpreted as aspects of the same discovery.[72]

However, the scope of the approach I favor, despite its similarities with recent treatments of scientific discovery, is more restricted. Whereas the intricacies of the debate on scientific realism and related philosophical issues suggest that an agnostic perspective is best suited for reconstructing the "discovery" of unobservable entities, it is not thereby implied that the discovery of observable entities and phenomena should be treated in a similar agnostic fashion. In this case the traditional category of discovery might be retained. It might be possible to specify when, say, a new planet has been discovered, without relying on the notion of consensus within the relevant scientific community.

My appeal to the consensus of the scientific community is likely to be interpreted as a social constructionist position.[73] The approach I outlined is constructionist, in that it incorporates a view of concept formation as a gradual process of construction. However, it is not social constructionist in the usual sense, according to which the emergence of consensus is the outcome of professional interests, the distribution of power within the scientific community, and so on. Rather, I think that this process of consensus formation is usually driven by good, epistemic reasons. Thus, the aim of my approach is to steer through the dire straits of the realism-constructionism debate and make possible historical narratives that would be acceptable to audiences of different philosophical persuasions, realists and antirealists alike.[74]

Finally, the approach advocated here does not run a risk associated with more sweeping historicist claims. A reconstruction of past scientific developments in terms of criteria taken from the historical actors themselves turns out to be problematic when the criteria in question are not universally

72. See C. Smith, *The Science of Energy: A Cultural History of Energy Physics in Victorian Britain* (Chicago: Univ. of Chicago Press, 1999).

73. See, e.g., P. Achinstein, "Who Really Discovered the Electron?" in Buchwald and Warwick (eds.), *Histories of the Electron*, pp. 403–424, on p. 410.

74. Cf. M. Friedman, "On the Sociology of Scientific Knowledge and Its Philosophical Agenda," *Studies in History and Philosophy of Science*, 29 (1998): 239–271.

shared.[75] If, for instance, the actors involved in an ontological debate, where the existence of an entity is contended, accept different existence proofs, then it is impossible to offer an actor-oriented reconstruction of that debate "without making some actors' voices inaudible."[76] This difficulty does not plague the approach I recommend, which regards consensus formation as an essential aspect of scientific discovery. No episode where controversy persists can be interpreted as constituting a discovery.

Chapters 3 and 4 will provide concrete illustrations of the theoretical issues discussed so far and will attempt a revisionist account of the "discovery" of the electron.

## 5. Scientific Realism: The Charybdis of Meaning Change

If the above section achieved its purpose, it should be evident that one cannot adequately discuss the issue of scientific discovery independently of the realism debate.[77] This is a quite complex debate that has taken place along several dimensions. In this book I will be exclusively concerned with the phenomenon of meaning change and its alleged subversion of a realist position. When Kuhn and Paul Feyerabend pointed out in the early 1960s that the meaning of scientific terms changes over time, it seemed that the instability of scientific concepts had as an immediate corollary the collapse of scientific realism. Given the then prevalent belief (inspired by Gottlob Frege) that the meaning of a term is specified by a set of conditions that are necessary and sufficient for the correct application of that term, the slightest change in those conditions (meaning change) would imply that the term as previously used was vacuous; that is, it referred to nothing at all.[78]

Hilary Putnam, among others, was alarmed by the antirealist implications of that view of meaning and developed an alternative theory whose purpose was to sustain a central realist intuition, namely, that the development of scientific knowledge amounts to learning more and more about the *same* things.

75. This difficulty has been pointed out by Steven Shapin. See his "Discipline and Bounding: The History and Sociology of Science as Seen Through the Externalism-Internalism Debate," *History of Science,* 30 (1992): 333–369, esp. pp. 353–354. Despite his warnings against an overenthusiastic historicism, however, Shapin espouses a moderately historicist outlook.

76. Ibid., p. 353.

77. This section is extremely schematic and merely aims at introducing the problem of the implications of meaning change for scientific realism. A detailed analysis of that problem is given in chapter 9.

78. Hence Kuhn's recent remark that "the history of science is the history of developing vacuity." T. S. Kuhn, "Possible Worlds in History of Science," in Sture Allén (ed.), *Possible Worlds in Humanities, Arts and Sciences* (Berlin: Walter de Gruyter, 1989), pp. 9–32, on p. 32.

According to Putnam's theory the reference of a term is an independent component of the term's meaning and is not affected by any changes the other components might undergo. In order for his theory to get off the ground there must be a way to fix a term's referent without relying on the descriptive part of the term's meaning. This limitation makes his conception of meaning applicable only to proper names that denote individual observable objects (e.g., the Statue of Liberty) and, perhaps, to natural kind terms whose referents are classes of observable objects (e.g., cats). On the other hand, it is not clear how this view of meaning would handle terms that denote unobservable entities (e.g., quarks), since we have no access to those entities independently of the descriptions provided by our theories (for details see chapter 9).

Ian Hacking tried to remedy this defect of Putnam's theory and make it a potent tool for the construction of a realist epistemology concerning those unobservable entities which can be manipulated in the laboratory. Hacking's view will be examined and partially rejected in chapter 9. I think, however, that the Putnam-Hacking approach toward meaning and realism is promising and deserves to be further developed. In particular, there is a way to separate the referential component of meaning from the rest even for the terms, or at least some of the terms, that denote unobservable entities. When presented with such a term, one has to identify the experimental situations that are taken to manifest the presence of (are causally attributed to) its referent, the corresponding unobservable entity. Thus, the experimental situations associated with a term provide a way to track its referent.[79]

As I will show in chapter 4, a great deal of theoretical and experimental knowledge is involved in interpreting an experimental situation as the manifestation of a particular entity. What binds all those, prima facie, different situations together and why are they attributed to the same entity? I can think of two possibilities. First, those situations may share some common feature, which indicates that a single entity is present in all of them. For example, it had been known since the seventeenth century that the processes of combustion, calcination (oxidation), and respiration took place only in the presence of atmospheric air. That fact was explained by the hypothesis that air was necessary for absorbing phlogiston, the entity that was given off in all those processes. Second, from the quantitative features of an experimental situation it

---

79. This proposal is not inspired by a belief that unobservable entities are accessed more readily in an experimental context. In fact, I do not believe this to be the case. As elsewhere in the book, my aim is to provide a viable option to the realist without, however, claiming that the option in question should be adopted by the antirealist.

may be possible to infer the value of some property of the entity involved in it. If that value turns out to be the same in different experimental situations, then this agreement binds them together as effects of the same entity. Furthermore, the value in question provides a way of identifying the entity in novel experimental situations. For example, at the turn of the nineteenth century several experimental phenomena (the Zeeman effect, cathode rays, β-rays, etc.) were attributed to the presence and action of hypothetical charged particles. It was not a priori evident that the particles involved in all those different phenomena were the same. Physicists were led to that conclusion when it turned out that the particles that were responsible for the Zeeman effect had approximately the same charge-to-mass ratio with the particles which constituted cathode rays. From that point on, the charge-to-mass ratio of the electron functioned as a criterion for identifying it in novel experimental situations (see chapter 4 for details).

To be a realist about a particular entity one has to establish the referential continuity of the corresponding term. That is, one has to show that over the history of the term's meaning the experimental situations that were assumed to be the observable manifestations of its referent have remained stable or, at least, exhibited a cumulative development (i.e., the previous set of situations associated with the term was a subset of the current set). This expansion may happen in three ways. First, new situations may get attributed to a familiar entity, without any modification in the theoretical description of that entity. In that case the meaning of the corresponding term will remain unaffected. Second, the accommodation of new situations as manifestations of a familiar entity may require the attribution of additional properties to that entity, which, moreover, do not conflict with its previously established properties. In that case, the meaning of the term will expand in a cumulative way and its reference will remain the same. Third, the accommodation of novel situations may require the attribution of novel properties to the entity, which, however, are incompatible with some of its previously accepted properties. The latter will, in the process, get rejected. In that case, the meaning of the term will change in a noncumulative fashion. Despite that change, the term may still refer to the same entity, provided that the experimental situations previously associated with the term continue to be attributed to its referent.

Thus, the question whether realism is a defensible attitude toward a certain entity requires a historical reconstruction of the evolution of the corresponding term's meaning (which requires considering the experimental situations that track its reference). Unless the historical and philosophical community undertakes successfully the enormous historiographical task of showing the referential continuity of most scientific terms, any realist position will make sense only with respect to specific entities (local realism).

As I will show toward the end of the book, with respect to the concept of the electron one can provide a realist reading of its historical development from the late nineteenth century till the mid-1920s. Despite the considerable change that the concept underwent during that period, the reference of the term "electron" remained stable. In other words, more and more experimental situations came to be interpreted as the observable manifestations of the presence and action of electrons (e.g., cathode rays, the Zeeman effect, β-rays, the photoelectric effect, thermionic emission, cloud chamber tracks, and Robert Millikan's oil-drop experiments). After the "discovery" of the electron, there was never a case where a certain experimental situation that had been previously thought to manifest the presence of electrons ceased to be so regarded. I want to stress, however, that referential continuity of this sort is not a sufficient condition for a realist position about the electron, but only a necessary one. One cannot exclude the possibility of an alternative theoretical account of the experimental situations in question that would dispense with electrons.

## Chapter 2 | Why Write Biographies of Theoretical Entities?

The issue of realism also comes up in relation with a methodological approach that I developed in the course of writing this book. This approach amounts to considering theoretical entities as active agents whose internal dynamic transcends the beliefs, abilities, and wishes of human actors and acts as a constraint on the development of scientific knowledge. Even though they are the products of scientific construction, they have a certain independence from the intentions of their makers; that is, they have a life of their own.[1]

As I mentioned in the introduction, I use the term "theoretical entity" as a shorthand expression for "representation of an unobservable entity." In particular, with the phrase "the electron qua theoretical entity" I refer to the representation of the electron. So when I stress the agency of theoretical entities, I locate it in *the representations* of the corresponding unobservable entities. That agency becomes manifest in several ways, both negative and positive.[2] First, at every stage of the career of theoretical entities, the scientists who are using them as theoretical tools represent the corresponding unobservable entities as having definite properties and obeying certain laws. As a result, theoretical

---

1. Cf. Popper's remark: "As with our children, so with our theories, and ultimately with all the work we do: our products become largely independent of their makers. We may gain more knowledge from our children or from our theories than we ever imparted to them." K. R. Popper, *Unended Quest: An Intellectual Autobiography* (La Salle, Ill.: Open Court, 1976), p. 196.

2. As I pointed out in the introduction, "agent" and "agency" need not be associated with intentional action.

entities resist all attempts at theoretical manipulation, for example, attempts to enrich them in order to account for novel phenomena, whose (usually unintended) consequence is the negation of some of those properties and the violation of some of those laws.[3] For example, before the development of the old quantum theory of the atom, electrons were represented as charged particles that obey the laws of classical electrodynamics. Confronted with the experimental results of Hans Geiger and Ernest Marsden on the scattering of α-particles from matter, Rutherford proposed a nuclear model of the atom, which portrayed electrons in constant circular motion around a nucleus. From my perspective, Rutherford attempted to manipulate the electron qua theoretical entity, that is, to enrich the representation of the electron in order to accommodate the novel results. The electron qua theoretical entity, however, resisted his attempt to manipulate it. According to the laws of classical electrodynamics, any accelerated charge would emit radiation. Thus, an electron orbiting the nucleus would radiate its energy and spiral into the nucleus (remember that circular motion is accelerated motion). On the other hand, the available experimental information from spectroscopy, which also constrained the representation of the electron's behavior within the atom, indicated that no such collapse of the electron into the nucleus takes place. So, given that experimental constraint, Rutherford's attempt to enrich the representation of the electron led to a violation of the classical laws that were supposed to govern its behavior. Bohr's proposal of the old quantum theory of the atom in 1913 was, in part, an attempt to come to terms with this paradox (see chapter 5).[4]

As indicated by the above example, the *recalcitrance of theoretical entities* is partly due to constraints arising from experiment.[5] Theoretical entities embody, appropriately translated, quantitative and qualitative features of the experimental situations that are deemed to be manifestations of the corresponding unobservable entities. In that sense, theoretical entities are constructions from experimental data.[6] After the initial stage of their construction, their

3. Cf. Gérard Jorland's comment that "ideas are objects that one cannot manipulate at will." G. Jorland, "The Coming into Being and Passing Away of Value Theories in Economics (1776–1976)," in L. Daston (ed.), *Biographies of Scientific Objects* (Chicago: Univ. of Chicago Press, 2000), pp. 117–131, on p. 128.

4. Here I am indebted to the analysis of K. Gavroglu and Y. Goudaroulis, *Methodological Aspects of the Development of Low Temperature Physics: Concepts out of Contexts* (Dordrecht: Kluwer, 1989), pp. 17, 25–30.

5. I owe the expression "recalcitrance of theoretical entities" to Simon Schaffer.

6. Cf. Hanson's remark that "[t]he idea of . . . atomic particles is a conceptual construction 'backwards' from what we observe in the large." N. R. Hanson, *The Concept of the Positron: A Philosophical Analysis* (Cambridge: Cambridge Univ. Press, 1963), p. 47.

further development is constrained by the experimental information built into them, in two ways. First, the experimentally inferred characteristics of an entity usually provide a stable backdrop for further investigations of its nature. Although the interpretation of those characteristics is theory dependent, their quantitative magnitude transcends any particular theory. For example, at the end of the nineteenth century quantitative information from certain experimental situations—for example, the magnitude of the magnetic splitting of spectral lines—was used to determine the charge-to-mass ratio of the electron and was, thereby, incorporated in its representation. From that point on, the specification of other properties of the electron—say, its size—conformed to its previously established charge-to-mass ratio (see chapter 4). Even though the interpretation of "charge" and "mass" varied considerably according to the specific theoretical context (e.g., J. J. Thomson's "charge," circa 1897, and Bohr's "charge," circa 1913, were understood very differently), the value of $e/m$ remained relatively stable across changes in theoretical perspective.

Second, when an experimental situation is attributed to an unobservable entity, some elements of the former have counterparts in the representation of the latter. To put it another way, observable features of the experimental situation are correlated with aspects of the putative behavior of the entity. Some correlation of this kind has to obtain, as long as the given situation is attributed to the entity in question. This requirement constrains and guides any changes the representation of the entity may undergo. For example, in Lorentz's theory of electromagnetic phenomena various characteristics of spectral lines (their frequency, intensity, and polarization) had counterparts in the representation of the electron (its frequency, amplitude, and direction of vibration respectively). On the other hand, in Bohr's 1913 theory of atomic structure only one feature of spectral lines (their frequency) could be correlated with the behavior of the electron (its transitions between different energy levels). In the subsequent development of the old quantum theory, however, the other characteristics of spectral lines were also linked with the quantum properties and behavior of the electron (see chapters 5, 6, and 8).[7]

Toward the end of that theory several physicists (most notably Wolfgang Pauli and Werner Heisenberg) and chemists (e.g., G. N. Lewis) argued that a

7. Cf. P. Galison, "Context and Constraints," in J. Z. Buchwald (ed.), *Scientific Practice* (Chicago: Univ. of Chicago Press, 1995), pp. 13–41; and A. Pickering, "Beyond Constraint: The Temporality of Practice and the Historicity of Knowledge," in Buchwald (ed.), *Scientific Practice,* pp. 42–55. Note that Galison and Pickering understand constraints very differently. According to Pickering, constraints emerge during scientific practice, whereas for Galison constraints are preexisting determinants of practice. My own view, supported by several cases discussed in this book, is closer to Galison's.

theoretical entity should not contain any superfluous features, that is, features with no observable counterparts. Read from my perspective, those scientists aimed at the construction of theoretical entities exclusively from empirical material. Their positivistic agenda was the main motivation behind the overthrow of the idea of electronic orbits.[8] Those orbits were considered unobservable, in the sense that they did not give rise to any observable effects. Because of their unobservability they were deemed redundant elements of the electron's representation and were, therefore, banished. Heisenberg's matrix mechanics and Pauli's representation of the electron dispensed with those elements.[9]

Whether or not the aims and agenda of the creators of quantum mechanics were justified, it is clear that experimental data do not uniquely determine the representations of unobservable entities, despite being embodied in them. We have here a typical underdetermination problem. That is, one cannot exclude the possibility that several incompatible representations can be constructed from the same experimental data. For instance, from 1916 till 1926 the fine structure of the hydrogen spectral lines was believed to reflect the relativistic behavior of the electron. In 1926 an alternative interpretation, which explained the phenomenon as the observable manifestation of spin and the quantum-mechanical behavior of the electron, was proposed. Those two different mechanisms of producing the phenomenon of fine splitting were quantitatively equivalent; that is, they led to the same predictions for the electron's energy levels. To put it another way, the same phenomenon was correlated at different times with different aspects of the electron's putative properties and behavior. Both of those aspects, as a result of "one of the fortunate coincidences in the history of physics,"[10] were qualitatively and quantitatively suited to the description of the phenomenon.

Notwithstanding the problem of underdetermination, experimental results, interpreted in terms of an unobservable entity, pose constraints on any attempt to manipulate its representation. Any attempt to attribute additional properties to the entity or modify the theoretical account of its behavior is

8. The origins of this agenda have not been adequately explored. It seems clear, however, that it was inspired by a positivistic outlook. For instance, Pauli's insistence on the elimination of unobservable elements from physical theory was probably stimulated by the philosophy of his godfather, Ernst Mach.

9. Cf. M. Beller, "Matrix Theory before Schrödinger: Philosophy, Problems, Consequences," *Isis*, 74 (1983): 469–491.

10. R. Kronig, "The Turning Point," in M. Fierz and V. F. Weisskopf (eds.), *Theoretical Physics in the Twentieth Century: A Memorial Volume to Wolfgang Pauli* (New York: Interscience Publishers, 1960), pp. 5–39, on p. 8.

subject to those constraints. If they are violated, some further adjustment is necessary so as to restore the coherence of the entity's representation.

Several other examples that exhibit the recalcitrance of theoretical entities will be developed in the course of this book: First, the introduction of quantum numbers (in addition to the one proposed by Bohr in 1913) had the unintended consequence that the allowed number of quantum transitions was multiplied beyond the number that was dictated by spectral data. In other words, the empirical constraints on the representation of the behavior of the electron within the atom had been violated. The subsequent proposal and theoretical justification of selection rules by means of the correspondence principle was an attempt to restrict the theorized freedom of the electron so as to obey those constraints (see chapter 6). Second, when Samuel Goudsmit and George Uhlenbeck portrayed the electron as a tiny magnet whose magnetic disposition was a result of its internal rotation, it turned out that any point on its surface would travel with a speed greater than the velocity of light. Indeed, this undesirable consequence, among other reasons, had prevented Ralph Kronig, who had also developed a similar model of the electron, from publishing his results. This paradox was solved later by denying a presupposition of the original spin concept, namely, that "spin" represented a literal rotation of the electron; instead it was interpreted as a quantum-mechanical property that had no classical correlate. Thus, the fact that the electron was supposed to obey the laws of special relativity functioned as a heuristic constraint on the development of quantum theory (see chapter 8).

Theoretical entities are active in a second, positive sense. They are usually proposed for specific theoretical purposes, and the properties of the corresponding unobservable entities are specified to the extent required by these purposes. However, as soon as they are introduced, further questions about the unobservable entities in question arise and provide heuristic guidance to the practicing scientist. For instance, when the electron became part of the scientific worldview, several of its properties were not explicitly specified. In particular, its charge-to-mass ratio and its dimensions were left unspecified. Several questions thus arose that were prompted by the incomplete description of the electron. In that sense, the electron qua theoretical entity "posed" questions about the "nature" of its unobservable counterpart. There was pressure to answer such questions only when they were implicated in the description of an experimental situation. The charge-to-mass ratio of the electron, for instance, was determined only after Zeeman's experiments on the magnetic splitting of spectral lines. Before those experiments there was no reason to specify that property because it was not implicated in any of the phenomena that contemporary theories of electrons were constructed to explain. Zeeman's attempt to estimate that property led, in turn, to the refinement of his original experimental discovery (for details see chapter 4).

Questions about the electron's dimensions exerted a formative influence on the development of physics. Classical electrodynamics precluded the electron from being a point particle. A point particle would have an infinite self-energy (a clearly absurd consequence). Thus, it followed that the electron was an extended particle. The question then arose, in the context of Lorentz's 1904 electromagnetic theory, What keeps it together, given that its negatively charged parts would pull it apart? To overcome this problem, Henri Poincaré proposed in 1905–1906 a cohesive force whose magnitude balanced those repulsive forces. However, as Lorentz showed in 1909, Poincaré's proposal did not suffice to explain fully the electron's stability, and some further adjustment was necessary.[11] The problem of the electron's self-energy remained unsolved for many years. Its eventual solution (renormalization) was also deplored by some physicists (e.g., Dirac).

There was another dimension to the heuristic role of the electron's representation. In certain theoretical contexts that representation contained simplified elements. When difficulties appeared, a plausible move toward their elimination was to remove those elements from it. For instance, Bohr assumed in 1913 that the electron's orbits inside the atom were circular and that its velocity was much smaller than the velocity of light. As soon as the problem of fine structure confronted the theory, a plausible problem-solving strategy was to dispense with those simplifying assumptions and see whether this realistic maneuver provided a solution to the problem. Indeed, the problem in question was resolved by developing a more accurate representation of the electron's motion within the atom, that is, by taking into account that the electronic orbits were, in general, elliptical (see chapters 5 and 6 for details).

It is these two aspects of theoretical entities—their tendency to resist manipulation and the heuristic resources they embody—that make them active participants in the development of science. Furthermore, the fact that they have an experimentally determined component, which is relatively stable, makes them independent to some extent of the vicissitudes of theory and gives them a life of their own.[12] If theoretical entities have a life of their own, then they can legitimately become the subject of biographies. A biographical

---

11. For a detailed exposition of this episode see A. I. Miller, *Albert Einstein's Special Theory of Relativity: Emergence (1905) and Early Interpretation (1905–1911)* (New York: Springer, 1998).

12. In that sense, they are "transtheoretical." Cf. D. Shapere, *Reason and the Search for Knowledge: Investigations in the Philosophy of Science,* Boston Studies in the Philosophy of Science 78 (Dordrecht: Reidel, 1984), p. 333.

approach in reconstructing their historical development follows their lifeline and highlights their active nature.[13] According to such an approach, the historian constructs narratives of past scientific episodes from the perspective of the theoretical entity in question.[14] First, he or she shows how it emerged out of conceptual and empirical problem situations and how the properties of its unobservable counterpart were determined so as to reflect experimental results. He or she then shows how the characteristics of the theoretical entity set limits on its manipulation for theoretical purposes and how they guided scientific practice. Since theoretical entities often cut across disciplinary boundaries, so will the narratives of their lives. The life of the electron qua theoretical entity, for instance, as I will show in chapter 7, cut across the boundaries of physics and chemistry.[15]

The episodes in which a theoretical entity participated can be told from its own perspective, regardless of whether it was the most important actor in them. Biographies aim at putting together a life, partly through a reconstruction of the developments in which the biographee participated. For example, any biographer of a scientist who participated in the development of quantum theory is compelled to discuss his involvement in that development. No assumptions need to be made about the significance of his contributions to the gradual construction of that theory. It does not matter, for the purposes of a scientist's biography, whether he was the most important actor in each of the episodes in which he participated. What matters is that the episodes under study affected his intellectual development and, more generally, his life.[16] In a similar fashion, a biographer of a theoretical entity has to discuss those episodes that affected its development. The aim of a narrative of this kind is to illuminate the gradual transformation of the theoretical entity, showing how that transformation was hindered or facilitated by its previously acquired character.

13. Cf. T. S. Kuhn, "Metaphor in Science," in A. Ortony (ed.), *Metaphor and Thought* (Cambridge: Cambridge Univ. Press, 1979), pp. 409–419, on p. 411. Note that Kuhn restricts the applicability of the notion of lifeline to individuals, whereas I think it is also useful for studying collective entities. For some thoughts that are relevant to this issue see D. Hull, "Historical Entities and Historical Narratives," in C. Hookway (ed.), *Minds, Machines and Evolution: Philosophical Studies* (Cambridge: Cambridge Univ. Press, 1984), pp. 17–42.

14. Cf. Rivka Feldhay's attempt to read a seventeenth-century treatise on the motion of the earth "from the perspective of the discourse of mathematical entities." R. Feldhay, "Mathematical Entities in Scientific Discourse: Paulus Guldin and his *Dissertatio De Motu Terrae*," in Daston (ed.), *Biographies of Scientific Objects*, pp. 42–66, on p. 66.

15. Cf. Hans-Jörg Rheinberger, "Cytoplasmic Particles: The Trajectory of a Scientific Object," in Daston (ed.), *Biographies of Scientific Objects*, pp. 270–294.

16. For some recent reflections on biography as a historiographical genre see M. Shortland and R. Yeo, *Telling Lives in Science: Essays on Scientific Biography* (Cambridge: Cambridge Univ. Press, 1996).

There are some interesting parallels between the history of a theoretical entity and the life of a person, which support the metaphorical use of the term "biography" in the former case. In the early stages of their history theoretical entities, like ordinary persons, are formed under various pressures from their environment—the theoretical, experimental, and sociocultural context in which they are embedded. Their capacity to resist the changes that are imposed on them is very limited, since their characteristics are not yet fixed. As they gradually reach maturity, however, their resistance to manipulation becomes more prominent and their personality becomes less flexible. The early period of the history of a theoretical entity, its infancy if you will, is crucial for establishing its identity. During the infancy of the representation of the electron, for instance, a key aspect of the electron's identity, its charge-to-mass ratio, was determined from the available experimental data. From that point on, that property functioned as the electron's "signature."[17] In other words, it became a way of identifying the presence of electrons in novel experimental situations. Furthermore, the robustness of that ratio reinforced the autonomy of the electron qua theoretical entity. It was an aspect of the representation of the electron that survived radical changes in electron theory.

One could use expressions like "birth," "character formation," and "death" to trace the career of theoretical entities. The *birth* of a theoretical entity amounts to the emergence of an explanatory representation of some unobservable entity, in response to various problem situations, empirical and/or conceptual. An outcome of that birth, as I indicated in the previous chapter, is that various phenomena are grouped together as manifestations of the unobservable entity in question. The metaphor of birth, usually associated with an event, may seem at odds with my claim, in chapter 1, that the discovery of unobservable entities is an extended process. This apparent tension dissolves when we take into account a point I already stressed: the notion of birth, and related biographical metaphors, concern not unobservable entities per se but their representations. A representation could very well have a beginning in time, even though the belief in the existence of the corresponding entity was gradually established.

After the birth of a theoretical entity, its *character,* that is, the properties attributed to the corresponding unobservable entity, the laws that it is supposed to obey as well as the experimental constraints governing the representation of its behavior, is gradually formed. This character is subject to

17. I borrow this term from B. Lelong, "Paul Villard, J. J. Thomson, and the Composition of Cathode Rays," in J. Z. Buchwald and A. Warwick (eds.), *Histories of the Electron: The Birth of Microphysics* (Cambridge, Mass.: MIT Press, 2001), pp. 135–167, on p. 146.

change; however, its various aspects are (more or less) resistant to change. As we will see in the following chapters, in the case of the electron, some of its properties that enabled its identification in novel experimental situations remained relatively stable throughout the development of its representation; whereas some of the laws that were supposed to govern its behavior changed radically. As long as a theoretical entity remains a fruitful investigative tool and the phenomena associated with it remain intact, it remains alive. *Death* usually comes when the theoretical entity ceases to aid theoretical practice and runs out of explanatory power; as a result, the bonds that tied together the phenomena associated with it start to dissolve, even when the existence of the phenomena themselves is not questioned.[18] Typically, the death of a theoretical entity is accompanied by the birth of another entity, which takes over part of the explanatory role of its predecessor. I will not elaborate further on the death of theoretical entities, since it is not relevant to the story of the electron qua theoretical entity, which more than a hundred years after its birth continues to be a central element of the ontology of physics and chemistry.

The main historiographical advantage of this approach is that theoretical entities become explanatory resources for the historian. To explain the outcome of an episode in which a theoretical entity participated, one has to take into account the entity's contribution (both positive and negative) to the outcome of that episode. If, on the other hand, one neglects the entity's active participation, the understanding of the episode will be in some respects flawed. For instance, in reconstructing the genesis of Bohr's atomic theory one has to consider, besides the resources that he brought to bear on the construction of that theory, the resistance of the electron qua theoretical entity toward his attempts to manipulate it to further his theoretical aims (see chapter 5). A disregard of that resistance would diminish our understanding of the construction of Bohr's theory.

This biographical approach suggests an enrichment of the notion of the problem situation that was outlined in chapter 1. There are various problems that are created by scientists' attempts to manipulate the representation of an unobservable entity to further their theoretical purposes. Those problems emerge as a result of the representation's resistance to the attempted manipulation. This resistance takes the form of an incoherence of the representation in question, which is due to the incompatibility between previously

---

18. Cf. S. Toulmin, "Do Sub-microscopic Entities Exist?" in E. D. Klemke et al. (eds.), *Introductory Readings in the Philosophy of Science,* 3rd ed. (Amherst, NY: Prometheus Books, 1998), pp. 358–362, on p. 360.

established aspects of the entity in question (e.g., empirically confirmed properties or patterns of behavior) and the novel properties that the scientists attempt to force on it. The problem thus created gives rise to a demand for a solution, that is, a maneuver that would eliminate it and restore the representation's coherence.[19]

Furthermore, this approach differs from more conventional attempts to trace the development of concepts denoting unobservable entities, insofar as ordinary narratives do not portray those concepts as active agents. Traditional reconstructions of the development of concepts do not take into account that the characteristics of an established concept are heavily implicated in its further evolution. The future form of a concept depends not only on contextual (theoretical, experimental, cultural) pressures but also on its already established features. These features, especially the ones established through experiment, both guide and frustrate any attempt to adapt the concept to suit a changing environment. Furthermore, their stability endows the concept with a relative independence from theory change and provides a way of identifying the corresponding unobservable entity across theoretical frameworks.

This points to an additional historiographical advantage of the biographical approach over more traditional conceptual histories. In view of the ubiquitous presence of conceptual change in science, one could argue that historians should avoid framing their narratives around concepts, because it would be impossible, given conceptual variance, to construct a meaningful and coherent narrative. This point has been forcefully made by Quentin Skinner in the context of the history of ideas: "[A]s soon as we see there *is* no determinate idea to which various writers contributed, but only a variety of statements made with the words by a variety of different agents with a variety of intentions, then what we are seeing is equally that there *is* no history of the idea to be written, but only a history necessarily focused on the various agents who used the idea, and on their varying situations and intentions in using it."[20] If, however, as I argued above and in the previous chapter, there is a way to identify a theoretical entity and trace its lifeline by focusing on the experimental situations associated with it and on the experimentally determined properties of the corresponding unobservable entity, Skinner's objections to the traditional history of ideas do not apply to the biographical approach I advocate.

19. Cf. Hacking's suggestion that *"[c]oncepts have memories."* I. Hacking, *Historical Ontology* (Cambridge, Mass.: Harvard Univ. Press, 2002), p. 37.

20. Q. Skinner, "Meaning and Understanding in the History of Ideas," *History and Theory*, 8, no. 1 (1969): 3–53, on p. 38.

Two remarks are in order here. First, I use the term "biography" in a metaphorical sense, to highlight some important aspects of the history of the representation of the electron. Every metaphor, however, has its limits. The history of the electron qua theoretical entity resembles a biography in some respects, but not in others. In particular, I do not want to attribute intentionality to the representation of the electron, or to imply that it had wishes or other anthropomorphic features. In this respect, my project differs significantly from the approach adopted by Bruno Latour, who obliterates completely the difference between human and nonhuman agents and attributes "purpose, will and life to inanimate matter, and . . . human interests to the nonhuman." [21] Even though, as we will see in the chapters that follow, several scientists often used anthropomorphic language to describe the behavior of the electron, the biographical approach I sketched above is not meant to put special emphasis on that aspect of the history of the electron qua theoretical entity. Furthermore, most of those anthropomorphic descriptions concerned the *electron itself,* whereas my use of the biography metaphor aims at capturing the active nature of the *representation of the electron.*

Second, even though this approach has been developed in the context of writing the history of the electron's representation, I see no reason why it would not be applicable to the historical development of other theoretical entities (e.g., representations of protons or quarks). The characteristics of the electron qua theoretical entity that rendered it an active agent are common to many, if not all, theoretical entities which exhibit a similar behavior. It remains to be seen in actual historical case studies whether other theoretical entities can also be usefully studied from a biographical perspective. In fact, the term "biography" was used, some time ago, in connection with the history of ideas. Its use, however, was quite casual and no attempt was made to associate it with a particular historiographical approach. [22] Furthermore, recently there has been an upsurge of biographies of inanimate entities. [23] Again, these

21. For a cogent critique of Latour's "hylozoism," see S. Schaffer, "The Eighteenth Brumaire of Bruno Latour," *Studies in History and Philosophy of Science,* 22 (1991): 174–192; the quote is from p. 182. There are further differences between Latour's approach and mine, which I discuss below.

22. See J. H. Hexter, *More's Utopia: The Biography of an Idea* (New York: Harper and Row, 1965; 1st ed., 1952); and G. Holton (ed.), *The Twentieth-Century Sciences: Studies in the Biography of Ideas* (New York: Norton, 1972).

23. See, e.g., R. D. Friedel and P. Israel with B. S. Finn, *Edison's Electric Light: Biography of an Invention* (New Brunswick: Rutgers Univ. Press, 1986); M. S. Malone, *The Microprocessor: A Biography* (Santa Clara, Calif.: Telos, 1995). R. P. Crease, *Making Physics: A Biography of Brookhaven National Laboratory, 1946–1972* (Chicago: Univ. of Chicago Press, 1999); wP. Ball, *A Biography of Water: Life's Matrix* (New York: Farrar, Straus and Giroux, 2000); D. Bodanis, $E = mc^2$:

works, some of which are quite engaging popular treatments of their subject matter, offer little or no commentary on the appropriation of the term "biography," instead of the more traditional "history."

There are, however, two recent important books which offer interesting reflections on "biography" as a historiographical genre pertaining to "things" or "objects." I refer to Hans-Jörg Rheinberger's "biography of things" and Lorraine Daston's "biography of scientific objects."[24] Rheinberger stresses some of the themes I discussed above, that is, the heuristic character and recalcitrance of what he calls "epistemic things," material entities "embodying concepts."[25] The heuristic role of an epistemic thing is shown by the fact that it "is first and foremost a question-generating machine."[26] As for the recalcitrance of epistemic things, Rheinberger, following Michael Polanyi, elevates it to a criterion of reality: the "reality of epistemic things lies in their resistance, their capacity to turn around the (im)precisions of our foresight and understanding."[27] I see two significant differences between Rheinberger's historiographical and philosophical analysis and my own biographical perspective. The first is that he is concerned with material entities, emerging and developing in the context of experimental systems, whereas I am interested in theoretical entities and focus mostly, though not exclusively, on their role in theoretical practice. Second, and more important, I do not think that the capacity of theoretical entities to resist has to be due to the reality of the corresponding unobservable entities; but more on this below.

Daston's collection contains biographies of a very wide spectrum of scientific objects: for example, of a category of classification ("preternatural" objects), a mathematical entity (the center of gravity of the earth), an everyday object (dreams), an idea (economic value), a hidden fictitious entity (the ether), and an experimental object (cytoplasmic particles). I have discussed the merits and limitations of that book elsewhere.[28] Here I want to indicate where its methodology differs from my own biographical approach. The ul-

*A Biography of the World's Most Famous Equation* (New York: Walker, 2000); J. Emsley, *The Shocking History of Phosphorus: A Biography of the Devil's Element* (London: Macmillan, 2000); and C. Seife, *Zero: The Biography of a Dangerous Idea* (New York: Penguin, 2000). I would like to thank Hasok Chang for bringing some of these books to my attention.

24. See Hans-Jörg Rheinberger, *Toward a History of Epistemic Things: Synthesizing Proteins in the Test Tube* (Stanford: Stanford Univ. Press, 1997); the quoted phrase is from p. 4; and Daston (ed.), *Biographies of Scientific Objects*.

25. Rheinberger, *Toward a History of Epistemic Things*, p. 8.

26. Ibid., p. 32.

27. Ibid., p. 23.

28. See my "Towards a Historical Ontology?" *Studies in History and Philosophy of Science*, 34, no. 2 (2003): 431–442.

timate aim of Daston and some of her collaborators is to historicize ontology, that is, to portray the existence of scientific entities as relative to a particular historical context of beliefs and practices. On their view, the birth of a new representation of an unobservable entity together with the theoretical and laboratory practices associated with it means that the entity in question literally comes into being. This could be the case if the representation and the practices were, somehow, constitutive of the entity. If there were "no reality without representation," changes in representation would amount to changes in reality.[29] Existence and reality would lose their absolute character, and scientific entities would become "relatively real and relatively existent."[30]

This radical ontological claim creates severe methodological problems. For one thing, if entities were constituted by their representations, we could not attribute existence to an entity before the emergence of its representation. That would limit substantially the scope of explanations based on scientific entities. We could not appeal to the effects of an entity to explain any event that took place before that entity came into existence. For instance, all the scientific explanations of events that took place before the late nineteenth century would be invalidated if they employed electrons, since there were no electrons in the universe before their birth in the laboratories of Continental and British physicists.

Furthermore, the unwillingness to distinguish an entity from its representation implies that if different scientists employ a scientific term, but associate different representations with it, they refer to different entities. This multiplication of entities would have two historiographical consequences. First, it would undermine the biographical approach, since no sense could be made of an entity's identity and ipso facto of its biography. Prosopography (collective biography) would be a more appropriate tool for studying the history of the family of entities associated with a single term.[31] In the case of the electron, for example, there would be no legitimate sense in which all the scientists who used the term "electron" between, say, 1895 and 1925 were "talking about the same thing." This lack of "sameness" would threaten the cohesion of the story of the electron and would throw doubt on the value of a biographical

29. The quote is from B. Latour, *Pandora's Hope: Essays on the Reality of Science Studies* (Cambridge, Mass.: Harvard Univ. Press, 1999), p. 304.

30. B. Latour, "On the Partial Existence of Existing *and* Nonexisting Objects," in Daston (ed.), *Biographies of Scientific Objects*, pp. 247–269, on p. 257.

31. I would like to thank Klaus Hentschel for pointing out the potential value of prosopography in this context.

project devoted to its reconstruction.[32] Second, one could not make sense of the scientists' practice, where it presupposes the stability of an entity's identity. This can be shown by means of an example. In Bohr's 1913 theory of the atom, the orbiting electron, as represented within the theory, did not radiate, in violation of the laws of classical electrodynamics. On the other hand, it was known that the free electron, as portrayed in Lorentz's classical electromagnetic theory, radiated when it underwent accelerated motion. Apparently there was a conflict between the two cases, which motivated the subsequent development of the quantum theory. That conflict, however, makes sense only on the assumption that the "Bohr electron" and the "Lorentz electron" referred to the same entity. If the Bohr electron and the Lorentz electron were completely different entities, then there would be no conflict or inconsistency. Thus, the physicists' search for new laws governing the electron, laws that would unify its behavior inside and outside the atom, is intelligible only on the assumption that they identified the Bohr electron with the Lorentz electron.[33]

In view of these difficulties I do not want to obliterate the distinction between an entity and its representation, at least with respect to entities like the electron, which, unlike their representations, are not historical.[34] While the beliefs about scientific entities and the human practices associated with them change, this does not warrant any radical ontological claims about the historicity of the entities themselves. What changes over time is how these entities are represented or manipulated and not necessarily the entities per se.[35] So this book is a biography of the representation of the electron and not of the electron itself. It is the representation that is active and makes possible a biographical project of the kind I suggest.

It should be emphasized that in attempting to write the biography of a theoretical entity one does not necessarily adopt from the outset a realist perspective. Rather, one puts aside (temporarily) ontological questions, to the extent that they did not form part of the problem situations out of which the entity in question emerged. One can write a biography of an imaginary per-

32. Note that if one did not distinguish the electron from its representation, the story of the electron would merge with the story of its representation(s).

33. Cf. H. Putnam, *Representation and Reality* (Cambridge, Mass: MIT Press, 1988), esp. pp. 10ff.

34. Cf. I. Hacking, *The Social Construction of What?* (Cambridge, Mass.: Harvard Univ. Press, 1999), esp. pp. 28ff.

35. Of course, there are cases, in the social sciences, for example, where the idea of historicizing ontology is very plausible. See Hacking, *Historical Ontology*.

son (e.g., Hamlet) or of a person whose existence is controversial (e.g., Homer, whose existence remains an open question). In both cases one may examine how the persons in question were represented in the relevant historical sources, or how their representations changed over time, without confronting the question of their existence.[36] Furthermore, the fact that the electron qua theoretical entity constrained and guided the development of physics can be accommodated within an antirealist outlook. The resistance exhibited by the representation of the electron need not come from its real counterpart in nature; rather, it comes from its being embedded in a framework of beliefs and implicit expectations. As we will see, the resistance in question was revealed in unsuccessful attempts to account for (old and novel) experimental situations in terms of the electron. To that effect, additional properties had to be attributed to the electron. It turned out, however, that these properties did not fit within its previously established representation. As a result, the representation of the electron lost its coherence. The emergence of incoherence was a mark of the resistance of the electron qua theoretical entity.

The story that will unfold in the following chapters suggests that the electron qua theoretical entity was active whether its counterpart in nature existed or not. That activity was due to the features embodied in the electron's representation. Whether that representation had a referent "out there" was immaterial for the role that the electron qua theoretical entity played in theory construction.

Moreover, the electron's role in experimental practice—for example, the experimental determination of its properties—can also be understood from an antirealist point of view. Even though I cannot give a detailed argument here, the following example will suffice to illustrate this point. The observation of the behavior of cathode rays under the influence of electric and magnetic fields, interpreted in terms of classical mechanics and electromagnetic theory, determined the value of the mass-to-charge ratio of the electron.[37] This measurement, as van Fraassen has argued, can be given an antirealist construal (for details see chapter 3). The antirealist, however, might have difficulty explaining the relative stability of that value.

---

36. In the former case the representation and the fictitious entity "merge into one thing" (I owe this phrase to an anonymous referee). In the latter case, however, the subject of the biography is, strictly speaking, the entity (person) per se and not its representation.

37. It is worth noting that the order of magnitude of that value remained stable throughout (and in spite of) the evolution of the concept associated with the term "electron."

In writing the biography of a person it is not necessary to have access to the subject's writings. The biography can be written even if none of those writings have survived. In such a case, the historical sources for the construction of the biography would be everything that has been written (by that person's contemporaries or by later generations) about the person in question. For the biography of the electron qua theoretical entity, one does not have direct access to the electron's "writings." Even though some experimental situations (e.g., tracks in a cloud chamber) were attributed to the presence of electrons, they do not provide unmediated access to those entities. An antirealist could claim that those "writings" might be due to a different entity or, perhaps, to no entity at all (i.e., that they are "brute facts" that do not reveal an unobservable reality). Thus, my sources will quite naturally be the physicists' and chemists' reflections about it.

However, I will also try to reconstruct some aspects of the "point of view" of the electron qua theoretical entity. In a biography the historian attempts to reconstruct both the individual's perspective about the events in which he or she participated as well as that of the community. In the case of the representation of the electron, its point of view amounts to the (negative and heuristic) constraints that it posed on scientific theorizing; it also appears in the various experimental situations in which, according to experimentalists, the electron manifested itself. It turns out that the electron qua theoretical entity had a "view" about the problem situations in which it participated. It prohibited certain theoretical moves and it suggested others.[38] Its negative role was evident whenever, as a result of a theoretical maneuver, some of the constraints on the electron's purported character and behavior were violated. This violation suggested, in turn, that the attempted maneuver had to be revised. On the other hand, its positive contribution to scientific practice took several forms. First, questions about its properties and behavior guided theoretical and experimental research (see chapter 4). Second, in some cases where empirical problems involving the electron appeared, the elimination of simplifications from its representation provided plausible problem-solving strategies (see chapter 6). Moreover, on the experimental side the electron produced certain "writings" that were deciphered by scientists. By "writings" I mean something wider than the traces of free electrons in cathode ray tubes or in cloud chambers. This term is meant to capture *all* the effects that were

38. Even though I cannot argue this point here, the active nature of the electron's representation reveals the limitations of the view that the products of scientific activity are social constructions.

attributed to the electron.[39] For instance, spectroscopic data were, in several cases, part of the electron's writings (see chapters 5, 6, 8).

The writings of the electron guided the articulation of its representation. For instance, once physicists attributed spectral lines to the action of electrons within the atom, the number, intensity, and polarization of those lines acted as positive constraints on the construction of an account of the properties and behavior of the electron within the atom. Thus, in accordance with the terminology introduced in the previous chapter, the reference of the term "electron" can be employed to reconstruct the experimental view of the electron qua theoretical entity. The rest of the meaning was negotiated between the representation of the electron and the scientists who used it as a theoretical tool.

The life of the representation of the electron can be seen in terms of the nature-nurture debate. The input of nature in the development of the "personality" attributed to the electron comes from the experimental detection of its properties. The influence of nurture comes from the theoretical practices of physicists and chemists in their attempts to resolve various problem situations. The confluence of nature and nurture resulted in the gradual construction of the "personality" attributed to the electron (the collection of its various properties). This gradual construction is, of course, an open-ended process, but it should be pointed out that throughout that process the "electron" has retained a common (to all stages) core of meaning (e.g., having negative charge[40] or a certain charge-to mass ratio). It is this aspect of the construction that supports, albeit not conclusively, a realist attitude toward the electron. To the extent that this study is a realist reading of the history of the representation of the electron, my realist stance comes from the historical record and is not imposed a priori on it (see chapter 9).

39. It is enough for my purposes that the scientists themselves believed that those effects were manifestations of the electron. Of course, the question of whether they were justified in doing so remains. Here, as elsewhere in the book, my focus is on how the electron was represented and not on the electron per se.

40. The only exception was Zeeman, who originally thought that the electron was positively charged. This exception, however, is insignificant, because Zeeman's belief was due to a straightforward mistake, which he soon realized and corrected. See chapter 4, p. 84.

# Chapter 3 | Rethinking "the Discovery of the Electron"

## 1. What Is Wrong with the Received View?

In the history and philosophy of science literature one reads
repeatedly that J. J. Thomson discovered the electron in
1897, while he was experimenting on cathode rays at the
Cavendish Laboratory.[1] On this widely held view, Thom-

1. See, e.g., P. Achinstein, *Particles and Waves* (New York: Oxford
Univ. Press, 1991), pp. 286–287, 299; M. Chayut, "J. J. Thomson: The
Discovery of the Electron and the Chemists," *Annals of Science,* 48 (1991):
527–544, see p. 531; P. Galison, *How Experiments End* (Chicago: Univ. of
Chicago Press, 1987), p. 22; R. Harré, *Great Scientific Experiments* (New
York: Oxford Univ. Press, 1983), pp. 157–165; J. Heilbron, *Historical Stud-
ies in the Theory of Atomic Structure* (New York: Arno Press, 1981), pp. 1,
14; E. N. Hiebert, "The State of Physics at the Turn of the Century," in
M. Bunge and W. R. Shea (eds.), *Rutherford and Physics at the Turn of the
Century* (New York: Dawson and Science History Publications, 1979),
pp. 3–22, see p. 7; M. Jammer, *The Conceptual Development of Quantum
Mechanics* (New York: McGraw-Hill, 1966), p. 121; H. Kragh, "Concept
and Controversy: Jean Becquerel and the Positive Electron," *Centaurus,*
32 (1989): 203–240, see p. 205; W. McGucken, *Nineteenth-Century Spectroscopy: Development of the Understanding of Spectra, 1802–1897* (Baltimore: Johns Hopkins Univ. Press, 1969), pp. xi, 209; A. I. Miller, "Have
Incommensurability and Causal Theory of Reference Anything to Do
with Actual Science?—Incommensurability, No; Causal Theory, Yes,"
*International Studies in the Philosophy of Science,* 5, no. 2 (1991): 97–108,
see p. 102; A. Pais, *Niels Bohr's Times, in Physics, Philosophy, and Polity*
(New York: Oxford Univ. Press, 1991), pp. 105–106; M. Paty, "The Scientific Reception of Relativity in France," in T. F. Glick (ed.), *The Comparative Reception of Relativity,* Boston Studies in the Philosophy of Science 103 (Dordrecht: Reidel, 1987), pp. 113–167, see p. 125; L. Pyenson,
"The Relativity Revolution in Germany," in Glick (ed.), *The Comparative
Reception of Relativity,* pp. 59–111, see p. 71; and A. N. Stranges, *Electrons
and Valence: Development of the Theory, 1900–1925* (College Station: Texas
A and M Univ. Press, 1982), pp. xi, 32. This list is by no means exhaustive.

son's measurement of the mass-to-charge ratio of the electron in 1897 and his subsequent measurement of its charge in 1899 eliminated all doubt about the existence of this new subatomic entity. In other words, the introduction and legitimization of the electron hypothesis has been portrayed as a discovery episode that can be assigned a more or less precise date. This is, of course, the typical physics textbook account, but it is also shared by many historians and philosophers of science. Here is a typical passage portraying Thomson's work as an almost heroic achievement: "In 1897 Joseph James Thomson discovered the electron in cathode rays. Others had intimated the discovery before Thomson, but the director of the Cavendish Laboratory at Cambridge had the acuity to identify the electron as the object radiating spectral lines, the constituent element of atoms, and the smallest unit of electrical charge. Thomson's discovery was the storming of the Bastille; it provided a definitive break with the old regime of physics." [2] From now on I will refer to this view of the discovery of the electron as the *received view*.[3] Another indication of the widespread acceptance of that view is that in 1997 there were many conferences on the occasion of the centennial of Thomson's work and special issues of journals were devoted to the one hundredth anniversary of the discovery of the electron.[4]

The received view is problematic, on both historiographical and philosophical grounds. On the historiographical side, it downplays several British and Continental developments that were quite decisive for the gradual acceptance of the electron as a universal, subatomic constituent of matter. On

2. Pyenson, "The Relativity Revolution in Germany," p. 71.

3. It should be pointed out that in some of the literature that credits the discovery of the electron to J. J. Thomson, there is ample information suggesting that the proposal and acceptance of the electron hypothesis were neither exclusively linked with J. J. Thomson nor confined to 1897. Furthermore, recently some historians have distanced themselves from the received view. See, e.g., O. Darrigol, "Aux confins de l'électrodynamique maxwelliene: Ions et électrons vers 1897," *Revue d'Histoire des Sciences*, 51, no. 1 (1998): 5–34; I. Falconer, "Corpuscles to Electrons," in J. Z. Buchwald and A. Warwick (eds.), *Histories of the Electron: The Birth of Microphysics* (Cambridge, Mass.: MIT Press, 2001), pp. 77–100; G. Gooday, "The Questionable Matter of Electricity: The Reception of J. J. Thomson's 'Corpuscle' among Electrical Theorists and Technologists," in Buchwald and Warwick (eds.), *Histories of the Electron*, pp. 101–134; B. Lelong, "Paul Villard, J. J. Thomson, and the Composition of Cathode Rays," in Buchwald and Warwick (eds.), *Histories of the Electron*, pp. 135–167; and G. E. Smith, "J. J. Thomson and the Electron, 1897–1899," in Buchwald and Warwick (eds.), *Histories of the Electron*, pp. 21–76.

4. See, e.g., "100 Years of Elementary Particles," *Beam Line*, 27, no. 1 (Spring 1997); also available at http://www.slac.stanford.edu/pubs/beamline; "The Electron: Discovery and Consequences," special issue, *European Journal of Physics*, 18, no. 3 (May 1997); "J. J. Thomson's Electron," special issue, *Physics Education*, 32, no. 4 (July 1997); "The Centenary of the Electron," special issue, *Physics Today*, Oct. 1997; and "100 Years of the Electron," special issue, *Physics World*, Apr. 1997.

the philosophical side, it presupposes a realist perspective toward unobservable entities and requires a theory of scientific discovery that would support such a perspective. As far as I can tell, no such adequate theory has been developed (see below).

In most of the works that propagate the received view, the discovery of the electron is credited to J. J. Thomson quite casually and no attempt is made to support this claim through a specific philosophical or historiographical model of scientific discovery. Very few authors seem to realize that "discovery" is an evaluative category with realist presuppositions. An antirealist would not accept that the electron has been discovered, since the antirealist favors an agnostic stance toward the existence of unobservable entities. Realist philosophers, for their part, as I pointed out in chapter 1, would have to suggest what constitutes an adequate demonstration for the existence of such entities and then justify, on that basis, the attribution of the electron's discovery to J. J. Thomson. There have been such proposals, in the philosophical and historical literature. I think, however, that these proposals beg the question.

Consider, first, a recent paper by the philosopher Peter Achinstein, in which he articulates necessary and sufficient conditions for the concept of scientific discovery and employs them to credit J. J. Thomson with the discovery of the electron. One of those conditions is epistemic: "[W]e can say that someone is in an *epistemic state necessary for discovering X* if that person knows that X exists."[5] From an antirealist point of view, however, one can never be in such an epistemic state with respect to unobservable entities, since the lack of direct access to those entities renders unattainable knowledge of their existence. To bypass this problem, Achinstein points out that "to discover X, you don't need to observe X directly. It suffices to observe certain causal effects of X that can yield knowledge of X's existence."[6] Again, this would hardly satisfy an antirealist, who denies that observations of the purported effects of an unobservable entity can yield knowledge of its existence.

This difficulty also comes up in Achinstein's discussion of the discovery of the electron: "The issue, as I have defined it, is simply this: even though others had provided some experimental evidence for the existence of charged particles as the constituents of cathode rays, were Thomson's experiments the first to *conclusively* demonstrate this? Were they the first on the basis of which knowledge of their existence could be correctly claimed? If so, he discovered

---

5. P. Achinstein, "Who Really Discovered the Electron?" in Buchwald and Warwick (eds.), *Histories of the Electron*, pp. 403–424, on p. 406.

6. Ibid., p. 406.

the electron. If not, he didn't."[7] This passage shows clearly the realist presuppositions of Achinstein's analysis. If we deny that it is possible to *conclusively* demonstrate "the existence of charged particles as the constituents of cathode rays," then neither Thomson nor any other scientist could have discovered the electron.

Part of the trouble with Achinstein's account is that he does not distinguish between discoveries in microphysics and discoveries of directly observable entities, like "the naturalist finding a new bug beneath a rock."[8] We should treat these two cases differently, I think, for two reasons: first, because the realism debate has focused on the existence of unobservable entities, with both sides sharing a belief in the existence of observable objects, and second, because in the philosophical analysis of the discovery of unobservable entities one has to face a difficulty that does not appear in the case of observables. The discovery of an observable entity might simply involve its direct observation and does not require that all, or even most, of the discoverer's beliefs about it be true. For example, to discover "that there is a person in the ditch, . . . not every belief about that person needs to be true or known to be true."[9] This is not the case, however, when it comes to unobservable entities where direct physical access is, in principle, unattainable. The lack of independent access to such an entity makes problematic the claim that the discoverer's beliefs about it need not be true.[10]

Another attempt to justify the attribution of the electron's discovery to J. J. Thomson is made by the historian Isobel Falconer, who has employed, to that effect, Hacking's criterion of what constitutes unobservable reality.[11] A discussion of her proposal not only will illuminate Thomson's achievement but will also reveal the limitations of Hacking's criterion as a historiographical tool.[12]

Falconer challenged traditional interpretations of Thomson's discovery

---

7. Ibid., p. 418.

8. N. R. Hanson, *The Concept of the Positron: A Philosophical Analysis* (Cambridge: Cambridge Univ. Press, 1963), p. 165.

9. Achinstein, "Who Really Discovered the Electron?" p. 416. Nevertheless, *some* beliefs about the discovered entity have to be true. For example, in the case of the discovery of a person in the ditch we have to know that what we have discovered is a person and not, say, a stone (or, in general, an I-know-not-what).

10. This is an aspect of the problem of identification, which I analyzed in chapter 1. I will try to come to terms with this problem in chapter 9.

11. See I. Falconer, "Corpuscles, Electrons and Cathode Rays: J. J. Thomson and the 'Discovery of the Electron,'" *British Journal for the History of Science,* 20 (1987): 241–276.

12. I should note that in her more recent work Falconer has distanced herself from the received view of the discovery of the electron (see her "Corpuscles to Electrons").

that portrayed "this discovery . . . [as] the outcome of a concern with the nature of cathode rays which had occupied Thomson since 1881 and had shaped the course of his experiments during the period 1881–1897."[13] Instead she argued that "[a]n examination of his work shows that he paid scant attention to cathode rays until late 1896."[14] Furthermore, "[t]he cathode ray experiments in 1897 were not the origin of the corpuscle [which has been renamed electron] hypothesis; instead they acted as a focus around which Thomson synthesized ideas he had previously developed."[15]

However, she did not deny a central presupposition of the received view, namely, that the discovery of the electron was a temporally nonextended event, which can be credited to a single individual. Even though the "corpuscle *hypothesis*" did not originate with Thomson's experiments with cathode rays, the *discovery* of corpuscles (i.e., the experimental demonstration of their existence) was the outcome of these experiments.

> Arriving at the theoretical concept of the electron was not much of a problem in 1897. Numerous such ideas were "in the air." What Thomson achieved was to demonstrate their validity experimentally. Regardless of his own commitments and intentions, it was Thomson who began to make the electron "real" in Hacking's sense of the word. . . . He pinpointed an experimental phenomenon in which electrons could be identified and methods by which they could be isolated, measured and manipulated. This was immensely significant for the development of the electron theory which hitherto had been an abstract mathematical hypothesis but now became an empirical reality.[16]

In terms of the methodological issues discussed in chapter 1, Falconer attempts to reduce the discovery process to the precise moment when experimental verification took place, thus equating discovery with the ability to isolate, measure, and manipulate. From my perspective this amounts to equating discovery with justification.

If, however, as I have argued in chapter 1, the context of discovery comprises both the context of generation and the context of justification, then the physicists who had formulated all those "ideas in the air" should also be considered to have taken part in the discovery of the electron. Furthermore, even if Falconer's attempt were sound from a methodological point of view, there

---

13. Falconer, "Corpuscles, Electrons and Cathode Rays," p. 241.
14. Ibid.
15. Ibid., p. 255.
16. Ibid., p. 276.

would still be several problems with her attribution of the discovery of the electron to J. J. Thomson. First, he was not the only one who provided an experimental demonstration of all those ideas in the air. Several months before Thomson, Pieter Zeeman had done the same with respect to Lorentz's theory of "ions" and Larmor's theory of "electrons."[17] Second, it is not clear in what sense Thomson pinpointed "methods by which . . . [electrons] could be *isolated, measured* and *manipulated*" (emphasis added). The actual isolation of the electron was accomplished several years later by Millikan and even then there were grave doubts that the electron had, in fact, been isolated.[18] Moreover, Thomson's measurement of the electron was by no means his exclusive achievement. Zeeman, several months before Thomson, had made an estimate of $e/m$ (the charge-to-mass ratio of an "ion"), and Emil Wiechert, as well as Walter Kaufmann, had also measured $e/m$, independently of and simultaneously with Thomson. Finally, Thomson was not the only one who could manipulate electrons. All those who experimented with cathode rays were able to manipulate them in various ways. For example, they could deflect them by means of magnetic fields. That is, from our perspective, given that they manipulated cathode rays and that cathode rays are streams of electrons, it follows that they manipulated electrons. And this brings me back to Hacking's criterion of what constitutes unobservable reality.

It is evident that Falconer has employed Hacking's criterion as a means for justifying attributing the discovery of the electron to J. J. Thomson. The validity of this attribution depends, therefore, on whether the manipulation of unobservable entities in the laboratory provides sufficient grounds for believing in their existence. To see the limitations of Hacking's criterion, consider Thomson's experiments with cathode rays. Since one could describe these experiments in terms of cathode rays as opposed to electrons, the act of manipulation could be described without even mentioning the entities that, according to present-day physics, were manipulated.[19] Moreover, an antirealist could give an even less theory-laden description, by avoiding the term "cathode rays" and using instead the phenomenological expression "spot on a

17. These theories, along with Zeeman's work, will be discussed below.

18. See G. Holton, "Subelectrons, Presuppositions and the Millikan-Ehrenhaft Dispute," in his *The Scientific Imagination: Case Studies* (Cambridge: Cambridge Univ. Press, 1978), pp. 25–83. Cf. A. Franklin, "Millikan's Published and Unpublished Data on Oil Drops," *Historical Studies in the Physical Sciences*, 11 (1981): 185–201.

19. This is not just an abstract possibility. In 1900 "Villard [a French opponent of the atomic hypothesis] described J. J. Thomson's work in considerable detail, but all references to atoms (and all the more to corpuscles) were removed." Lelong, "Paul Villard, J. J. Thomson, and the Composition of Cathode Rays," p. 149.

phosphorescent screen." The only thing that we know, the antirealist would argue, is that by activating an electromagnet Thomson could move a spot on a phosphorescent screen. Since an act of manipulation can be described without mentioning the unobservable entity that is (supposedly) manipulated, this act does not, by itself, imply the existence of the entity in question. Thus, given that experiments can be described in phenomenological terms, manipulability cannot be employed, to the satisfaction of an antirealist, for existential inferences. Whereas for Hacking manipulability justifies existence claims, for the antirealist it is the other way around: It is the belief in the existence of, for example, electrons, prior to the act of manipulation, that allows us to interpret that act as a manipulation of electrons (as opposed to something else).[20] In fact, as Falconer points out, well before his classic experiments with cathode rays, Thomson had put forward the hypothesis of "subatomic particles [that] were very small and . . . could have an independent existence, i.e., that the atom could be split up into them."[21]

Since Hacking's criterion does not provide adequate grounds for a realist position on unobservable entities, it cannot be employed to justify discovery claims. Thus, Falconer's claim that the discovery of the electron was Thomson's exclusive experimental achievement is undermined.[22] In other respects, however, her article is an excellent reconstruction of Thomson's theoretical and experimental contribution to the process that culminated to the consolidation of the belief that "electrons" denote real entities. In reconstructing that process below I will draw on her analysis.

Abraham Pais has also argued that "Thomson should be considered the sole discoverer of the first particle [the electron]."[23] His approach exemplifies what I called in chapter 1 the scientist's perspective on the issue of scientific discovery. A discussion of his claim, which he backs with a considerable

20. Cf. P. K. Feyerabend, "Das Problem der Existenz theoretischer Entitäten," in E. Topitsch (ed.), *Probleme der Wissenschaftstheorie* (Vienna: Springer-Verlag, 1960), pp. 35–72, esp. p. 64. This is just one problematic aspect of Hacking's entity realism. For further criticism of his view see chapter 9.

21. Falconer, "Corpuscles, Electrons and Cathode Rays," p. 264. Thomson made this proposal in a letter to Kelvin on 10 Apr. 1896.

22. In correspondence, Falconer has suggested to me that she agrees with my "objections to using Hacking's criteria for reality to justify 'discovery.'" But she still "think[s] it can be used as an analytical tool, to ask what did this or that physicist contribute to our understanding of the electron; did he help to give it manipulative reality for the physicist?" I do not object to using Hacking's manipulability criterion in this way, provided that it gives a good descriptive account of how scientists construct "existence proofs" for unobservable entities. As I will argue in chapter 9, however, it also faces difficulties in this respect.

23. A. Pais, *Inward Bound* (New York: Oxford Univ. Press, 1988), p. 78.

amount of historical information, reveals in a concrete manner the limitations of this approach.

Even though he attributes the discovery of the electron to J. J. Thomson, he disputes the traditional claim that this discovery took place in 1897:

> It is true that in that year Thomson made a good determination of $e/m$ for cathode rays, an indispensable step toward the identification of the electron, but he was not the only one to do so. Simultaneously Walter Kaufmann had obtained the same result. It is also true that in 1897 Thomson, less restrained than Zeeman, Lorentz, and (as we shall see) Kaufmann, correctly conjectured that the large value for $e/m$ he had measured indicated the existence of a new particle with a very small mass on the atomic scale. However, he was not the first to make that guess. Earlier in that year, Emil Wiechert had done likewise, on sound experimental grounds, even before Thomson and Kaufmann had reported their respective results. Nevertheless, it is true that Thomson should be considered the sole discoverer of the first particle, since he was the first to measure not only $e/m$ but also (within 50 per cent of the correct answer) the value of $e$, thereby eliminating all conjectural elements—but that was in 1899.[24]

According to Pais, then, the most central features of the modern concept of the electron are the values of its charge and its mass. An important step in measuring these two values was the measurement of the charge-to-mass ratio, achieved independently by Zeeman, Kaufmann, Thomson, and Wiechert. This step, according to Pais, was not sufficient for establishing the existence of the electron. One had also to conjecture that the unexpectedly large value of $e/m$ was due to the very small mass of a new subatomic particle and, more important, to confirm experimentally that conjecture. Even though Wiechert had also made that conjecture, Thomson was the only one who both made it and demonstrated it experimentally. By measuring the charge of the electron in 1899 he was able to deduce from that measurement and the already known charge-to-mass ratio the mass of the electron.

As I have argued in chapter 1, there is a serious difficulty that undermines Pais's approach. The charge and mass of the electron are only two of the sev-

---

24. Ibid. Some historians have also advocated a similar view. Barbara Turpin, for instance, has suggested that "it would be more correct to set the date of the discovery of the electron as 1899 rather than 1897, the date most scholars have adopted." B. M. Turpin, *The Discovery of the Electron: The Evolution of a Scientific Concept, 1800–1899* (Ph.D. dissertation, Univ. of Notre Dame, 1980), p. 202.

eral properties associated with the electron of present-day physics. It is not at all clear why the theoretical and experimental detection of just two of these properties constitutes the discovery of the electron. On the one hand, one could claim that the measurement of the charge-to-mass ratio was by itself the key aspect of that discovery. Stuart Feffer, for instance, has suggested that "[t]he finding that this ratio was much larger than anticipated was the key experimental ingredient in the discovery of the electron."[25] On the other hand, one could argue that Thomson's measurement of the electron's charge and mass is not sufficient for earning him the status of the "discoverer of the electron." According to such an account, one should also have detected experimentally the intrinsic magnetic disposition of the electron (spin) or its wave properties in order to be considered its discoverer. Laszlo Tisza, for example, believes that the electron was discovered in the late 1920s when Clinton J. Davisson and Lester H. Germer and, independently, George P. Thomson detected its wave properties. From the measurement of the electron's wavelength it became possible to calculate its momentum. That, according to Tisza, rendered electrons directly detectable.[26]

Besides this methodological difficulty, there are further historiographical and philosophical problems that undermine Pais's claim. To begin with, it is not true that Thomson's measurement of the charge of the electron eliminated "all conjectural elements." If that measurement had led the scientific community to a consensus on the reality of electrons, this would have immediate repercussions for the atomism debate. Someone who believes in the existence of subatomic constituents of all atoms believes ipso facto in the reality of atoms. The atomism debate, however, remained open till shortly after 1910[27]—a fact that clearly contradicts the view that Thomson had already proved beyond doubt in 1899 the existence of electrons.[28] Furthermore, there is specific evidence that Thomson's measurements had not convinced everybody of the reality of electrons. Max Planck, for instance, "confessed that as

25. S. M. Feffer, "Arthur Schuster, J. J. Thomson, and the Discovery of the Electron," *Historical Studies in the Physical and Biological Sciences*, 20, no. 1 (1989): 33–61, on p. 33.

26. I would like to thank Professor Tisza for discussing with me some of the ideas I develop in this chapter.

27. "For the majority of the scientific community the atomic debates were resolved about 1911, the year of the first Solvay Conference." M. J. Nye, "The Nineteenth-Century Atomic Debates and the Dilemma of an 'Indifferent Hypothesis,'" *Studies in History and Philosophy of Science*, 7, no. 3 (1976): 245–268, on pp. 267–268. See also M. J. Nye, *Molecular Reality* (New York: Elsevier, 1972).

28. I discuss further the entanglement of the "discovery" of the electron with the atomic debates in the following section.

late as 1900 [one year after Thomson's "discovery" of the electron] he did not fully believe in the electron *hypothesis*" (emphasis added).[29]

Thus, from the perspective of the historical actors Thomson's measurements did not eliminate "all conjectural elements." However, one might want to disregard that perspective and argue that any doubts about the existence of electrons were unwarranted after Thomson had performed his measurements. From a philosophical point of view, I do not see how such an ahistorical judgment could be justified. Even Hacking, who is one of the most eminent advocates for entity realism, has admitted that "[o]nce upon a time it made good sense to doubt that there are electrons. Even after Thomson had measured the mass of his corpuscles, and Millikan their charge, doubt could have made sense."[30]

Furthermore, from an antirealist perspective the above ahistorical judgment leaves much to be desired. Even though the word "measurement" is laden with realist presuppositions (if we measure something, then, no doubt, this something exists), a realist stance is not a prerequisite for making sense of measurement. The antirealist has the option of interpreting the measurement of various parameters as a process of "theory construction by other means." The function of measurement, on this view, is to specify certain parameters of a theory, which were not specified when the theory was initially proposed, in such a way as to make the theory under consideration empirically adequate.[31] Paradoxical as it sounds, the measurement of the properties associated with an unobservable entity does not imply that this entity exists. To appreciate the plausibility of this view, one needs only to be reminded that late-eighteenth-century chemists were measuring, or so they thought, the weight of phlogiston.[32]

29. Holton, "Subelectrons, Presuppositions and the Millikan-Ehrenhaft Dispute," p. 42.

30. I. Hacking, *Representing and Intervening* (Cambridge: Cambridge Univ. Press, 1983), p. 271. It should be noted that this excerpt is incompatible with Hacking's criterion, since Thomson and Millikan *could* manipulate the electron in various ways. It is true that they could not manipulate it in as many ways as modern experimenters can, since the latter's capacity to manipulate it is based on properties (e.g., spin) that were not known to the former. However, if Thomson's and Millikan's ability to manipulate electrons did not establish their existence, it is not clear why our current ability to manipulate them guarantees their reality. Since manipulability is a matter of degree, one needs to know the extent of manipulation that constitutes an adequate "existence proof."

31. See B. C. van Fraassen, *The Scientific Image* (New York: Oxford Univ. Press), pp. 73–77.

32. Note that my suggestion in chapter 2 that theoretical entities are constructions from experimental data is compatible with this antirealist construal of measurement. From my perspective, the measurement of unobservable properties is a process of "translation," whereby quantitative information from experimental situations is incorporated within the theoretical entity associated with them.

## 2. Early-Twentieth-Century Views of the Acceptance of the Electron Hypothesis

My criticism of the received view aimed at showing that the portrayal of the introduction and acceptance of the electron as a sudden discovery episode is too simplistic. As I will show in the following chapter, the acceptance of the electron as a new and fundamental constituent of matter was an evolving process, lasting from the early 1890s till the early twentieth century, that involved scientists working in different areas, from the discharge of electricity through gases to spectroscopy and electromagnetic theory. In this section I will present evidence from early-twentieth-century sources indicating that the received view is a later construction, which is at odds with the views of contemporary scientists on the early history of the electron hypothesis. I will also discuss briefly an important aspect of that history, namely, its interconnection with the debate over the existence of atoms.

The perspective of contemporary scientists, as opposed to later scientists writing surveys of the history of that period, supports my suggestion that the electron was not "discovered" but was gradually incorporated in the ontology of physics and chemistry. Several contemporary scientists who were involved in the development of the theory of electrons did not think that the electron had been discovered by J. J. Thomson, or any other scientist for that matter.[33] Lorentz, for instance, in a commentary on Zeeman's discovery, pointed out that "[w]e are all well aware of the extent to which this hypothesis of ions contained in all imaginable bodies was made probable by a number of phenomena of another nature, and particularly by electrical discharge in rarefied gases."[34] Here Lorentz refers to Thomson's experiments with cathode rays and points out that those experiments rendered probable the "hypothesis of ions."[35] His tentative language is significant; Thomson's investigations are

33. Cf. Achinstein's remark that, on my view, "Thomson discovered the electron if he is generally regarded by physicists as having done so. The physicists who so regard him may have different reasons for doing so, but these reasons do not make him the discoverer: simply their regarding him as such does. Even if the reasons are false (in some 'absolute,' non-consensual sense), he is still the discoverer, unless the physics community reaches a different consensus" ("Who Really Discovered the Electron?" p. 411). In chapter 1 I suggested as a pragmatic (historiographical) criterion of discovery the formation of consensus regarding the existence of an entity, not the formation of consensus about the discoverer of that entity. Furthermore, from the evidence I present below it is clear that the view of Thomson as the discoverer of the electron was *not* the view of the scientific community at the time that the presumed discovery took place.

34. H. A. Lorentz, "Théorie des phénomènes magnéto-optiques récemment découverts," *Rapports présentés au Congrès International de Physique,* vol. 3 (Paris, 1900), p. 2; quoted in N. Robotti and F. Pastorino, "Zeeman's Discovery and the Mass of the Electron," *Annals of Science,* 55 (1998): 161–183, on p. 182.

35. Lorentz sometimes used the terms "ion" and "electron" interchangeably.

not portrayed as a decisive discovery, but just as additional evidence for the existence of "ions."

Furthermore, in 1901 a volume of papers was published to celebrate the twenty-fifth anniversary of Lorentz's graduation. Thomson was one of the contributors to that volume, and Lorentz sent him a laudatory letter to thank him for his contribution. Lorentz characterized Thomson's achievement as having "taken so prominent a part in the investigation of the subject [of the theory of electrons]."[36] Thus, Lorentz regarded Thomson as a significant contributor to the development of the theory of electrons and not as the discoverer of these entities. Furthermore, in his Nobel Lecture Lorentz mentioned Thomson's name only once, in connection with the latter's secondary confirmation that the mass of the electron was very small compared to the mass of the hydrogen atom. Lorentz pointed out that this result had been previously obtained by himself, on the basis of Zeeman's measurements of $e/m$ together with an analysis of the phenomenon of dispersion (see p. 85 below).[37]

In 1901 the physicist Arthur W. Rücker (president of the University of London), in his presidential address to the British Association for the Advancement of Science, made a considerable effort to support the atomic theory. Here is what he had to say about Thomson: "The dream that matter of all kinds will some day be proved to be fundamentally the same has survived many shocks. . . . Sir Norman Lockyer has long been a prominent exponent of the view that the spectra of the stars indicate the reduction of our so-called elements to simpler forms, and now Professor J. J. Thomson believes that we can break off from an atom a part, the mass of which is not more than one-thousandth of the whole, and that these corpuscles, as he has named them, are the carriers of the negative charge in an electric current."[38] Again, there is no mention of any discovery in this passage, and this becomes even more significant if we take into account that Rücker was not hostile to the atomic theory. There is only a reference to Thomson's "belief," a belief that is placed on a par with Lockyer's previous speculations on the complexity of chemical atoms.

In 1902 Rutherford, in a review of the evidence that supported "the existence of bodies smaller than atoms," did not attribute the electron's discovery

---

36. H. A. Lorentz to J. J. Thomson, 1 Feb. 1901, Cambridge Univ. Library, Add 7654/L61.

37. See H. A. Lorentz, "The Theory of Electrons and the Propagation of Light," Nobel Lecture, 11 Dec. 1902, in *Nobel Lectures: Physics, 1901–1921* (Amsterdam: Elsevier, 1967), pp. 14–29, esp. p. 24.

38. A. Rücker, "Address of the President of the British Association for the Advancement of Science," *Science,* 14, no. 351 (20 Sept. 1901): 425–443, on p. 438.

to any particular scientist. Rather, his view was that "DURING the last few years considerable evidence has been obtained of the production, under various conditions, of bodies which behave as if their mass was only a small fraction of the mass of the chemical atom of hydrogen. As far as we know at present, these minute particles are always associated with a negative electric charge. For this reason they have been termed 'electrons.'" [39] Thus, according to Rutherford, the belief in the existence of the electron was the outcome of a gradual process of accumulating evidence, as opposed to an event. Even though Thomson contributed significantly to that process, he did not produce all that evidence by himself. More important, as late as 1904 the existence of the electron was not universally accepted: "The physical existence of electrons is now accepted by many scientific men." [40]

Owen Richardson, who had been Thomson's student and was an expert on thermionic emission and electron theory, shared a similar perspective: "The electron theory may now be said to have developed far beyond the region of hypothesis. Discovery after discovery *during the last fifteen years* has established indubitably the existence of a negative electron whose properties are independent of the matter from which it originates" (emphasis added). [41] So, according to Richardson, it was a fifteen-year process that established the existence of the electron, and not any isolated experiment or measurement.

Millikan also did not think that Thomson's experiments were decisive with respect to the electron's discovery. As he said in 1924, "J. J. Thomson and after him other experimenters . . . [provided] an indication of an affirmative answer to the . . . question above [is there a primordial subatom out of which atoms are made?]—an indication which was strengthened by Zeeman's discovery in 1897." [42] Even though Millikan got the chronology wrong (Zeeman's results had been obtained in the fall of 1896, well before Thomson's experiments), his assertion is significant. Neither Thomson nor Zeeman discovered the electron; rather, their work provided an *indication* of its reality. Some doubts remained with respect to its existence, doubts that were re-

39. E. Rutherford, "The Existence of Bodies Smaller than Atoms," *Transactions of the Royal Society of Canada,* 2nd series, section 3, 8 (1902): 79–86; repr. in *The Collected Papers of Lord Rutherford of Nelson,* vol. 1 (London: George Allen and Unwin, 1962), pp. 403–409, on p. 403.

40. Ibid. (in *The Collected Papers of Lord Rutherford of Nelson*), p. 409.

41. O. W. Richardson, *The Electron Theory of Matter* (Cambridge: Cambridge Univ. Press, 1914), p. 3. Cf. a recent paper on Richardson: "He began his physics education in the year of the discovery of the electron [1897]." O. Knudsen, "O. W. Richardson and the Electron Theory of Matter, 1901–1916," in Buchwald and Warwick (eds.), *Histories of the Electron,* pp. 227–253, on p. 228.

42. R. A. Millikan, *The Electron: Its Isolation and Measurement and the Determination of Some of Its Properties,* 2nd ed. (Chicago: Univ. of Chicago Press, 1924; 1st ed., 1917), pp. 42–43.

moved by Millikan's own experimental work. As he claimed in his Nobel Lecture, "The most direct and unambiguous proof of the existence of the electron will probably be generally admitted to be found in an experiment which for convenience I will call the oil-drop experiment." [43] The self-serving character of this assertion notwithstanding, it points the way to an adequate understanding of the process that established the reality of the electron. That process was an extended one, stretching from Faraday to Millikan, and culminating in the oil-drop experiments.

All of these excerpts point to a similar way of understanding the early history of the representation of the electron, namely, as a gradual process of accumulation of evidence to the effect that a novel subatomic entity, the electron, exists. Furthermore, the reception of the electron hypothesis was a very complex process that depended on several factors: the problems that a scientist was working on, his disciplinary affiliation, and his philosophical commitments. For some scientists, the electron qua subatomic particle had "alchemical" connotations, which were clearly undesirable. As we will see, for that reason George Francis FitzGerald rejected the representation of the electron as a subatomic, material particle. Some chemists were also skeptical for similar reasons. Mendeleev, for instance, who was a believer in the atomic theory, rejected the "vague hypothesis of electrons." [44] His refusal to accept the existence of electrons was also due to their alchemical implications. If electrons were real, then atoms would be divisible and that would bring the specter of the transmutation of elements. [45]

Moreover, Henry E. Armstrong, in the 1899 British Association for the Advancement of Science meeting at Dover, expressed his skepticism toward the electron qua detachable constituent of the atom. [46] Ten years later, in his presidential address to the chemistry section of the 1909 BAAS meeting, he voiced once again his strong reservations toward the existence of elec-

---

43. R. A. Millikan, "The Electron and the Light-Quant from the Experimental Point of View," Nobel Lecture, 23 May 1924, in Nobel Lectures: Physics, 1922–1941 (Amsterdam: Elsevier, 1965), pp. 54–66, on p. 55.

44. D. Mendeleev, An Attempt towards a Chemical Conception of the Ether (London: Longmans, Green and Co., 1904); quoted in R. Kargon, "Mendeleev's Chemical Ether, Electrons, and the Atomic Theory," Journal of Chemical Education, 42, no. 7 (1965): 388–389, on p. 388.

45. For Mendeleev's aversion to the possibility of the transmutation of elements, see M. D. Gordin, "Making Newtons: Mendeleev, Metrology, and the Chemical Ether," Ambix, 45, no. 2 (July 1998): 96–115, esp. pp. 99–100. Cf. also H. Kragh, "The Electron, the Protyle, and the Unity of Matter," in Buchwald and Warwick (eds.), Histories of the Electron, pp. 195–226, on p. 209.

46. See Gooday, "The Questionable Matter of Electricity," p. 114.

trons.[47] These examples reinforce the view that the reception of the electron hypothesis was not a linear and straightforward process. I could go on with similar examples. The fact of the matter is that we lack a systematic study of that reception in different contexts.[48]

Finally, the gradual acceptance of the electron hypothesis was entangled with the debate over the existence of atoms.[49] The belief in the existence of the electron qua subatomic particle presupposed a conviction in the existence of atoms. If one believes in subatomic constituents of all atoms, then one has to believe in atoms. Conversely, one who rejects the atomic hypothesis has to reject ipso facto the electron hypothesis. The logical point I just made may seem like an exercise in a priori history. However, it is borne out in the writings of atomists and antiatomists alike. A straightforward connection between the atomism debate and the electron hypothesis was perceived by Lorentz, who suggested that the grounds for skepticism toward atoms could also throw doubt on the existence of electrons: "[T]he theory of electrons is to be regarded as an extension to the domain of electricity of the molecular and atomistic theories that have proved of so much use in many branches of physics and chemistry. Like these, it is apt to be viewed unfavourably by some physicists, who prefer to push their way into new and unexplored regions by following those great highways of science which we possess in the laws of thermodynamics, or who arrive at important and beautiful results, simply by describing the phenomena and their mutual relations by means of a system of suitable equations."[50] Some of the main opponents of the atomic hypothesis were unfavorably disposed toward the electron. Pierre Duhem, for instance, detested atoms and electrons alike.[51] The same was true of Ernst Mach.[52]

47. See H. E. Armstrong, presidential address, *British Association for the Advancement of Science, Transactions of Section B,* 1909 (delivered on 26 Aug.), pp. 420–454, on p. 432. I would like to thank Kostas Gavroglu for bringing Armstrong's address to my attention.

48. In a recent paper, Graeme Gooday examined the reception of the electron among electrical engineers. Gooday's analysis provides additional support for the approach I have advocated here. See Gooday, "The Questionable Matter of Electricity," esp. pp. 108–109, 111.

49. This entanglement has not been adequately addressed in the historical literature. An exception is Benoit Lelong, who studied how the skepticism of some French scientists toward atoms shaped their experimental investigation of cathode rays. See his "Paul Villard, J. J. Thomson, and the Composition of Cathode Rays," pp. 144–145, 147, 149.

50. H. A. Lorentz, *The Theory of Electrons,* 2nd ed. (Leipzig: Teubner, 1916; 1st ed., 1909), p. 10.

51. See P. Duhem, *The Aim and Structure of Physical Theory* (Princeton: Princeton Univ. Press, 1991), p. 304; and Nye, *Molecular Reality,* p. 166.

52. See E. Mach, *Knowledge and Error: Sketches on the Physchology of Enquiry* (Dordrecht: Reidel, 1976; 1st German ed., 1905), p. 77.

Here it is worth pointing out that Thomson's experiments with cathode rays were perceived by some antiatomists as a reductio ad absurdum of the atomistic doctrine. When the dispute between Millikan and Felix Ehrenhaft about the reality of the electron was under way, one of Mach's followers, Anton Lampa, wrote to Mach: "If the provisional measurements should be confirmed which Ehrenhaft carried out when I was now in Vienna as part of his continuing research on the charges on colloidal particles, then the electron would be divisible. . . . It would be just too beautiful if now the electron were to undergo the same fate as the atom did as a result of cathode rays."[53] So, from the point of view of an antiatomist the experimental investigation of cathode rays may have undermined the atomic theory, which postulated the existence of indivisible units of matter. What remains to be done is to explore in detail the attitudes of the antiatomic opposition toward the electron. This task, however, is beyond the scope of this chapter.[54]

At this point, one may object that I am confusing two issues: the acceptance of the electron qua constituent of all atoms and the acceptance of the electron qua atom of electricity. It may be argued that a prior belief in the existence of atoms was not necessary for resolving the issue of the existence of the electron qua atom of electricity.[55] However, as we will see below in Hermann von Helmholtz's atomistic interpretation of Faraday's work on electrolysis, the belief in the atomicity of matter was necessary for the hypothesis of the atomic structure of electricity to get off the ground. Furthermore, the establishment of the atomic nature of electricity was also a gradual process, stretching from Faraday to Millikan, whose achievement consisted in showing that the atomic structure of electricity was not a statistical delusion. Millikan's balanced-drop method "made it possible to examine the properties of individual isolated electrons and to determine whether different ions actually carry one and the same charge. That is to say, it now became possible to determine whether electricity in gases and solutions is actually built up out of electrical atoms, each of which has exactly the same value, or whether the electron which had first made its appearance in Faraday's experiments on solutions . . . is after all only a *statistical mean* of charges which are themselves

53. Lampa to Mach, 1 May 1910; quoted in Holton, "Subelectrons, Presuppositions and the Millikan-Ehrenhaft Dispute," p. 82.

54. See my "The Discovery of the Electron and the Atomism Debate" (unpublished manuscript).

55. Note that Thomson originally proposed the "corpuscle" as a subatomic particle; this turned out to be the most controversial of his claims about the constituents of cathode rays.

greatly divergent."[56] If the electron were a *statistical mean,* then it would not have a real existence, in the sense that the "average man" does not have a real existence. It is just a statistical construct.

I hope that by now the notion of the "discovery of the electron" has become a problem demanding explanation, as opposed to an explanatory resource. How did it happen that a complex, multidimensional, and long process was collapsed to a discovery episode? How did it happen that, whereas for turn-of-the-century physicists and chemists the electron had not been discovered by any particular scientist, in the 1920s and 1930s Thomson was hailed as the "discoverer" of the electron? I will not try to address these questions here, partly because I am not in a position to provide satisfactory answers to them.[57] Rather, I will proceed with my own account of the birth and infancy of the representation of the electron, an account that embodies the historiographical approach I sketched in the previous chapter.

56. Millikan, *The Electron,* p. 58.

57. There is some scholarship that broaches these questions, but much still remains to be learned. See, e.g., Falconer, "Corpuscles to Electrons"; and Gooday, "The Questionable Matter of Electricity."

## Chapter 4 | The Birth and Infancy of the Representation of the Electron

### 1. Introduction

In this chapter I discuss the early life of the electron qua theoretical entity. Sections 2–6 contain a reconstruction of the birth and infancy of the representation of the electron. The aim of these sections is twofold: first, to unfold some of the main aspects of the process that led to the consolidation of the electron in the ontology of physics, and thereby to highlight the complexity of that process, and second, to show how the early characteristics of the electron qua theoretical entity, which turned out to be crucial for the subsequent development of its character, were formed. The chapter ends with some reflections on the historiographical and philosophical issues raised in the previous chapters, in light of the historical material discussed in the previous sections.

### 2. The Birth of the Term "Electron"

The name "electron" was introduced by George Johnstone Stoney in 1891 to denote an elementary quantity of electricity.[1] At the Belfast meeting of the British Association for

---

1. See G. J. Stoney, "On the Cause of Double Lines and of Equidistant Satellites in the Spectra of Gases," *Scientific Transactions of the Royal Dublin Society,* 2nd series, 4 (1888–1892): 563–608, see p. 583. Cf. J. G. O'Hara, "George Johnstone Stoney, F.R.S., and the Concept of the Electron," *Notes and Records of the Royal Society of London,* 29 (1975): 265–276; and J. G. O'Hara, "George Johnstone Stoney and the Conceptual Discovery of the Electron," in *Stoney and the Electron: Papers from a Seminar Held on November 20, 1991, to Commemorate the Centenary of the Naming of*

the Advancement of Science in 1874 Stoney had already suggested that "[n]a-ture presents us in the phenomenon of electrolysis, with a single definite quantity of electricity which is independent of the particular bodies acted on."[2] In 1891 he proposed, "[I]t will be convenient to call [these elementary charges] *electrons.*"[3] Stoney's electrons were permanently attached to atoms; that is, they could "not be removed from the atom," and each of them was "associated in the chemical atom with each bond." Furthermore, their oscil-lation within molecules gave rise to "electro-magnetic stresses in the sur-rounding aether."[4]

Even though Stoney coined the term "electron," the representation asso-ciated with that term had several ancestors.[5] Key aspects of that representa-tion, most notably the notion of the atomicity of charge, considerably pre-ceded his proposal. In the period between 1838 and 1851 a British natural philosopher, Richard Laming, conjectured "the existence of sub-atomic, unit-charged particles and pictured the atom as made up of a material core sur-rounded by an 'electrosphere' of concentric shells of electrical particles."[6] On the Continent several physicists had made similar suggestions. Those physi-cists attempted to explain electromagnetic phenomena by action-at-a-distance forces between electrical particles. As an example of the Continental approach to electrodynamics consider Wilhelm Weber's electrical theory of matter and ether.[7] Weber's theory originated in 1846 and continued to evolve till the time of his death (1891). According to the initial version of that theory, electricity consisted of two electrical fluids (positive and negative). The interactions of these fluids were governed by inverse square forces, which were functions of

---

*the Electron* (Dublin: Royal Dublin Society, 1993), pp. 5–28. The introduction of a new term is an event that can be easily identified and, thus, provides a convenient starting point for a bio-graphical narrative whose subject is the corresponding representation. The appearance of a new term also signals the birth of a novel concept, whose identity has not yet solidified. Thus, it is not surprising that in its subsequent development the concept may merge with other re-lated concepts. As we will see below, this is what happened in the case of the electron.

2. Stoney's paper was first published in 1881. See G. J. Stoney, "On the Physical Units of Na-ture," *Scientific Proceedings of the Royal Dublin Society,* new series, 3 (1881–1883): 51–60, on p. 54.

3. Stoney, "On the Cause of Double Lines," p. 583.

4. Ibid.

5. Note that the biographical approach can also come to grips with the "prehistory" of the electron's representation.

6. Kragh, "Concept and Controversy: Jean Becquerel and the Positive Electron," *Centau-rus,* 32 (1989): 203–240, on p. 205.

7. For an extended discussion of Weber's program see M. N. Wise, "German Concepts of Force, Energy, and the Electromagnetic Ether, 1845–1880," in G. N. Cantor and M. J. S. Hodge (eds.), *Conceptions of Ether: Studies in the History of Ether Theories, 1740–1900* (Cambridge: Cam-bridge Univ. Press, 1981), pp. 269–307, esp. pp. 276–283.

their relative velocity and their relative acceleration. In the 1870s Weber attempted to construct an electrical theory of matter, based on discrete electrical particles, suggesting "that each of the identical ponderable atoms combining to form chemical elements was a neutral system consisting of a negative, but highly massive, central particle with a positive satellite of much smaller mass." [8] Thus, Weber's "electrical particles" shared some aspects with the subsequently introduced "electrons": both denoted particles of electricity and subatomic constituents of all atoms.

Another central aspect of the representation of the electron, the notion that electricity (like matter) has an atomic structure, emerged as a result of Faraday's experiments on electrolysis and his enunciation of the "second law of electrochemistry." [9] This law stated that when the same quantity of electricity passed through different electrolytic solutions, the amounts of the decomposed substances were proportional to their chemical equivalents. What Faraday meant by "chemical equivalents" is revealed by the following passage from his *Experimental Researches in Electricity:* "the equivalent weights of bodies are simply those quantities of them which contain equal quantities of electricity, or have naturally equal electric powers; it being the ELECTRICITY which *determines* the equivalent number, *because* it determines the combining force. Or if we adopt the atomic theory or phraseology, then the atoms of bodies which are equivalents to each other in their ordinary chemical action, have equal quantities of electricity naturally associated with them. But I must confess I am jealous of the term *atom;* for though it is very easy to talk of atoms, it is very difficult to form a clear idea of their nature." [10] For someone like Faraday who was "jealous of the term *atom*" the regularities in the phenomena of electrolysis need not have implied the atomic character of electricity. The inference of the atomicity of charge from "Faraday's law" required a prior conviction of the atomic nature of matter and was clearly drawn by Helmholtz in 1881. In his Faraday Lecture to the Fellows of the Chemical Society Helmholtz argued that "the most startling result of Faraday's law is perhaps this. If we accept the hypothesis that the elementary substances are composed of atoms, we cannot avoid concluding that electricity also, positive as well as negative, is divided into definite elementary portions, which behave

8. Ibid., p. 282. Cf. R. McCormmach, "H. A. Lorentz and the Electromagnetic View of Nature," *Isis,* 61 (1970): 459–497, on p. 472.

9. See L. P. Williams, *Michael Faraday: A Biography* (New York: Da Capo Press, 1987), pp. 256–257.

10. M. Faraday, *Experimental Researches in Electricity,* 3 vols. (London, 1839–1855), vol. 1, p. 256, par. 869.

like atoms of electricity."[11] So the hypothesis of the atomicity of matter made possible the hypothesis of the atomic nature of electricity. If one did not accept the existence of atoms of matter, the existence of atoms of electricity did not follow from the regularities in electrolysis.

As I mentioned above, Helmholtz's conclusion had already been drawn by Stoney in 1874. Its significance is shown by the fact that Stoney was eager to reaffirm his priority over Helmholtz when he discovered that it was (inadvertently) challenged. In a letter to the editors of the *Philosophical Magazine* on 4 September 1894 he disputed a claim that had appeared in the same month's issue, according to which "Von Helmholtz, on the basis of Faraday's Law of Electrolysis, *was the first* to show in the case of electrolytes that each valency must be considered charged with a minimum quantity of electricity, . . . which like an electrical atom is no longer divisible."[12] In rebutting this claim Stoney emphasized that

> I had already twice pointed out this remarkable fact: first, at the Belfast meeting of the British Association in August 1874, in a paper "On the Physical Units of Nature," in which I called attention to this minimum quantity of electricity as one of three . . . physical units, the absolute amounts of which are furnished to us by Nature. . . . This same paper was again read before the Royal Dublin Society on the 16th of February, 1881, and is printed both in the Proceedings of that meeting and in the Phil. Mag. of the following May. . . . In this paper an estimate was made of the actual amount of this most remarkable fundamental unit of electricity, for which I have since ventured the name *electron*.[13]

Even though Stoney's reaffirmation of his priority over Helmholtz betrays the importance of the proposal of the atomicity of electricity, it lacks, like most questions of priority, any further historiographical significance. What is important is that this proposal and its reformulation by Stoney in terms of the electron hypothesis played a significant role in a different con-

11. H. von Helmholtz, "On the Modern Development of Faraday's Conception of Electricity," *Journal of the Chemical Society,* 39 (1881): 277–304, on p. 290. Note, however, that, pace Helmholtz, a belief in the atomicity of matter did not necessarily lead to a conviction of the atomic nature of electricity. Some atomists (e.g., Maxwell and Kelvin) doubted the atomicity of electricity. See J. C. Maxwell, *A Treatise on Electricity and Magnetism,* 2 vols. (Oxford: Clarendon Press, 1873), vol. 1, pp. 312–313.

12. Quoted in G. J. Stoney, "Of the 'Electron,' or Atom of Electricity," *Philosophical Magazine,* 5th series, 38 (1894): 418–420, on p. 418.

13. Ibid., pp. 418–419.

text, that of electromagnetic theory. In the 1890s two of the main electromagnetic theories were those of H. A. Lorentz and Joseph Larmor. The entities postulated by those theories ("ions" and "electrons," respectively) were crucially transformed, as a result of an experimental discovery (the magnetic splitting of spectral lines) by the Dutch physicist Pieter Zeeman. His discovery not only provided evidence for the existence of those entities, but also led to a specification of two of their key properties, their charge-to-mass ratio and the sign of their charge. Shortly after Zeeman's discovery, those entities were identified and denoted by a single name, "electrons." Before we discuss these developments, however, we should take a look at Zeeman's discovery itself.

### 3. The Discovery of the Zeeman Effect:
### The First Experimental Manifestation of the Electron

Pieter Zeeman (1856–1943) began to study magneto-optic phenomena in 1890, as Lorentz's assistant at the University of Leiden.[14] The first phenomenon he investigated was the Kerr effect—the rotation of the plane of polarization of light upon reflection from a magnetized substance. The investigation of this phenomenon was also the subject of his doctoral dissertation, which he completed in 1893 under the supervision of Kamerlingh Onnes.[15] In the course of that research he made an unsuccessful attempt to detect the influence of a magnetic field on the sodium spectrum.[16] Several years later, inspired by reading "Maxwell's sketch of Faraday's life" and finding out that "Faraday thought of the possibility of the above mentioned relation [between magnetism and light]," he thought "it might yet be worthwhile to try the ex-

---

14. For a more detailed account of Zeeman's path to his discovery see my "The Discovery of the Zeeman Effect: A Case Study of the Interplay between Theory and Experiment," *Studies in History and Philosophy of Science,* 23 (1992): 365–388. Cf. also J. B. Spencer, *An Historical Investigation of the Zeeman Effect (1896–1913)* (Ph.D. dissertation, Univ. of Wisconsin, 1964); and K. Hentschel, "Die Entdeckung des Zeeman-Effekts als Beispiel für das komplexe Wechselspiel von wissenschaftlichen Instrumenten, Experimenten und Theorie," *Physikalische Blätter,* 52 (1996): 1232–1235. Another recent study, based on an examination of Zeeman's laboratory notebooks, is A. J. Kox, "The Discovery of the Electron," pt. 2, "The Zeeman Effect," *European Journal of Physics,* 18 (1997): 139–144.

15. See J. B. Spencer, "Zeeman, Pieter," in C. C. Gillispie (ed.), *Dictionary of Scientific Biography,* 16 vols. (New York: Charles Scribner's Sons, 1970–1980), vol. 14, pp. 597–599.

16. See P. Zeeman, "On the Influence of Magnetism on the Nature of the Light Emitted by a Substance," pt. 1, *Communications from the Physical Laboratory at the University of Leiden,* 33 (1896): 1–8, on p. 1; trans. from *Verslagen van de Afdeeling Natuurkunde der Koninklijke Akademie van Wetenschappen te Amsterdam,* 31 Oct. 1896, p. 181. There is no unpublished record of this early attempt. See Kox, "The Discovery of the Electron," pp. 139–140.

periment again with the excellent auxiliaries of the spectroscopy of the present time."[17] This time the experiment turned out to be a success.[18]

Zeeman placed the flame of a Bunsen burner between the poles of an electromagnet and held a piece of asbestos impregnated with common salt in the flame. After turning on the electromagnet, the two D-lines of the sodium spectrum, which had been previously narrow and sharply defined, were clearly widened. In shutting off the current the lines returned to their former condition. Zeeman then replaced the Bunsen burner with a flame of "lightgas" fed with oxygen and repeated the experiment. The spectral lines were again clearly broadened. Replacing the sodium with lithium, he observed the same phenomena.

Zeeman was not convinced that the observed widening was due to the action of the magnetic field directly upon the emitted light. The effect could be caused by an increase of the radiating substance's density and temperature. As noted by Zeeman, a similar phenomenon had been reported by Ernst Pringsheim in 1892.[19] Since the magnet caused an alteration of the flame's shape, a subsequent change of the flame's temperature and density was also possible. To exclude this possibility, Zeeman tried another, more complicated experiment. He put a porcelain tube horizontally between the poles of the electromagnet, with the tube's axis perpendicular to the direction of the magnetic field (see fig. 4.1). Two transparent caps were attached to each terminal of the tube, and a piece of sodium was introduced into the tube. Simultaneously the tube's temperature was raised by the Bunsen burner. At the same time the light of an electric lamp was guided by a metallic mirror to traverse the entire tube.

In the next stage of the experiment the sodium, under the action of the Bunsen flame, began to gasify. The absorption spectrum was obtained by means of the Rowland grating, and finally the two sharp D-lines of sodium were observed. The heterogeneity of the density of the vapor at different

17. Zeeman, "On the Influence of Magnetism," pt. 1, p. 3. Those "excellent auxiliaries" included a concave Rowland grating that had recently been acquired by the Physical Laboratory of the University of Leiden. The exceptional resolving power of that instrument played a crucial role in Zeeman's discovery. See Arabatzis, "The Discovery of the Zeeman Effect," p. 372; and Hentschel, "Die Entdeckung des Zeeman-Effekts," p. 1233.

18. There is a record of this experiment in Zeeman's laboratory notebook, dated 2 Sept. 1896. See Kox, "The Discovery of the Electron," p. 140. The description that follows is from Zeeman's published paper, which appeared on 31 Oct. 1896. Kox's analysis of Zeeman's notebooks shows that Zeeman's publications gave a very faithful account of his actual research.

19. See E. Pringsheim, "Kirchhoff'sches Gesetz und die Strahlung der Gase," *Wiedmannsche Annalen der Physik*, 45 (1892): 428–459, on pp. 455–457.

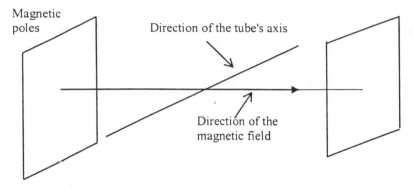

**Fig. 4.1** This diagram represents Zeeman's experiment for detecting the influence of a magnetic field on the absorption spectrum of sodium.

heights of the tube produced a corresponding asymmetry in the lines' width, making them thicker at the top. By activating the electromagnet the lines became broader and darker. When it was turned off the lines recovered their initial form.

Zeeman, however, was still skeptical about whether the experiment's purpose, to demonstrate the direct effect of magnetism on light, had been accomplished. The temperature difference between the upper and lower parts of the tube was responsible for the heterogeneity of the vapor's density. The vapor was denser at the top of the tube and, since their width at a given height depended on the number of incandescent particles at that height, the spectral lines were thicker at the top. It was conceivable that the activation of the magnetic field could give rise to differences of pressure in the tube of the same order of magnitude as and in the opposite direction to those produced by the differences of temperature. If this were the case, the action of magnetism would move the denser layers of vapor toward the bottom of the tube and would alter in this way the width of spectral lines without interacting directly with the light that generated the spectrum.

To exclude the possibility of these phenomena, which would undermine the experiment's aim, Zeeman performed a more refined experiment. He used a smaller tube and heated it with a blowpipe to eliminate disturbing temperature differences. Moreover, he rotated the tube around its axis and thus achieved equal densities of sodium vapor at all heights. The D-lines were now uniformly wide along their whole length. The subsequent activation of the electromagnet resulted in their uniform broadening.

Zeeman was by then nearly convinced that the outcome of his experiments was due to the influence of magnetism directly upon the light emitted or absorbed by sodium: "The different experiments . . . make it more and more probable, that the absorption—and hence also the emission—lines of an incandescent vapor, are widened by the action of magnetism." [20] The sentence immediately following is instructive with respect to the theoretical significance of Zeeman's experimentation: "Hence if this is really the case, then by the action of magnetism in addition to the free vibrations of the *atoms, which are the cause of the line spectrum,* other vibrations of changed period appear" (emphasis added).

It is evident that Zeeman identified the origin of spectral lines with the vibration of atoms. H. A. Lorentz, Zeeman's mentor and collaborator, had developed a theory of electromagnetic phenomena that accounted for the emission of light in this way. As the above excerpt indicates, Lorentz's theory could be used to provide a theoretical understanding of Zeeman's experimental discovery. As it turned out, that theory guided Zeeman's subsequent experimental researches and was, in turn, shaped by them. Through that interplay of theory and experiment, Lorentz's "ions" were consolidated and refined, and in the process transformed into "electrons." [21] Let us examine more closely the state of Lorentz's theory at that time.

## 4. Lorentz's "Ion": A Somewhat Startling Hypothesis

In 1878 Lorentz had already suggested that the phenomenon of dispersion could be explained by assuming that molecules are composed of charged particles that may perform harmonic oscillations.[22] In 1892 he developed a unification of the Continental and the British approaches to electrodynamics that incorporated those particles. From the British approach he borrowed the notion that electromagnetic disturbances travel at the speed of light. That is, his theory was a field theory that dispensed with action at a distance. From the Continental approach he borrowed the conception of electric charges as

20. Zeeman, "On the Influence of Magnetism," pt. 1, p. 8.

21. Here I should point out that the attribution of the novel phenomenon to the entities postulated by Lorentz and its subsequent use for the articulation of the representation of those entities were made possible, in the first place, by Zeeman's remarkable experimental knowledge and skill. These qualities were displayed in the strategies he devised for distinguishing possible artifacts of the experimental apparatus from clues to the phenomenon under investigation. The outcome of those strategies was a strong argument that the experimental results obtained were a direct manifestation of a deeper reality, and not due to some superficial features of the experimental situation.

22. See H. A. Lorentz, "Concerning the Relation between the Velocity of Propagation of Light and the Density and Composition of Media," in his *Collected Papers,* ed. P. Zeeman and A. D. Fokker, 9 vols. (The Hague: Martinus Nijhoff, 1935–1939), vol. 2, pp. 1–119.

ontologically distinct from the field. Whereas in Maxwell's theory charges were mere epiphenomena of the field, in Lorentz's theory they became the sources of the field.[23]

The aim of Lorentz's combined approach, in 1892, was to analyze electromagnetic phenomena in moving bodies. That analysis required a model of the interaction between matter and ether. The notion of "charged particles," an ancestor of the electron qua theoretical entity, provided him with a means of handling this problem.[24] The interaction in question could be understood if one reduced all "electrical phenomena to . . . [the] displacement of these particles."[25] The movement of a charged particle altered the state of the ether, which, in turn, influenced the motion of other particles. Furthermore, macroscopic charges were "constituted by an excess of particles whose charges have a determined sign, [and] an electric current is a true stream of these corpuscles."[26] This proposal was similar to the familiar conception of the passage of electricity through electrolytic solutions and metals. Thus, Lorentz constructed the representation of "charged particles" in such a way that several characteristics of macroscopic phenomena had counterparts in the properties and behavior of those entities. For instance, the charge of macroscopic bodies corresponded to the charge of those particles, and the unobservable counterpart of electric currents became the motion of those particles.

It is worth pointing out that in the last section of his 1892 paper Lorentz deduced a formula for the velocity of light in moving media that had been derived by Augustin-Jean Fresnel on the assumption that the ether was dragged

23. H. A. Lorentz, "La théorie électromagnétique de Maxwell et son application aux corps mouvants," in his Collected Papers, vol. 2, pp. 164–343, esp. p. 229. Cf. R. McCormmach, "Einstein, Lorentz, and the Electron Theory," Historical Studies in the Physical Sciences, 2 (1970): 41–87. In a letter to Rayleigh on 18 Aug. 1892 Lorentz characterized the "ions" as "a supposition which may appear somewhat startling but which may, as I think, serve as a working hypothesis." Quoted in N. J. Nersessian, Faraday to Einstein: Constructing Meaning in Scientific Theories (Dordrecht: Martinus Nijhoff, 1984), p. 108.

24. See T. Hirosige, "Origins of Lorentz' Theory of Electrons and the Concept of the Electromagnetic Field," Historical Studies in the Physical Sciences, 1 (1969): 151–209, on pp. 178–179, 198; and Nersessian, Faraday to Einstein, p. 98.

25. "[L]es phénomènes électriques sont produits par le déplacement de ces particules." Lorentz, "La théorie électromagnétique de Maxwell," p. 228.

26. "[U]ne charge électrique est constituée par un excès de particules dont les charges ont un signe déterminé, un courant électrique est un véritable courant de ces corpuscules." Lorentz, "La théorie électromagnétique de Maxwell," pp. 228–229. Cf. J. L. Heilbron, A History of the Problem of Atomic Structure from the Discovery of the Electron to the Beginning of Quantum Mechanics (Ph.D. dissertation, Univ. of California, Berkeley, 1964), p. 98.

by moving matter. Lorentz's derivation, however, discarded that assumption and capitalized on the influence of light on moving charged particles. The latter were forced to vibrate by the ethereal waves constituting light and gave rise to a complex interaction that produced the effect named after Fresnel. Lorentz's analysis enhanced considerably the credibility of his theory and facilitated the acceptance of his "charged particles" as real entities.[27]

In 1895 he explicitly associated those particles with the ions of electrolysis.[28] The transformation of "ions" to "electrons" took place as a result of Zeeman's experimental discovery, which after its initial stage was dominated by Lorentz's theory. To understand how this transformation took place it is necessary to examine Lorentz's theoretical analysis of Zeeman's initial results and its role in guiding further Zeeman's experimental research. The first form of that analysis is recorded in Zeeman's second paper on his celebrated discovery.[29] Zeeman initially thought that Lorentz's theory could provide an explanation of his experimental results. Thus, he asked Lorentz to provide a quantitative treatment of the influence of magnetism on light:

> Prof. Lorentz to whom I communicated these considerations, at once kindly informed me of the manner, in which according to his theory the motion of an ion in a magnetic field is to be calculated, and pointed out to me that, if the explanation following from his theory was true, the edges of the lines of the spectrum ought to be circularly polarized. The amount of widening might then be used to determine the ratio of charge and mass to be attributed in this theory to a particle giving out the vibrations of light.
>
> The above mentioned extremely remarkable conclusion of Prof. Lorentz relating to the state of polarization in the magnetically widened line, I have found to be fully confirmed by experiment.[30]

As I mentioned above, the emission of light, according to Lorentz, was a direct result of the vibrations of small electrically charged particles ("ions"),

27. Cf. N. J. Nersessian, "Hendrik Antoon Lorentz," in *The Nobel Prize Winners: Physics* (Pasadena, CA: Salem Press, 1989), pp. 35–42, esp. p. 39.

28. H. A. Lorentz, "Versuch einer Theorie der elektrischen und optischen Erscheinungen in bewegten Körpern," in his *Collected Papers*, vol. 5, pp. 1–137, esp. p. 5.

29. See P. Zeeman, "On the Influence of Magnetism on the Nature of the Light Emitted by a Substance," pt. 2, *Communications from the Physical Laboratory at the University of Leiden, 33* (1896): 9–19; translated from *Verslagen van de Afdeeling Natuurkunde der Koninklijke Akademie van Wetenschappen te Amsterdam,* 28 Nov. 1896, p. 242.

30. Ibid., p. 12.

which exist in all material bodies. In the absence of a magnetic field an ion would oscillate about an equilibrium point under the action of an elastic force. The influence of a magnetic field would alter the mode of vibration of the ion. Suppose that an ion is moving in the $xy$-plane under the action of a uniform magnetic field, which is parallel to the $z$-axis. The equations of motion are

$$m\frac{d^2x}{dt^2} = -k^2x + eH\frac{dy}{dt}$$

and

$$m\frac{d^2y}{dt^2} = -k^2y - eH\frac{dx}{dt},$$

where $e$ and $m$ are the charge and the mass of the ion, respectively, and $H$ is the intensity of the magnetic field. The first term on the right side of the equations denotes the elastic force and the second term represents the force due to the magnetic field (the "Lorentz force"). Assuming that

$$x = a\,e^{st} \quad \text{and} \quad y = \beta\,e^{st},$$

we get

$$m\,s^2\,a = -k^2\,a + e\,H\,s\,\beta$$

and

$$m\,s^2\,\beta = -k^2\,\beta - e\,H\,s\,a.$$

In the absence of a magnetic field ($H = 0$), we can easily obtain the period of vibration of the ion:

$$s = i\frac{k}{\sqrt{m}} = i\frac{2\pi}{T} \Rightarrow T = \frac{2\pi\sqrt{m}}{k}.$$

When a magnetic field is present the period becomes

$$s \cong i\frac{k}{\sqrt{m}}\left(1 \pm \frac{eH}{2k\sqrt{m}}\right) \Rightarrow T' \cong \frac{2\pi\sqrt{m}}{k}\left(1 \pm \frac{eH}{2k\sqrt{m}}\right).$$

It follows that

$$(4.1) \qquad \frac{T' - T}{T} = \frac{eH}{2k\sqrt{m}} = \frac{e}{m}\frac{HT}{4\pi}.$$

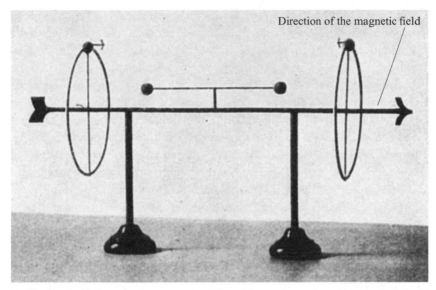

Direction of the magnetic field

**Fig. 4.2** "A model for the simple electronic motions." (From P. Zeeman, *Researches in Magneto-Optics: With Special Reference to the Magnetic Resolution of Spectrum Lines* [London: Macmillan, 1913], p. 32.)

The physical implications of this analysis are as follows:[31] In the general case, the oscillation of the ion has an arbitrary direction in space. In the absence of a magnetic field the motion of the ion can be resolved into three components: a linear oscillation and two circular oscillations in a plane perpendicular to the first. All three oscillations have the same frequency, and the two circular ones have opposite directions (see fig. 4.2). When a magnetic field is present, the oscillations along the direction of the field remain unaltered. But one of the circular components is accelerated, while the other is retarded. Thus, under the influence of magnetism the charged particle will yield three distinct frequencies. If the particle is observed along the direction of the field, a doublet of lines will be seen. Each line represents circularly polarized light. If it is observed in a direction perpendicular to the field, a triplet of lines will be seen. The middle component represents plane-polarized light, its plane of polarization being parallel to the field. The two outer components also represent plane-polarized light, but their plane of polarization is perpendicular to the field.

All these theoretical expectations were subsequently confirmed by experiments designed specifically to detect them. In the same paper that contained

31. Ibid., p. 16.

Lorentz's analysis Zeeman confirmed that the polarization of the edges of the broadened lines followed the theoretical predictions. Lorentz considered the confirmation of his predictions "direct proof for the existence of ions."[32] Furthermore, Zeeman estimated the order of magnitude of the ratio $e/m$. As we saw, the change in the period of vibration of an ion due to the influence of a magnetic field depends on $e/m$ (see equation [4.1] above). Thus, the widening of spectral lines, which is a reflection of the alteration in the mode of vibration of an ion, is proportional to the ionic charge-to-mass ratio. According to Zeeman's approximate measurements a magnetic field of 10,000 Gauss produced a widening of the D-lines equal to 2.5 percent of their distance. From the observed widening of the spectral lines, Zeeman calculated (using equation [4.1]) $e/m$, which turned out to be unexpectedly large ($10^7$ e.m.u.). As he recalled, when he announced the result of his calculation to Lorentz, the latter's response was: "That looks really bad; it does not agree at all with what is to be expected."[33]

Here we see one of the characteristics of theoretical entities I discussed in chapter 2, namely, their relative independence from the intentions of their makers. Once an experimental situation (Zeeman's) was interpreted as a manifestation of an unobservable entity (Lorentz's ion), some features of that situation were used for the determination of some of the properties of the entity in question. Spectral lines were attributed to the vibrations of ions, and thereby the features of the former were linked with the properties and behavior of the latter, in three ways. First, the frequencies of spectral lines corresponded to the frequencies of vibration of the ions. Second, the polarization of spectral lines was linked with the direction and mode of vibration of the

32. We know that from an entry in Zeeman's diary, dated 23 Nov. 1896. See Kox, "The Discovery of the Electron," p. 142.

33. Cited in Zeeman, "Faraday's Researches on Magneto-Optics and Their Development," *Nature,* 128 (1931): 365–368, on p. 367. Even though Lorentz associated his "ions" with the ions of electrolysis, he did not specify, to the best of my knowledge, their charge-to-mass ratio. Thus, there is no conclusive evidence that he assumed his "ions" to be of an order of magnitude comparable with an atom's. Further indirect evidence is found in O. Lodge, "The History of Zeeman's Discovery and Its Reception in England," *Nature,* 109 (1922): 66–69, on p. 67. There we read that Larmor, *"like everyone else at that time, . . .* considered that the radiating body must be an atom or part of an atom with an $e/m = 10^4$" (emphasis added). Moreover, Zeeman, while discussing various aspects of his discovery, remarked that "[t]he value found [for the charge-to-mass ratio of the "ion"] is about 1500 times that of the corresponding value which can be derived for hydrogen from the phenomena of electrolysis. *This was something entirely new in 1896"* (emphasis added). P. Zeeman, *Researches in Magneto-Optics: With Special Reference to the Magnetic Resolution of Spectrum Lines* (London: Macmillan, 1913), pp. 39–40.

ions. Third, the magnitude of the magnetic splitting of spectral lines was co-ordinated with their charge-to-mass ratio, assuming that the ions were subject to an elastic force, the Lorentz force, and Newton's second law of motion. On the basis of this coordination, Zeeman's experimental results could be used to determine one of the properties of Lorentz's ion, namely, its charge-to-mass ratio. In that way, those results were embodied in the representation of the ion, which in the process turned out to be very different from what its creator had intended. Thus, the fact that the ion qua theoretical entity had an exper-imentally determined component made it partly independent of Lorentz's theoretical framework. A mark of that independence was the violation of his expectations about the charge-to-mass ratio of the ion.

It should be noted that this was the first estimate of that ratio that indi-cated that the "ions" did not refer to the well-known ions of electrolysis, but corresponded instead to extremely minute subatomic particles. J. J. Thom-son's measurement of the mass-to-charge ratio of the particles that consti-tuted cathode rays was announced several months later and was in close agreement with Zeeman's result.[34] It is worth pointing out that the priority of Zeeman over Thomson was not always acknowledged. Oliver Lodge, for in-stance, claimed that Zeeman's results were obtained after Thomson's mea-surements.[35] Not surprisingly, Zeeman did not appreciate that remark. In a letter to Lodge, praising "[y]our book on Electrons" and thanking him for be-ing "kind enough to send me a copy," he defended his priority over Thomson:

> May I make a remark concerning the history of the subject? On p. 112 of your book you mention that the small mass of the electron was deduced from the ra-diation phenomena in the magnetic field, the result "being in general confor-mity with J. J. Thomson's direct determination of the mass of an electron *some months previously.*" I think, my determination of $e/m$ being of order $10^7$ has been previous to all others in this field. My paper appeared in the "Verslagen" of the Amsterdam Academy of October and November 1896. It was translated in the "Communications from the Leyden Laboratory" and then appeared in the Phil. Mag. for *March* 1897. Prof. Thomson's paper on cathode rays appeared in the Phil. Mag. for *October* 1897. (Emphasis in the original.)[36]

34. See J. J. Thomson, "Cathode Rays," *Proceedings of the Royal Institution,* 15 (1897): 419–432.

35. O. Lodge, *Electrons, or the Nature and Properties of Negative Electricity* (London: George Bell and Sons, 1906), p. 112.

36. Zeeman to Lodge, 3 Aug. 1907, University College Library, Lodge Collection, MS. Add 89/116.

Even though Zeeman neglected to mention that an early report of Thomson's measurements appeared in April 1897,[37] his complaint was justified. Thomson's supposed priority, however, continued to be promoted. In 1913, for instance, Norman Campbell erroneously suggested that Thomson's measurement of the charge-to-mass ratio of cathode ray particles preceded Zeeman's estimate of $e/m$.[38] Millikan also spread the same mistaken view.[39]

The splitting of lines was initially observed by Zeeman in 1897.[40] Instead of sodium he had used cadmium. Its indigo line was found to split into a doublet or triplet depending on whether the light was emitted in a direction parallel or perpendicular to the magnetic field. This stage of Zeeman's experimentation was dominated completely by the theoretical insight of Lorentz. Lorentz's theoretical anticipations led to new aspects of the novel phenomenon. The refinement of the experiment, however, soon led to theoretical advances. For instance, from the direction of polarization of the higher-frequency component of the doublet Zeeman inferred that the charge of the "ions" was negative.[41] Moreover, he gave a more accurate value of $e/m$ ($1.6 \times 10^7$), and finally, by considering this unexpectedly large ratio, he was able to distinguish the "ions" from the electrolytic ions. Here we see the heuristic significance of Lorentz's "ions" for the refinement of Zeeman's experimental results, which was due to the latter's attempt to answer specific questions about the ions.

To summarize here, Zeeman's experimental discovery occupies a prominent place in the early life of the electron qua theoretical entity for three reasons. First, it provided direct empirical support for Lorentz's postulation of the ion-electron. As Zeeman remarked, it "furnishes, as it occurs to me, direct

37. Thomson, "Cathode Rays," *Proceedings of the Royal Institution.*

38. See N. R. Campbell, *Modern Electrical Theory,* 2nd ed. (Cambridge: Cambridge Univ. Press, 1913), pp. 148–149. This mistake became very common at the time. See, e.g., A. D. Cole, "Recent Evidence for the Existence of the Nucleus Atom," *Science,* new series, 41, no. 1046 (15 Jan. 1915): 73–81, on p. 74.

39. See R. A. Millikan, *The Electron: Its Isolation and Measurement and the Determination of Some of Its Properties,* 2nd ed. (Chicago: Univ. of Chicago Press, 1924; 1st ed., 1917), pp. 42–43.

40. P. Zeeman, "Doublets and Triplets in the Spectrum Produced by External Magnetic Forces," *Philosophical Magazine,* 5th series, 44 (1897): 55–60, 255–259.

41. It should be noted that Zeeman initially reported that these polarization results led to the conclusion that the "ions" were positively charged. See Zeeman, "On the Influence of Magnetism," pt. 2, p. 18. However, he soon corrected his erroneous statement in his following paper. See Zeeman, "Doublets and Triplets," p. 58. My attention was drawn to Zeeman's mistake by S. Endo and S. Saito, "Zeeman Effect and the Theory of Electron of H. A. Lorentz," *Japanese Studies in History of Science,* 6 (1967): 1–18.

experimental evidence for the existence of electrified ponderable particles (electrons) in a flame." [42] Apparently, "observations of . . . its [the electron's] direct effects caused, or are among the things that caused . . . [Zeeman] to believe that . . . [the electron] exists, and among . . . [his] reasons for believing that . . . [the electron] exists is that . . . its direct effects have been observed." [43] Second, it led to an approximately correct value of a central property of the electron, namely, its charge-to-mass ratio. The large value of that ratio indicated that Lorentz's "ions" were different from the ions of electrolysis and, thus, led to a revision of the taxonomy of the unobservable realm. Whereas before Zeeman's experiments the term "ions" denoted the ions of electrolysis as well as the entities producing electromagnetic phenomena, after those experiments the extension of the term was restricted to the ions of electrolysis. That is why Lorentz started using the expression "light-ions" to refer to the entities of his electromagnetic theory [44] and later adopted the term "electrons." [45] Third, Zeeman's results in conjunction with Lorentz's analysis of optical dispersion led to an estimate of the mass of the electron. In particular, using his equations for dispersion, Lorentz expressed the mass of the "light-ion" as a function of $e/m$. By substituting Zeeman's estimate of that ratio, he obtained a value of the mass in question that was approximately 350 times smaller than the mass of the hydrogen atom. [46]

Thus, as a result of Zeeman's discovery, the assumption that the radiating particles were as massive as hydrogen ions was abandoned and Lorentz's the-

42. Zeeman to Lodge, 24 Jan. 1897, University College Library, Lodge Collection, MS. Add. 89/116. In a subsequent letter he clarified his previous remark: "I have called electrons ponderable particles; I wished to express that they must possess inertia." Zeeman to Lodge, 28 Jan. 1897, ibid.

43. P. Achinstein, "Who Really Discovered the Electron?" in J. Z. Buchwald and A. Warwick (eds.), *Histories of the Electron: The Birth of Microphysics* (Cambridge, Mass.: MIT Press, 2001), pp. 403–424, on p. 406. It seems to me that all of Achinstein's components of discovery (ontological, epistemic, and priority) can be found in Zeeman's case. Achinstein, however, thinks that Zeeman's "results [as well as the other experimental results explained by Larmor's and Lorentz's theories] were not sufficiently strong to justify a knowledge-claim about the electron's existence" (ibid., p. 408).

44. See, e.g., H. A. Lorentz (1898), "Optical Phenomena Connected with the Charge and Mass of the Ions," pts. 1 and 2, in his *Collected Papers,* vol. 3, pp. 17–39, on p. 24.

45. He began using this term in 1899. See C. Jungnickel and R. McCormmach, *Intellectual Mastery of Nature: Theoretical Physics from Ohm to Einstein,* 2 vols. (Chicago: Univ. of Chicago Press, 1986), vol. 2, p. 233.

46. Lorentz, "Optical Phenomena Connected with the Charge and Mass of the Ions," on pp. 24–25. Cf. C. L. Maier, *The Role of Spectroscopy in the Acceptance of an Internally Structured Atom, 1860–1920* (Ph.D. dissertation, Univ. of Wisconsin, 1964), p. 298.

ory of ions was subsequently transformed into his theory of electrons. In terms of my biographical perspective, the representation of the electron was born out of the representation of the ion, through the mediation of Zeeman. On the basis of his experiments, the ion qua theoretical entity spawned the electron qua theoretical entity, defying the expectations of Lorentz, the ion's maker. Zeeman's discovery had a similar effect on the transformation of the ion's British counterpart—Joseph Larmor's electron.[47]

## 5. Larmor's "Electron"

In 1894 Joseph Larmor appropriated Stoney's "electron," "at the suggestion of G. F. FitzGerald,"[48] to resolve a problem situation that had emerged in the context of the Maxwellian research tradition.[49] Larmor's adoption of the electron represented the culmination (and perhaps the abandonment) of that tradition. A central aspect of the research program initiated by Maxwell was that it avoided microscopic considerations altogether and focused instead on macroscopic variables (e.g., field intensities). This macroscopic approach ran into both conceptual and empirical problems. Its main conceptual shortcoming was that it proved unable to provide an understanding of electrical conduction. Its empirical defects were numerous: "It could not explain the low opacity of metal foils, or dispersion, or the partial dragging of light waves by moving media, or a number of puzzling magneto-optic effects."[50] It was in response to these problems that Larmor started to develop a theory whose aim was to explain the interaction between ether and matter.

The first stage in that development was completed with the publication of part 1 of "Λ Dynamical Theory of the Electric and Luminiferous Medium" in

47. It is worth noting that Larmor acknowledged "Lorentz's priority about electrons which he introduced in 1892 very candidly." Larmor to Lodge, 7 Feb. 1897, Univ. College London, Lodge Collection, MS. Add 89/65 (ii).

48. J. Larmor, *Mathematical and Physical Papers*, 2 vols. (Cambridge: Cambridge Univ. Press, 1929), vol. 1, p. 536, footnote added in the 1929 edition but not in the original paper.

49. This tradition has been thoroughly studied by Jed Buchwald, Bruce Hunt, and Andrew Warwick. See J. Z. Buchwald, *From Maxwell to Microphysics: Aspects of Electromagnetic Theory in the Last Quarter of the Nineteenth Century* (Chicago: Univ. of Chicago Press, 1985); B. J. Hunt, *The Maxwellians* (Ithaca: Cornell Univ. Press, 1991); and A. Warwick, "On the Role of the FitzGerald-Lorentz Contraction Hypothesis in the Development of Joseph Larmor's Electronic Theory of Matter," *Archive for History of Exact Sciences,* 43, no. 1 (1991): 29–91. For what follows, I am indebted to their analysis.

50. Hunt, *The Maxwellians*, p. 210.

August 1894.[51] Its initial version was submitted to the *Philosophical Transactions* on 15 November 1893 and was revised considerably in the months that preceded its publication under the critical guidance of FitzGerald. What is crucial for my purposes is that the published version concluded with a section, added on 13 August, titled "Introduction of Free Electrons."[52]

According to Larmor's representation of field processes, "[T]he electric displacement in the medium is its absolute rotation . . . at the place, and the magnetic force is the velocity of its movement."[53] In order for a medium to be able to sustain electric displacement it must have rotational elasticity. In the original formulation of his theory conductors were conceived as regions in the ether with zero elasticity, since Larmor had "assumed that the electrostatic energy is null inside a conductor."[54] Conduction currents were regarded, in Maxwellian fashion, as mere epiphenomena of underlying field processes and were represented by the circulation of the magnetic field in the medium encompassing the conductor.

To explain electromagnetic induction, Larmor had to find a way in which a changing electric displacement would change that circulation. If conductors were totally inelastic, a changing displacement in their vicinity could not affect them.[55] Therefore, Larmor had to endow conductors with the following peculiar feature: they were supposed to contain elastic zones that were affected by displacement currents and were the vehicle of electromagnetic induction. This implied that in conductors the ether had to be ruptured, a consequence strongly disliked by Larmor. This problem could be circumvented, however, if one assumed that the process of conduction amounted to charge convection.[56] As he remarked, "If you make up the world out of monads, electropositive and electronegative, you get rid of any need for such a barbarous makeshift as rupture of the aether. . . . A monad or an atom is what a geometer would call a 'singular point' in my aether, i.e. it is a singularity naturally arising out of its constitution, and not something foreign to it from outside."[57]

51. J. Larmor, "A Dynamical Theory of the Electric and Luminiferous Medium," pt. 1, *Philosophical Transactions of the Royal Society of London A*, vol. 185 (1894): 719–822; repr. in his *Mathematical and Physical Papers*, vol. 1, pp. 414–535.

52. Ibid. (in *Mathematical and Physical Papers*), pp. 514–535.

53. Ibid., p. 447.

54. Ibid., p. 448.

55. Ibid., p. 462.

56. Cf. Buchwald, *From Maxwell to Microphysics*, p. 161.

57. Larmor to Lodge, 30 Apr. 1894, University College Library, Oliver Lodge Collection, MS. Add 89/65(i); also quoted in Buchwald, *From Maxwell to Microphysics*, pp. 152–153.

There was another conceptual problem related to the phenomenon of electromagnetic induction. Larmor had initially appropriated William Thomson's conception of atoms as vortices in the ether and suggested that magnetism was due to closed currents within those atoms (already postulated by André-Marie Ampère).[58] FitzGerald pointed out, however, that currents of this kind would not be affected by electromagnetic induction, since the ether could not get a hold on them. To solve this problem, Larmor suggested that the currents in question were unclosed. In connection with this issue FitzGerald sent a letter to Larmor which provided the inspiration for the introduction of the electron:[59]

> I don't see where you *require* a discrete structure except that you *say* that it is required in order to make the electric currents unclosed, yet I think that electrolytic and other phenomena prove that there is this discrete structure and you *do* require it, where you *don't* call attention to it, namely where you speak of a rotational strain near an atom. You *say* that electric currents are unclosed vortices but I can't see that this *necessitates* a *molecular* structure because in the matter the unclosedness might be a continuous peculiarity so far as I can see. That it is molecular is due to the molecular constitution of matter and not to any necessity in your theory of the ether.[60]

FitzGerald's point was that the discrete structure of electricity was an independently established fact that did not follow from Larmor's theory, but had to be added to it.

In a few months Larmor reconstructed his theory on the basis of FitzGerald's suggestion. Currents were now identified with the transfer of free charges ("monads"), which were also the cause of magnetic phenomena. Those charges had the ontological status of independent entities and ceased to be epiphenomena of the field. Furthermore, material atoms were represented as stable configurations of those entities. In Larmor's words, "[T]he core of the vortex ring [constituting an atom] . . . [is] made up of discrete electric nuclei or centres of radial twist in the medium. The circulation of these nuclei along the circuit of the core would constitute a vortex. . . . [I]ts strength is now subject to variation owing to elastic action, so that the motion is no longer purely cyclic. A magnetic atom, constructed after this type, would be-

---

58. See Larmor, *Mathematical and Physical Papers*, vol. 1, p. 467. Cf. Hunt, *The Maxwellians*, p. 218.

59. Cf. Buchwald, *From Maxwell to Microphysics*, pp. 163–164.

60. FitzGerald to Larmor, 30 Mar. 1894, Royal Society Library, Joseph Larmor Collection, 448. This excerpt is also reproduced in Buchwald, *From Maxwell to Microphysics*, p. 166.

have like an ordinary electric current in a non-dissipative circuit. It would, for instance, be subject to alteration of strength by induction when under the influence of other changing currents, and to recovery when that influence is removed." [61] Thus, the problem that FitzGerald had brought up disappeared, since the ether could now get a hold on the core of the vortex ring and the atomic currents could be influenced by electromagnetic induction.

In July 1894 FitzGerald suggested the word "electron" to Larmor, as a substitute for the familiar "ion." In FitzGerald's words, Stoney "was rather horrified at calling these ionic charges 'ions.' He or somebody has called them 'electrons' and the ion is the atom not the electric charge." [62] This was the first hint of the need for a distinction between the entities introduced by Larmor and the well-known electrolytic ions. This distinction was obscured, however, by the fact that the effective mass of Larmor's electrons was of the same order of magnitude as the mass of the hydrogen ion. In this respect the subsequent discovery of the Zeeman effect was crucial, since it indicated that the electron's mass was three orders of magnitude smaller than the ionic mass (see below for details).

Larmor's "electrons" were conceived as permanent structures in the ether with the following characteristics:

> [A]n electron has a vacuous core round which the radial twist is distributed. . . .
> It may be set in radial vibration, say pulsation, and this vibrational energy will
> be permanent, cannot possibly be radiated away. All electrons being alike have
> the same period: if the amplitudes and phases are also equal for all at any one
> instant, they must remain so. . . . Thus an electron has the following properties,
> which are by their nature permanent
>
> > (i) its strength [= electric charge]
> > (ii) its amplitude of pulsation
> > (iii) the phase of its pulsation.
>
> These are the same for all electrons . . . The equality of (ii) and (iii) for all elec-
> trons may be part of the pre-established harmony which made them all alike at
> first,— or may, very possibly, be achieved in the lapse of aeons by the same kind
> of averaging as makes the equalities in the kinetic theory of gases. [63]

61. Larmor, *Mathematical and Physical Papers*, vol. 1, p. 515.

62. FitzGerald to Larmor, 19 July 1894; quoted in Hunt, *The Maxwellians*, p. 220.

63. Larmor to Lodge, 29 May 1895, Univ. College London, Lodge Collection, MS. Add 89/65 (i). Larmor's use of terms like "monad" and "pre-established harmony" indicates a Leibnizian element in his thought that has not been explored. Note that his suggestion of a pulsating electron appeared in print. See J. Larmor, "A Dynamical Theory of the Electric and Lumi-

Furthermore, he suggested that they were universal constituents of matter. He had two arguments to that effect. First, spectroscopic observations in astronomy indicated that matter "is most probably always made up of the same limited number of elements."[64] This would receive a straightforward explanation if "the atoms of all the chemical elements [were] to be built up of combinations of a single type of primordial atom."[65] Second, the fact that the gravitational constant was the same in all interactions between the chemical elements indicated that "they have somehow a common underlying origin, and are not merely independent self-subsisting systems."[66]

Larmor's electronic theory of matter received strong support from experimental evidence. First, it could explain the Michelson-Morley experiment. Inspired by Lorentz, Larmor managed to derive the so-called FitzGerald contraction hypothesis, which had been put forward to accommodate the null result of that experiment.[67] As he mentioned in a letter to Lodge, "I have just found, developing a suggestion that I found in Lorentz, that if there is nothing else than electrons—i.e. pure singular points of simple definite type, the only one possible, in the aether—then movement of a body, *transparent* or *opaque,* through the aether *does actually* change its dimensions, just in such way as to verify Michelson's second order experiment."[68] Second, as I mentioned in the previous section, Fresnel had suggested that the ether was dragged by moving matter and had derived from this hypothesis a formula for the velocity of light in moving media. Larmor's theory was able to reproduce Fresnel's result: "The application [of electrons] to the optical properties of moving media leads to Fresnel's well known formula."[69]

The introduction of the electron initiated a revolution that resulted in the abandonment of central features of Maxwellian electrodynamics. Although

niferous Medium," pt. 3, "Relations with Material Media," *Philosophical Transactions of the Royal Society of London A,* 190 (1897): 205–300; repr. in his *Mathematical and Physical Papers,* vol. 2, pp. 11–132, on p. 25.

64. Larmor, *Mathematical and Physical Papers,* vol. 1, p. 475.

65. Ibid. Cf. O. Darrigol, "The Electron Theories of Larmor and Lorentz: A Comparative Study," *Historical Studies in the Physical and Biological Sciences,* 24 (1994): 265–336, on p. 312.

66. Larmor, *Mathematical and Physical Papers,* vol. 1, p. 475.

67. See Warwick, "On the Role of the FitzGerald-Lorentz Contraction."

68. Larmor to Lodge, 29 May 1895, Univ. College Library, Lodge Collection, MS. Add 89/65 (i). This excerpt from Larmor's letter is reproduced in Warwick, "On the Role of the FitzGerald-Lorentz Contraction," p. 56.

69. J. Larmor, "A Dynamical Theory of the Electric and Luminiferous Medium," pt. 2, "Theory of Electrons," *Philosophical Transactions of the Royal Society of London A,* 186 (1895): 695–743; repr. in his *Mathematical and Physical Papers,* vol. 1, pp. 543–597; the quote is from p. 544. Cf. Darrigol, "The Electron Theories of Larmor and Lorentz," pp. 315–316.

in Larmor's theory, as in Maxwell's, the concept of charge was explicated in terms of the concept of the ether, there were significant differences between the two electromagnetic theories. In contrast to Maxwellian theory, which did not attribute independent existence to charges, in Larmor's theory the electron acquired an independent reality. Furthermore, the macroscopic approach to electromagnetism was jettisoned, and microphysics was launched. Conduction currents were represented as streams of electrons, and dielectric polarization was attributed to the polarizing effect of an electric field on the constituents of molecules. In a fashion similar to the representation of Lorentz's charged particles, the representation of Larmor's electrons was constructed in such a way that several characteristics of macroscopic phenomena had counterparts in the properties and behavior of those entities.

Larmor's "electrons" were transformed as a result of Zeeman's discovery. Before that discovery, Larmor thought that a magnetic widening of spectral lines would be beyond experimental detection. The widening in question was proportional to the charge-to-mass ratio of the electron and, on the assumption that "electrons were of mass comparable to atoms," he was led to "the improbability of an observable effect."[70] Larmor's reaction to an announcement of Zeeman's discovery in *Nature* shows that he immediately realized its far-reaching implications with respect to the characteristics of the electron.[71] In a letter to Lodge, asking him to confirm Zeeman's results, he writes: "There is an experiment of Zeeman's . . . which is fundamental + ought to be verified. . . . It demonstrates that a magnetic field can alter the free period of sodium vapor by a measurable amount. I have had the fact as I believe it is (on my views) before my mind for months . . . [but] it never occurred to me that it could be great enough to observe: and it needs a lot of proof that it is so."[72] Several days later he was even more skeptical about the possibility of observing the effect: "I don't expect you will find the effect all the same. The only theory I have about it is that it must be extremely small."[73] Lodge managed to reproduce Zeeman's results and informed Larmor of his success several weeks after Larmor's initial request: "Did I tell you that I had verified Zee-

70. Larmor, *Mathematical and Physical Papers*, vol. 1, p. 622, footnote added in the 1929 edition and not in the original paper. The accuracy of Larmor's retrospective remark is confirmed by contemporary evidence. See below.

71. For the announcement of Zeeman's discovery, see *Nature*, 55 (24 Dec. 1896): 192; cf. N. Robotti and F. Pastorino, "Zeeman's Discovery and the Mass of the Electron," *Annals of Science*, 55 (1998): 161–183, p. 172.

72. Larmor to Lodge, 28 Dec. 1896, University College Library, Oliver Lodge Collection, MS. Add 89/65 (ii).

73. Larmor to Lodge, 6 Jan. 1897, ibid.

man's result, to the extent of seeing the broadening of a Na line from a flame between magnetic poles. It is a *small* effect though." [74]

The implications of Zeeman's discovery were clear for Larmor: "[I]n an ideal simple molecule consisting of one positive and one negative electron revolving round each other, the inertia of the molecule would have to be considerably less than the chemical masses of ordinary molecules, in order to lead to an influence on the period, of the order observed by Dr. Zeeman." [75] Larmor's electron turned out to be as recalcitrant as Lorentz's ion. In the former case, as in the latter, once the effect discovered by Zeeman was attributed to the action of an unobservable entity (Larmor's electron and Lorentz's ion), the characteristics of that effect constrained the representation of the entity in question. In particular, the magnitude of the Zeeman effect implied, contrary to Larmor's original assumption, that the mass of the electron was much smaller than the "chemical masses of ordinary molecules." From my biographical perspective, this amounts to an early manifestation of the agency of the electron qua theoretical entity. The representation of the electron had to be different from what its creator, Larmor, had intended. According to his original theory, the electron, because of its big mass, was supposed to remain relatively unaffected by the action of a magnetic field. As the magnitude of the Zeeman effect indicated, however, the frequency of vibration of the electron was altered significantly after the application of a magnetic field. This experimental result implied, in turn, a smaller value for the mass of the electron. Once more, the experimentally determined component of the representation of the electron made it partly independent of the theoretical framework in which it was embedded.

Furthermore, Zeeman's result and his subsequent estimate of $e/m$ enabled Larmor and Lodge to determine a property of the electron that had been left unspecified in Larmor's original theory, the electron's size. In 1894 Larmor had suggested that a "molecule may quite well be composed of a single positive . . . electron and a single negative . . . one revolving round each other," without, however, specifying the size of the electron relative to the molecule. [76] About a year later he conjectured privately that "if the core [of the electron] is as small compared with the sphere of action of the molecule

74. Lodge to Larmor, 6 Feb. 1897, Royal Society Library, Larmor Collection, 1244. The repetition of Zeeman's experiment was reported by Lodge on 11 Feb. 1897 in the *Proceedings of the Royal Society.* See O. Lodge, "The Influence of a Magnetic Field on Radiation Frequency," *Proceedings of the Royal Society of London,* 60 (1897): 513–514.

75. J. Larmor, "The Influence of a Magnetic Field on Radiation Frequency," *Proceedings of the Royal Society of London,* 60 (1897): 514–515, on p. 515.

76. Larmor, *Mathematical and Physical Papers,* vol. 1, pp. 515–516.

as the latter is compared with 1 cm, this will represent *ordinary mass* in all its physical and chemical aspects."[77] Thus, according to Larmor's early speculations, the electron's size was of the order of $10^{-16}$ cm.[78] The value of $e/m$ obtained by Zeeman together with the concept of electromagnetic mass made possible a more reliable estimate of the electron's size, which was at odds with Larmor's original conjecture. The concept of electromagnetic mass was introduced by J. J. Thomson in 1881. A charged spherical body would possess, besides its material mass, an additional inertia due to its charge. The value of that inertia would depend on $\mu e^2/\alpha$, where $\mu$ was the magnetic permeability of the ether and $\alpha$ the radius of the sphere.[79] Now assuming that the electron's mass was purely electromagnetic, one could calculate its size. Lodge performed the calculation and asked Larmor whether the result that he obtained was acceptable: "Zeeman's $e/m = 10^7$ means if $m = 2\mu e^2/3\alpha$ that $\alpha = 10^{-14}$. . . . [I]s this too small for an electron?"[80]

Larmor's reply is very revealing about the process that led to the construction of the representation of the electron:

> I don't profess to know à priori anything about the size or constitution of an electron except what the spectroscope may reveal. I do assert that a logical aether theory must drive you back on these electrons as the things whose mutual actions the aether transmits: but for that general purpose each of them is a point charge just as a planet is an attracting point in gravitational astronomy. But as regards their constitution am inclining to the view that an atom of $10^{-8}$ cm is a complicated sort of solar system of revolving electrons, so that the single electron is very much smaller, $10^{-14}$ would do very well—is in fact the sort of number I should have guessed.[81]

77. Larmor to Lodge, 29 May 1895, Univ. College London, Lodge Collection, MS. Add 89/65 (i); also quoted and discussed in S. B. Sinclair, "J. J. Thomson and the Chemical Atom: From Ether Vortex to Atomic Decay," *Ambix,* 34, pt. 2 (1987): 89–116, on p. 106.

78. In print, Larmor had attributed to the molecule dimensions of the order of $10^{-8}$ cm. See his "A Dynamical Theory of the Electric and Luminiferous Medium," pt. 1, p. 807. Cf. also Robotti and Pastorino, "Zeeman's Discovery and the Mass of the Electron," p. 171.

79. See J. J. Thomson, "On the Electric and Magnetic Effects Produced by the Motion of Electrified Bodies," *Philosophical Magazine,* 5th series, 11 (1881): 229–249, on p. 234.

80. Lodge to Larmor, 8 Mar. 1897, Royal Society Library, Larmor Collection, 1247. Lodge's calculation was probably prompted by a letter that he had received from FitzGerald (FitzGerald to Lodge, 6 Mar. 1897, Univ. College London, Lodge Collection, MS. Add 89/35 (iii)). This calculation appeared a few days later in the *Electrician.* See. O. Lodge, "A Few Notes on Zeeman's Discovery," *Electrician,* 12 (Mar. 1897): 643–644, on p. 644.

81. Larmor to Lodge, 8 May 1897, Univ. College London, Lodge Collection, MS. Add 89/65 (ii). Well before he wrote this letter, Larmor had found out that Lodge "had verified Zeeman's result" (see Lodge to Larmor, 6 Feb. 1897, Royal Society Library, Larmor Collec-

So originally the representation of the electron was arrived in an a priori fashion, that is, as a solution to theoretical problems. The remaining task was to determine the properties of the electron so as to accommodate the available empirical evidence. The size of the electron, for instance, was calculated by Lodge so as to "attain Zeeman's quantitative result."[82] The resulting representation had also to be coherent. The specification of, say, the electron's size from spectroscopic evidence had to be compatible with other indications about that size from the domain of atomic structure.

Larmor's detailed analysis of the Zeeman effect was completed by 8 November 1897.[83] Larmor considered "a single ion $e$, of effective mass M, describing an elliptic orbit under an attraction to a fixed centre proportional to the distance therefrom."[84] If a magnetic field was introduced, Larmor proved, by solving the corresponding equations of motion, that instead of the original frequency of vibration three distinct ones would appear: one of them would coincide with the original, whereas the other two would be shifted by an amount equal to $\pm eH/4\pi Mc^2$. A "striking feature" of Larmor's analysis was "that the modification thus produced is the same whatever be the orientation of the orbit with respect to the magnetic field."[85] This feature resulted from a general theorem that he had managed to prove a few weeks before he submitted his paper to the *Philosophical Magazine*. In his words, "[T]he following math prop is true:—Consider any system of (say) *negative* ions, with charges proportional to their effective masses, attracting each other according to some laws & attracted to fixed centres anywhere on the axes of the magnetic field: then their motion when the magnetic field is turned on relative to an observer fixed is the same as when it was off relative to an observer attached to

---

tion, 1244). Thus, he had no reason to doubt the validity of that result. Furthermore, he had realized early on its implications with respect to the magnitude of the electron's mass. Therefore, his allusion to Lodge's estimate of the size of the electron as "the sort of number I should have guessed" is not surprising and does not contradict my previous claim that, prior to Zeeman's discovery, Larmor had attributed to the electron a mass comparable to the mass of the hydrogen ion. Furthermore, $10^{-14}$ was not "the sort of number" that Larmor had guessed before Zeeman's discovery. Rather, as we saw above, his original conjecture for the size of the electron was $10^{-16}$. A size of that magnitude would imply an electron mass two orders of magnitude larger than Zeeman's experiments indicated.

82. Lodge, "A Few Notes on Zeeman's Discovery," p. 644. This, by the way, undermines Achinstein's suggestion that "claims made about the electron by . . . Larmor, and Lorentz, . . . were primarily theory-driven, not experimentally determined" (Achinstein, "Who Really Discovered the Electron?" p. 414).

83. J. Larmor, "On the Theory of the Magnetic Influence on Spectra; and on the Radiation from moving Ions," *Philosophical Magazine*, 5th series, 44 (1897): 503–512.

84. Ibid., p. 503.

85. Ibid., p. 504.

a frame rotating round the axis of the field H with ang. velocity $eH/Mc^2$ where $e/M$ is the constant charge/mass and $c$ is the velocity of radiation." [86] In this respect Larmor's analysis was superior to Lorentz's less general explanation of the results obtained by Zeeman. In other respects, such as the polarization of the emitted spectral lines, Larmor reached conclusions identical to those obtained by Lorentz (see above). Larmor's analysis, in conjunction with Zeeman's experiments, enabled the approximate estimate of $e/M$. As it turned out, "the effective mass of a revolving ion, supposed to have the full unitary charge or electron, is about $10^{-3}$ of the mass of the atom." [87]

As a result of Larmor's work and the support that it received from Zeeman's experiments, by 1898 the electron qua theoretical entity had become an essential ingredient of British scientific practice in the domain of electromagnetism.[88] By that time, it had also emerged, in the guise of J. J. Thomson's "corpuscle," out of a problem situation in the investigation of cathode rays. I will now proceed to a brief examination of Thomson's achievement.

### 6. Thomson's "Corpuscle": A "By No Means Impossible Hypothesis"

Thomson's contribution to the acceptance of the electron hypothesis was closely tied with the experimental and theoretical investigation of cathode rays.[89] These rays were detected in the discharge of electricity through gases at very low pressure. Their main observable manifestation was that they gave rise to a fluorescent spot when they collided with certain substances. There were two conflicting interpretations of the nature of cathode rays. According to the first view, mainly endorsed by British physicists, they were beams of charged particles. William Crookes, for instance, suggested in 1879 that they consisted of charged molecules. His proposal was based on two aspects of their behavior, namely, that they were emitted in a direction perpendicular to the cathode and that their trajectory was bent by a magnetic field.

The alternative view, favored by German physicists, was that cathode rays were another species of waves in the ether. Eugen Goldstein, for instance, gave the following argument to that effect. It was known that they traveled in straight lines and that their impact caused fluorescence. Since ultraviolet rays,

---

86. Larmor to Lodge, 12 Oct. 1897, University College Library, Lodge Collection, MS. Add. 89/65 (ii).

87. Larmor, "On the Theory of the Magnetic Influence on Spectra," p. 506.

88. Cf. Buchwald, *From Maxwell to Microphysics*, p. 172; and Hunt, *The Maxwellians*, pp. 220–221.

89. This characterization of Thomson's proposal of the corpuscle comes from FitzGerald. See his "Dissociation of Atoms," *Electrician*, 39 (1897): 103–104, on p. 104.

which were represented as waves in the ether, had similar characteristics, this suggested by analogy that cathode rays too were waves in that medium. There were other arguments in favor of this representation. In 1883 Heinrich Hertz had attempted to deflect them with an electric field, but failed to do so. If they were charged particles, their path should have been affected by the electric field.[90]

Experimental research revealed some further characteristics of cathode rays. It turned out that they could not penetrate certain materials that did not block the course of ultraviolet radiation. This weighed heavily against their being waves in the ether. Furthermore, Hertz established in 1892 that they could pass through thin layers of metals. This result could also be explained by those who conceived cathode rays as charged particles. J. J. Thomson, for instance, argued in 1893 that the material bombarded by cathode rays turned into a source of cathode rays itself.[91] Finally, in 1895 Jean Perrin showed that they were carriers of negative charge.[92]

Thus, there were conflicting pieces of evidence about the nature of cathode rays. On the one hand, the fact that they could not be deflected by an electric field and that they could pass through metals that were impenetrable to particles of atomic size suggested that they were waves in the ether. On the other hand, the fact that their trajectory was influenced by a magnetic field and that they were carriers of electricity supported their representation as charged particles.

This was the state of knowledge about cathode rays when Thomson put forward his corpuscle hypothesis in a lecture to the Royal Institution on 30 April 1897.[93] According to that hypothesis, cathode rays consisted of extremely small, subatomic particles. He inferred their small size from their rate of absorption in gases, which had been investigated experimentally by Philipp Lenard.[94] Thomson pointed out that if cathode rays were particles of atomic size, their rate of absorption would be much higher than that indicated by

90. See D. L. Anderson, *The Discovery of the Electron: The Development of the Atomic Concept of Electricity* (1964; repr., New York: Arno Press, 1981), pp. 27–30; I. Falconer, "Corpuscles, Electrons and Cathode Rays: J. J. Thomson and the 'Discovery of the Electron,'" *British Journal for the History of Science*, 20 (1987): 241–276, on p. 244; and Heilbron, *A History of the Problem of Atomic Structure*, pp. 61–62.

91. See Heilbron, *A History of the Problem of Atomic Structure*, p. 65.

92. See Falconer, "Corpuscles, Electrons and Cathode Rays," p. 244.

93. Thomson, "Cathode Rays," *Proceedings of the Royal Institution*. It is worth mentioning that no unpublished material documenting Thomson's path to that hypothesis still exists. Cf. Falconer, "Corpuscles, Electrons and Cathode Rays," p. 267.

94. P. Lenard, "Ueber die Absorption der Kathodenstrahlen," *Annalen der Physik und Chemie*, 56 (1895): 255–275.

Lenard's results.[95] Moreover, the unexpectedly small value of their mass-to-charge ratio further supported that inference. By that time he had not yet succeeded in detecting the influence of an electric field on cathode rays. However, he explained away his failure to do so, by assuming that the gas in the cathode ray tube was ionized by the rays and, thus, screened off the external electric field.[96]

In October 1897 Thomson published a more elaborate report of his experiments with cathode rays and the conclusions that he drew from them.[97] By that time he had managed to deflect the rays by means of an electric field, but "only when the vacuum was a good one."[98] Furthermore, by deflecting the rays by a magnetic field he observed that "the path of the rays is independent of the nature of the gas."[99] Those results threw light on the question of the nature of cathode rays. In Thomson's words, "As the cathode rays carry a charge of negative electricity, are deflected by an electrostatic force as if they were negatively electrified, and are acted on by a magnetic force in just the way in which this force would act on a negatively electrified body moving along the path of these rays, I can see no escape from the conclusion that they are charges of negative electricity carried by particles of matter."[100] Thus, Thomson inferred some of the qualitative characteristics of the constituents of cathode rays from their observable behavior. Those characteristics (material, negatively charged particles) were determined so as to reflect their observable counterparts (the direction of the electric and magnetic deflections of cathode rays). In that sense, the representation of the entities that constituted cathode rays was a construction from the available experimental data. The term "construction" is apt here, because it reminds us that the representation in question was not passively read from the experimental situation. Rather, it was the outcome of an active, creative reading of that situation. In fact, an alternative interpretation of the same experimental data was possible.

95. See Thomson, "Cathode Rays," *Proceedings of the Royal Institution*, pp. 430–431.

96. Cf. Falconer, "Corpuscles, Electrons and Cathode Rays," p. 266.

97. J. J. Thomson, "Cathode Rays," *Philosophical Magazine*, 5th series, 44 (1897): 293–316.

98. Ibid., p. 297. It is worth pointing out that a German physicist, Gustav Jaumann, had reported an electrostatic deflection of cathode rays in April 1896. See G. Jaumann, "Electrostatische Ablenkung der Kathodenstrahlen," *Annalen der Physik und Chemie*, 59 (1896): 252–266. Cf. G. Gooday, "The Questionable Matter of Electricity: The Reception of J. J. Thomson's 'Corpuscle' among Electrical Theorists and Technologists," in Buchwald and Warwick (eds.), *Histories of the Electron*, pp. 101–134, on p. 105. So, even in this respect, Thomson's achievement was not without precedent. Cf. I. Falconer, "Corpuscles to Electrons," in Buchwald and Warwick (eds.), *Histories of the Electron*, p. 92.

99. Thomson, "Cathode Rays," *Philosophical Magazine*, p. 301.

100. Ibid., p. 302.

As FitzGerald argued, there was "escape" from Thomson's conclusion. The latter's observations were consistent with the hypothesis that cathode rays were free electrons, that is, disembodied charges; but more on this below.

Having argued that cathode rays were charged material particles, Thomson then inquired further into their nature: "The question next arises, What are these particles? are they atoms, or molecules, or matter in a still finer state of subdivision?" [101] His mass-to-charge measurements aimed at elucidating this question.[102] The heuristic role of the theoretical entities ("charged material particles") that Thomson put forward as an explanatory representation of cathode rays is here manifest.

Let us see how he constructed a quantitative attribute of cathode ray particles $(m/e)$ from the experimental data at his disposal. First, the charge carried by cathode rays was $Q = ne$, where $n$ was the number of particles. This charge could be measured by collecting "the cathode rays in the inside of a vessel connected with an electrometer." [103] Second, their kinetic energy was $W = (1/2)nmv^2$. This energy could be measured by "the increase in the temperature of a body of known thermal capacity caused by the impact of these rays." [104] Third, the effect of a magnetic field on the trajectory of the rays was given by the following formula: $mv/e = H\rho = I$, where $\rho$ was "the radius of curvature of the path of these rays." [105] From these measurable magnitudes he could deduce $m/e$:

$$\frac{1}{2}\frac{m}{e}v^2 = \frac{W}{Q},$$

$$v = \frac{2W}{QI},$$

and

$$\frac{m}{e} = \frac{I^2Q}{2W}.$$

---

101. Ibid.

102. As John Heilbron has noted, in his October paper Thomson presented the arguments for his corpuscle hypothesis in the reverse order from that of his April account. That is, he introduced his measurements of $m/e$ as the main argument for the smallness of corpuscles and then presented Lenard's experiments on the absorption of cathode rays in gases as secondary confirmation. Heilbron, *A History of the Problem of Atomic Structure*, p. 81.

103. Thomson, "Cathode Rays," *Philosophical Magazine*, p. 302.

104. Ibid.

105. Ibid.

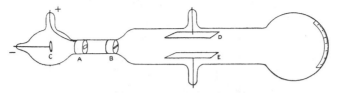

**Fig. 4.3** J. J. Thomson's cathode ray tube. "The rays from the cathode C pass through a slit in the anode A, which is a metal plug fitting tightly into the tube and connected with the earth; after passing through a second slit in another earth-connected metal plug B, they travel between two parallel aluminium plates [connected to the poles of a battery] . . . ; they then fall on the end of the tube and produce a narrow well-defined phosphorescent patch. A scale pasted on the outside of the tube serves to measure the deflexion of this patch." (From J. J. Thomson, "Cathode Rays," *Philosophical Magazine*, 5th series, 44 [1897]: 293–316, on p. 296.)

Thomson came up with a different method for measuring $m/e$ that capitalized on the deflection of the cathode rays by an electric field (see fig. 4.3). When the rays entered the region where the electric field was active, they would be accelerated in the direction of the field. The magnitude of that acceleration would be $Fe/m$, where $F$ was the intensity of the field. Assuming that the length of the region under the influence of the field was $l$, the time taken by the cathode rays to traverse that region would be $l/v$. Thus, when they exited that region their direction would be deflected by an angle

$$\theta = \frac{Fe}{m}\frac{l}{v^2}.$$

(more precisely, this would be the value of $\tan\theta$, but Thomson did not mention that). By a similar reasoning, Thomson calculated that angle when the rays were deflected by a magnetic field. The value that he obtained was

$$\phi = \frac{He}{m}\frac{l}{v}.$$

From these two formulas, he derived the mass-to-charge ratio:

$$\frac{m}{e} = \frac{H^2\theta l}{F\phi^2}.$$

By subjecting the cathode rays simultaneously to the action of an electric and a magnetic field and by adjusting the intensity of the former so that $\phi = \theta$, he obtained the value of

$$\frac{m}{e} = \frac{H^2 l}{F\theta}.$$

The results that he obtained with both of the above methods indicated "that the value of $m/e$ is independent of the nature of the gas."[106] That value (approximately $10^{-7}$) was a thousand times smaller than "the smallest value of this quantity previously known, and which is the value for the hydrogen ion in electrolysis."[107] It should be emphasized that in 1897 Thomson did not guess correctly one of the key properties of the electron, the value of its charge: "The smallness of $m/e$ may be due to the smallness of $m$ or the largeness of $e$, or to a combination of these two."[108] He preferred the last option. There was independent warrant for the conjecture that the size of cathode ray particles was very small. As I mentioned above, Lenard had investigated "the rate at which the brightness of the phosphorescence produced by these rays diminishes with the length of path travelled by the ray."[109] His results were incompatible with the view that cathode rays consisted of particles of atomic dimensions, since particles of this kind would have been slowed down by collisions with the molecules of the surrounding gas much more rapidly than indicated by observation. Furthermore, Thomson suggested that there was "some evidence that the charges carried by the corpuscles in the atom are large compared with those carried by the ions of an electrolyte."[110]

Thomson's methods for inferring the properties of the constituents of cathode rays illuminate (and are illuminated by) my contention that theoretical entities are, to some extent, constructions from experimental data. In all probability, he was convinced that cathode rays were charged particles before his 1897 experiments.[111] He then interpreted his experimental results in terms of his prior belief in the reality of those corpuscles. The qualitative and quantitative results of his experiments along with previously established information enabled him to construct the properties that those particles should have in order to produce the observed results. A prior conviction in the cathode rays' particulate nature along with Lenard's results on their rate of absorption in gases led to the conclusion that they were very small compared to the ions of electrolysis. Thomson's quantitative results (the amount of their deflection by electric and magnetic fields, the amount of heat generated by their impact, etc.) and the belief that they were subject to classical mechanics and electromagnetic theory enabled him to calculate their mass-to-charge ratio. Thus,

106. Thomson, "Cathode Rays," *Philosophical Magazine,* p. 310.
107. Ibid.
108. Ibid.
109. Ibid.
110. Ibid., p. 312.
111. Cf. Falconer, "Corpuscles, Electrons and Cathode Rays," p. 270.

the qualitative and quantitative elements of his experimental situation were embodied, appropriately translated, in the representation of the corpuscle. In subsequent investigations of the nature of the corpuscle, which ended in its identification with the electron, the experimentally obtained information that had been built into its representation remained relatively immune to shifts in theoretical perspective. This stability made the corpuscle qua theoretical entity partly independent of the evolving theoretical framework in which it was embedded.

Thomson also conjectured that the corpuscle was a universal constituent of all matter.[112] He had several arguments to that effect: The mass-to-charge ratio of cathode rays depended neither on the chemical composition of the gas within the cathode ray tube nor on the material of the tube's electrodes. Furthermore, as Lenard had demonstrated in 1895, their absorption by a substance depended only on its density and not on its chemical composition. If matter were composed of corpuscles, Lenard's observations would be readily explained.[113] Finally, by means of the corpuscle hypothesis one could explain away the measurements of atomic weights that were not exact multiples of the weight of the hydrogen atom.[114]

However, the seemingly "alchemical" connotations of Thomson's suggestion generated a great deal of resistance toward his corpuscle theory. Fitz-Gerald, for instance, interpreted Thomson's theory as implying the "presence of a possible method of transmutation of matter."[115] FitzGerald's aversion to alchemy led him to deny that the cathode ray particles were material constituents of atoms. He offered, instead, an alternative interpretation of them as free charges ("electrons") that did not require the divisibility of atoms and kept him away from "the track of the alchemists."[116]

112. Thomson's conjecture might seem incompatible with his suggestion that the charge of the corpuscle was larger than the elementary unit of electric charge, the unit carried by a hydrogen ion in electrolysis. An electrolytic ion was also supposed to consist of corpuscles, and it is not clear how it could carry a charge smaller than the charge carried by its constituents. For Thomson, however, this was not a problem. Considering the molecules of HCl as an example, he represented the "components of the hydrogen atom as held together by a great number of tubes of electrostatic force; the components of the chlorine atom are similarly held together, while only one stray tube binds the hydrogen atom to the chlorine atom"(Thomson, "Cathode Rays," *Philosophical Magazine,* p. 312). Thus, the above difficulty did not arise, since the unit of charge associated with each electrostatic tube was portrayed as one "stray" component of the charge associated with the unit of matter.

113. For details see Thomson, "Cathode Rays," *Philosophical Magazine,* pp. 310–312.

114. Cf. Heilbron, *A History of the Problem of Atomic Structure,* pp. 82–83.

115. FitzGerald, "Dissociation of Atoms," p. 104.

116. Ibid. Cf. P. Achinstein, *Particles and Waves* (New York: Oxford Univ. Press, 1991), p. 288; and Falconer, "Corpuscles, Electrons and Cathode Rays," p. 271.

William Sutherland made a similar proposal in 1899.[117] Thomson's reaction was negative. When he found out about Sutherland's suggestion, he prepared a rebuttal, in which he pointed out its shortcomings: "[T]he view that in the cathode rays we have electric charges without matter Electrons . . . I took at the beginning of my experiments, but exactly in the form (which I gather Mr. Sutherland does not adopt) that the atoms are themselves a collection of electrons . . . , in which form it only differs verbally from the form I used. [R]eplacing corpuscles by electrons, seems to be wanting in precision + definiteness as compared with the other view and beset with difficulties from which the other was free. . . . [T]he advantages of the hypothesis as far as I can see . . . are only that it does not involve the necessity of the atom being split up."[118] The main drawback of the hypothesis of free electrons, on the other hand, was that "it supposes that a charge of electricity can exist apart from matter of which there is as little direct evidence as of the divisibility of the atom."[119] As he had mentioned a few years earlier, "*An electric charge* is always found associated with matter."[120]

Thomson's objection to Sutherland's view was not applicable to Larmor's version of the electron hypothesis. According to Larmor, electricity did not exist apart from matter, because it constituted matter. Electrons were material particles, in the sense that their material aspect (inertia) was an epiphenomenon of their electric charge. Furthermore, material atoms themselves were composed of (positive and negative) electrons in dynamical equilibrium. That is why Thomson referred to Larmor's view as only "verbally" different from his own.[121]

---

117. W. Sutherland, "Cathode, Lenard and Röntgen Rays," *Philosophical Magazine,* 5th series, 47 (1899): 269–284. Cf. S. M. Feffer, "Arthur Schuster, J. J. Thomson, and the Discovery of the Electron," *Historical Studies in the Physical and Biological Sciences,* 20, no. 1 (1989): 33–61, p. 59.

118. Cambridge Univ. Library, Add 7654/NB 44. This notebook is not dated, but it must have been written sometime in early March 1899, since Sutherland's paper appeared in the March issue of the *Philosophical Magazine* and Thomson's published response was dated 11 Mar. 1899. See J. J. Thomson, "Note on Mr. Sutherland's Paper on the Cathode Rays," *Philosophical Magazine,* 5th series, 47 (1899): 415–416.

119. Cambridge Univ. Library, Add 7654/NB 44.

120. Cambridge Univ. Library, Add 7654/NB 40. The contents of this notebook, which is undated, "were probably written in August or September 1896." Falconer, "Corpuscles, Electrons and Cathode Rays," p. 258.

121. It is worth noting that Larmor had identified cathode rays with free electric charges as early as 1894: "What strikes me also is the fact that free electric charges of ionic or some such character can flash about space with velocity comparable with radiation, provided they are not bothered by inertia other than that of the medium around them: cf. discharge in vac-

Nevertheless, Thomson wanted to differentiate his "corpuscle" from the "electron" (even in Larmor's sense). His resistance to the latter can be explained by two factors: First, in 1897 he still subscribed to a Maxwellian conception of charge; that is, he considered charge an epiphenomenon of the interaction between ether and matter.[122] Since charges could not exist independently of matter, the "corpuscle" had to be a material particle that carried charge.[123] Thus, one of the reasons he avoided the term "electron" was that it denoted a disembodied charge, whereas his "corpuscle" was a material particle.[124] Second, the term "electron" denoted both negative and positive elementary charges. Whereas the former, according to Thomson, were real entities, the reality of the latter had not been established (as late as 1907). Since he did not want to inadvertently confer reality on the positive electron, he did not use that term.[125] These were not very significant differences, however, and eventually negative "electrons" and "corpuscles" came to be identified, even by Thomson himself, who managed to overcome his "parental affection" for the corpuscle.[126]

The acceptance of Thomson's proposal was, however, gradual. Whereas his success in deflecting cathode rays by means of an electric field established that they were charged particles, his suggestion that they were universal, subatomic constituents of matter was not accepted till, at least, 1899. In 1897 he had not shown that those entities were present in other phenomena besides

---

uum tubes." Larmor to FitzGerald, 14 June 1894, quoted in Buchwald, *From Maxwell to Microphysics,* p. 167.

122. Cf. Falconer, "Corpuscles, Electrons and Cathode Rays," pp. 261–262; and Feffer, "Arthur Schuster, J. J. Thomson," pp. 60–61.

123. Thomson's choice of the term "corpuscle" must have been deliberate. Since the seventeenth century, when Boyle and others had used this term in connection with the mechanical philosophy, corpuscles had been conceived as minute particles of matter.

124. See also J. J. Thomson, "Carriers of Negative Electricity," Nobel Lecture, 11 Dec. 1906, in *Nobel Lectures: Physics, 1901–1921* (Amsterdam: Elsevier, 1967), pp. 145–153, esp. p. 149.

125. See Kragh, "Concept and Controversy," p. 210; and H. Kragh, "The Electron, the Protyle, and the Unity of Matter," in Buchwald and Warwick (eds.), *Histories of the Electron,* pp. 195–226, on pp. 205–206.

126. In 1907 Thomson declared that one of the reasons he preferred the corpuscle was that "it is my own child, and I have a kind of parental affection for it" ( J. J. Thomson, "The Modern Theory of Electrical Conductivity in Metals," *Journal of the Institution of Electrical Engineers,* 38 [1906–1907]: 455–465; Thomson et al., "Discussion," *Journal of the Institution of Electrical Engineers,* 38 [1906–1907]: 465–468; the quote is on p. 467; cf. Gooday, "The Questionable Matter of Electricity," p. 122). Later, however, he disowned his "own child" and adopted the electron. In 1923, for instance, he published a book titled *The Electron in Chemistry* (Philadelphia: Franklin Institute).

the discharge of electricity through gases. Furthermore, he did not measure separately the charge and mass of the corpuscle, and, thus, the smallness of $m$ was not sufficiently established.[127]

By 1899 these difficulties had been alleviated. In 1898 he devised a method for measuring the charge of ions in gases that had been ionized by x-rays. The results of his measurements agreed with those for the charge carried by electrolytic ions. In 1899 he published measurements of the charge-to-mass ratio of the particles produced in the photoelectric effect as well as by thermionic emission. That ratio agreed with the corresponding ratio of cathode ray particles. Furthermore, he measured the charge of those particles by the new method that he had come up with, and he found that it coincided with the charge carried by hydrogen ions. This result along with the large charge-to-mass ratio implied the smallness of $m$. Finally, the reception of Thomson's proposal was facilitated by the theoretical and experimental developments that we examined in the previous sections—namely, the construction of Lorentz's and Larmor's theories and the discovery of the Zeeman effect.[128]

## 7. Concluding Remarks

By the turn of the century, the constituents of cathode rays had been identified with Lorentz's ions,[129] and both had already been associated with electrons by Larmor himself.[130] This identification strengthened the case for the existence of the electron, since it unified the theoretical and experimental evidence on which the claim for its existence rested. Furthermore, by that

127. Cf. Falconer, "Corpuscles, Electrons and Cathode Rays," p. 271; and Heilbron, *A History of the Problem of Atomic Structure,* p. 84.

128. J. J. Thomson, "On the Masses of the Ions in Gases at Low Pressures," *Philosophical Magazine,* 48 (1899): 547–567. Thomson's research between 1897 and 1899 is discussed in N. Robotti, "J. J. Thomson at the Cavendish Laboratory: The History of an Electric Charge Measurement," *Annals of Science,* 52 (1995): 265–284. Cf. also Falconer, "Corpuscles, Electrons and Cathode Rays," pp. 272–273; and Heilbron, *A History of the Problem of Atomic Structure,* pp. 84–85. Heilbron's thesis continues to be the most comprehensive discussion of the various developments that led to the acceptance of the electron hypothesis. However, in some of his statements he attributes the electron's discovery exclusively to Thomson, an attribution that is at odds with the rest of his analysis.

129. See, e.g., P. Zeeman, "Experimentelle Untersuchungen über Teile, welche kleiner als Atome sind," *Physikalische Zeitschrift,* 1 (1900): 562–565, 575–578, esp. pp. 577–578.

130. Larmor, "On the Theory of the Magnetic Influence on Spectra," esp. p. 506. Cf. Heilbron, *A History of the Problem of Atomic Structure,* pp. 103–104. I should add that this identification was by no means universally accepted circa 1900. The most notable exception was, of course, Thomson.

time the electron had become indispensable for the practice of physicists concerned with electromagnetic theory.[131] As we read in a textbook from that period, a "significant sign of its [the electron theory's] acceptance is the almost complete absence of attempts to formulate electrical theories not based upon electrons."[132] In that sense, it would remain a permanent part of the ontology of physics.

If one insists on using the expression "the discovery of the electron," one would have to use it as meaning the complex process that led to the consolidation of the belief that "electrons" denote real entities. In that sense, was the electron discovered by Thomson? The answer must be negative. His significant contribution to the acceptance of the belief in the reality of the electron was not sufficient for establishing that belief. Furthermore, most of his theoretical suggestions had been previously put forward by others, and many of the experimental results that he obtained were also independently produced by others.

The conception of the ion-electron as a universal constituent of material bodies had been developed earlier by Lorentz and Larmor. More important, the experimental evidence that was brought to bear on this issue was not exclusively produced by Thomson. Zeeman's discovery and the subsequent identification of β-rays, emitted in radioactive decay, with fast-moving electrons also indicated the ubiquity of the electron.[133]

Thomson's measurements of the mass-to-charge ratio of his corpuscle were preceded by Zeeman's measurement of the corresponding ratio of Lorentz's ions, and similar measurements were independently carried out by Walter Kaufmann and Emil Wiechert. In the beginning of 1897 the latter reported his measurements of the charge-to-mass ratio of cathode rays, which he obtained by subjecting them to a magnetic field and observing how it affected their trajectories. Since that ratio was several orders of magnitude larger than the corresponding ratio for the hydrogen ion, it followed that the constituents of cathode rays were very different from electrolytic ions. Wiechert, thus, identified those constituents with the atomic charges postu-

---

131. For an interesting reading of the emergence of microphysics from the point of view of the physicists' practices, see J. Z. Buchwald, "How the Ether Spawned the Microworld," in L. Daston (ed.), *Biographies of Scientific Objects* (Chicago: Univ. of Chicago Press, 2000), pp. 203–225.

132. E. E. Fournier D'Albe, *The Electron Theory: A Popular Introduction to the New Theory of Electricity and Magnetism*, 2nd ed. (London: Longmans, Green, and Co., 1907), p. xxi (from the author's preface).

133. Cf. Anderson, *The Discovery of the Electron*, p. 112.

lated by Helmholtz, and from the large value of $e/m$ he inferred the small value of the electromagnetic mass associated with them.[134]

Kaufmann also measured the charge-to-mass ratio of cathode rays, on the provisional assumption that they were charged particles, and ascertained that it was three orders of magnitude larger than the corresponding ratio of the ions of electrolysis. Furthermore, he pointed out that the value of this ratio depended neither on the chemical composition of the gas contained in the cathode ray tube nor on the material of that tube's electrodes. These results, according to Kaufmann, eliminated the possibility that cathode rays were composed of electrolytic ions. Ironically, he concluded that the identification of cathode rays with charged particles was untenable.[135]

One could suggest that Thomson should be considered the electron's discoverer because his "corpuscle" is closer to our electron than any of the entities put forward by his contemporaries. On such a view, it was Thomson who discovered the electron—not the "electron" of the late 1890s, which was "erroneously" thought to be a disembodied charge, but the subsequent electron, a material particle and the carrier of the unit of electric charge. The corpuscle's description (as a material particle, and as the carrier of the unit of negative charge) suffices to identify it with our electron, whereas the description of the late-nineteenth-century "electron" (as a disembodied charge, either positive or negative) is incompatible with our use of that term.

This suggestion also leaves much to be desired. Consider, first, the material origin of the corpuscle's mass. Even though the mass of Larmor's electrons was purely electromagnetic, they did not cease to be material particles. Their material aspect was inextricably tied with their electrical character. In Larmor's scheme matter (inertia) was an epiphenomenon of electricity. More important, a few years later, as a result of Kaufmann's experiments on the velocity dependence of the electron's mass, Thomson was converted to the view "that this mass arises entirely from the charge of electricity on the

134. See E. Wiechert, "Ueber das Wesen der Elektrizität," *Schriften der Physikalisch-Ökonomischen Gesellschaft zu Königsberg*, 38 (7 Jan. 1897): 3–12; Wiechert, "Experimentelles über die Kathodenstrahlen," *Schriften der Physikalisch-Ökonomischen Gesellschaft zu Königsberg*, 38 (7 Jan. 1897): 12–16. Cf. Heilbron, *A History of the Problem of Atomic Structure*, p. 96; P. F. Dahl, *Flash of the Cathode Rays: A History of J. J. Thomson's Electron* (Bristol: Institute of Physics Publishing, 1997), pp. 154–155; and J. F. Mulligan, "Emil Wiechert (1861–1928): Esteemed Seismologist, Forgotten Physicist," *American Journal of Physics*, 69, no. 3 (2001): 277–287, on p. 280.

135. W. Kaufmann, "Die magnetische Ablenkbarkeit der Kathodenstrahlen und ihre Abhängigkeit vom Entladungspotential," *Annalen der Physik und Chemie*, 61 (1897): 544–552 (received on 21 May 1897). Cf. Heilbron, *A History of the Problem of Atomic Structure*, p. 79; and Dahl, *Flash of the Cathode Rays*, pp. 156–158.

corpuscle."[136] His conversion would be an embarrassing incident if we attributed the discovery of the electron to him because of the materiality of his corpuscle.

Second, consider the negative sign of the corpuscle's charge. Can one attribute the electron's discovery to Thomson because "these physicists [Stoney, Larmor, and Lorentz] were referring to particles or charges or both that could be positive as well as negative, and not to the negative particles that comprise cathode rays"?[137] Again, the answer must be negative. Zeeman's experiments had also indicated that the charge of the ion cum electron was negative.

Finally, even if the "corpuscle" resembled our "electron" more than, say, Lorentz's "ions," this resemblance would still be slight, to say the least. First, the corpuscle was conceived as a structure in the ether, which has disappeared from the ontology of physics. Second, all of the bizarre properties now associated with electrons (quantum numbers, wave-particle duality, indeterminate position-momentum, etc.) were literally inconceivable in Thomson's time.

The radical difference between Thomson's "corpuscle" and our "electron" reminds us that the problem of identification, which I outlined in chapter 1, is lurking in the background. The problem has both a synchronic and a diachronic dimension. Lorentz's "ion," Larmor's "electron," and Thomson's "corpuscle" were associated with different representations, all of which are a far cry from our contemporary representation of the electron. In view of these differences, is it possible to maintain that the same entity is denoted by all of these terms? The key to a satisfactory answer to this question can be found in the linguistic practices of those physicists. As we have seen, Lorentz, Larmor, and even Thomson eventually adopted a single name, "electron," for the theoretical entities they had put forward. Apparently, they must have thought that those theoretical entities were representations of the same unobservable entity.[138] A prominent reason for their thinking so was that the charge-to-mass ratio of ions, electrons, and corpuscles turned out to be approximately the same. As a result of the stability of that quantity across different experimental contexts, several experimental situations (the Zeeman effect, cathode rays, thermionic emission, the photoelectric effect, β-rays, etc.) came to be considered observable manifestations of the same entity, the elec-

136. J. J. Thomson, *The Corpuscular Theory of Matter* (London: Charles Scribner's Sons, 1907), p. 28. Cf. Feffer, "Arthur Schuster, J. J. Thomson," p. 60; and Heilbron, *A History of the Problem of Atomic Structure*, p. 98.

137. Achinstein, *Particles and Waves*, p. 286.

138. The fact that Thomson, who was so fond of his "corpuscle," ultimately overcame his attachment and embraced the term "electron" provides further grist for my mill.

tron.[139] In the subsequent development of the representation of the electron, the value of its charge-to-mass ratio was refined (but not altered significantly) and the experimental situations associated with it expanded in a cumulative fashion (see chapter 9). Even when Kaufmann's investigations of β-rays showed that their charge-to-mass ratio varied according to their velocity, this effect did not call into question the identification of β-rays with fast-moving electrons.[140]

This continuity in the history of the representation of the electron binds together the various uses of the term "electron" (and related terms, like "corpuscle") and suggests a sense in which the scientists who used it were referring to the same "thing."[141] Note, however, that from this point of view Thomson's "corpuscle" is not closer to our "electron" than, say, Larmor's "electron." Both of these theoretical entities are associated with the same set of experimental situations and attribute to their referents the same charge-to-mass ratio.

Even though my approach to the problem of identification does not privilege Thomson's "corpuscle," it should be clear that my aim has not been to settle a priority question or to undermine his achievement. The point I have tried to convey is not that Larmor, Lorentz, Zeeman, Wiechert, and others should be given due credit for the discovery of the electron.[142] Such a sugges-

---

139. For the photoelectric effect see B. R. Wheaton, "Philipp Lenard and the Photoelectric Effect, 1889–1911," *Historical Studies in the Physical Sciences,* 9 (1978): 299–322, esp. pp. 310ff.; for β-rays see Dahl, *Flash of the Cathode Rays,* pp. 234ff. The significance of the "overdetermination" of $e/m$ for convincing physicists about the reality of the electron is stressed by John Norton in his "How We Know about Electrons," in R. Nola and H. Sankey (eds.), *After Popper, Kuhn and Feyerabend: Recent Issues in Theories of Scientific Method* (Dordrecht: Kluwer, 2000), pp. 67–97.

140. For a recent and detailed study of this episode see G. Hon, "Is the Identification of Experimental Error Contextually Dependent? The Case of Kaufmann's Experiment and Its Varied Reception," in J. Z. Buchwald (ed.), *Scientific Practice: Theories and Stories of Doing Physics* (Chicago: Univ. of Chicago Press, 1995), pp. 170–223.

141. This takes care of the worry that my "approach appears to be fraught with the difficulty of individualism when different scientists attach different concepts to the same word, as is the case with 'electron'" (Falconer, "Corpuscles to Electrons," p. 93). Furthermore, the common reference of the various tokens of the term "electron" reinforces a point I made in chapter 2—namely, that in tracing the career of the electron's representation a biographical approach is preferable to a prosopographical one.

142. Some physicists and historians have argued that they should. As early as 1901, Walter Kaufmann suggested that the existence of the electron had been established by Zeeman's discovery (see Falconer, "Corpuscles to Electrons," p. 82). More recently, according to "the opinion of Leiden physicists, as told to me by H. B. G. Casimir, . . . Lorentz was the 'discoverer' of the electron" (N. J. Nersessian, "'Why Wasn't Lorentz Einstein?' An Examination of the Scientific Method of H. A. Lorentz," *Centaurus,* 29 [1986]: 205–242, on p. 209). Cf. also A. Romer, "Zeeman's Discovery of the Electron," *American Journal of Physics,* 16 (1948): 216–223.

tion would face, of course, all the problems I have pointed out in connection with the attribution of the electron's discovery to Thomson. What I have tried to show, instead, is that there is something wrong, both historiographically and philosophically, with viewing the introduction and gradual consolidation of theoretical entities through the distorting lens of a simple-minded notion of "discovery." The electron was not the product of a sudden discovery. Rather, its representation emerged from several problem situations in the study of chemical phenomena (such as electrolysis), in the context of electromagnetic theory, and in the study of the discharge of electricity in gases. By 1900 those diverse situations had found a single solution in the representation of the electron as a subatomic, charged particle. Several historical actors provided the theoretical reasons and the experimental evidence that persuaded the physics community of the electron's reality. However, none of those people discovered it. The most that we can say is that one of those, say Thomson, contributed significantly to the acceptance of electrons as real entities.[143]

My objections to retaining a traditional notion of discovery with respect to the electron do not rule out the possibility of a distinction between the construction of a theoretical framework and the discovery of a real thing in nature. In the first stages of the construction process it is not clear, *even to the scientists who carry it out,* that a theoretical entity within that framework stands for a real, natural kind. At some point, however, the accumulation of evidence about the existence of the entity in question and the indispensability of its representation for theoretical and experimental practice establish the belief that a "real thing in nature has been discovered."[144] Both realists and antirealists can accept, I think, my approach to the "discovery" of unobservable entities and to the "discovery of the electron" in particular. A realist need not have any objections in principle to my point that an evolving process of evidence accumulation, as opposed to an event, led to the acceptance of the electron hypothesis. This would mark, for the realist, the discovery of a "real thing in nature." The antirealist, on the other hand, could very well accept

143. These points are in tune with some recent literature in the history and philosophy of physics. Margaret Morrison, for instance, has pointed out that "it is sometimes impossible to isolate an exact moment or specific procedure that transformed the entity or property from having a mere existence on paper to acquiring a more robust nature as something physically real" ("History and Metaphysics: On the Reality of Spin," in Buchwald and Warwick [eds.], *Histories of the Electron,* pp. 425–449, on p. 426). Cf. Peter Galison's remark that "the 'moment of discovery' of the electron is itself arguably undefinable" (*Image and Logic: A Material Culture of Microphysics* [Chicago: Univ. of Chicago Press, 1997], p. 230).

144. I am using here the phrase of an anonymous referee.

my account of that process and continue to doubt the existence of electrons, opting to maintain a critical distance from the judgment of the scientific community.[145]

Regardless of our philosophical gloss on the early history of the representation of the electron, we could agree that it occupies a prominent place in its biography. During that period some important traits of its character were formed in response to empirical and conceptual problem situations, in some cases even against the expectations of its creators. One of those traits, the charge-to-mass ratio ($e/m$) attributed to the electron, reflected the quantitative aspects of its experimental manifestations and survived, more or less intact, very radical changes in the understanding of its behavior. Furthermore, $e/m$ bound together several experimental situations, which provided the experimental constraints that guided the subsequent development of the representation of the electron. The robustness of that ratio and the cumulative expansion and refinement of those experimental situations gave to the electron qua theoretical entity a life of its own.

Its life in the early twentieth century was closely tied with experimental developments that stemmed from one of those situations, the Zeeman effect. This effect turned out to be much more recalcitrant to theoretical analysis than initially thought. New, more complicated patterns of magnetic splitting (the so-called anomalous Zeeman effect) that violated the predictions of Lorentz's and Larmor's theories were discovered. This remained a problem throughout the development of the "old" quantum theory and was resolved only in 1925 with the attribution of a novel property to the electron, the property of "spin."[146] In the meantime, the representation of the electron's behavior within the atom had been completely transformed. The laws that it was supposed to obey during the infancy of its representation (those of classical mechanics and electromagnetic theory) were gradually abandoned and new properties (e.g., quantum numbers) were attributed to it. What is important for my purposes is that these developments were predicated on the belief that certain experimental situations (most notably in spectroscopy) were the electron's "writings," that is, its observable manifestations. The physicists tried to construct a representation of the electron that would do

145. Cf. B. C. van Fraassen, *The Empirical Stance* (New Haven: Yale Univ. Press, 2002), p. 63.

146. For details see chapters 6 and 8. Cf. J. B. Spencer, "The Historical Basis for Interactions between the Bohr theory of the Atom and Investigations of the Zeeman Effect: 1913–1925," in *XII Congrès International d'Histoire des Sciences: Actes* (Paris: Blanchard, 1971), vol. 5, pp. 95–100.

justice to the complexity of its writings. In the process salient features of those writings (the frequencies of spectral lines, their intensity, their polarization, their patterns of magnetic splitting) obtained counterparts in the properties and behavior of the electron within the atom. It is time now to turn to these developments.

# Chapter 5  |  The Genesis of the Quantum Electron

## 1. Preliminary Remarks

In this chapter we will examine how the electron ceased to be represented as a deterministic mechanical particle, governed by classical mechanics and electromagnetic theory, and began instead to be portrayed as a peculiar entity, which in the so-called quantum transitions was endowed with "free will."[1] From the perspective of the physicists who studied its behavior, the electron started to acquire a life of its own. That was the first instance of the physicists' resort to anthropomorphic metaphors in their struggle to come to terms with the peculiarities of its behavior within the atom.[2]

The historiographical advantages of focusing on the electron, while attempting to reconstruct the development of the old quantum theory, have been recognized in some of the historical literature covering the period we are considering here. John Heilbron, for instance, has suggested that

---

1. Endowing the electron with free will was one of the possibilities envisaged by C. G. Darwin in 1919, when he suggested that physicists were confronted with three alternatives: "to make fundamental changes in our ideas of space and time, or to abandon the conservation of matter and electricity, *or even in the last resort to endow electrons with free will*" (emphasis added). Cited in J. Hendry, "Weimar Culture and Quantum Causality," *History of Science,* 18 (1980): 155–180, on p. 162. The subsequent development of physics realized the final alternative; but more on this below.

2. In the following chapters we will come across further instances of anthropomorphic descriptions of the electron. Note, however, that this aspect of its history should not be conflated with my attribution of agency to the representation of the electron. See the clarifying remarks in chapter 2, p. 46.

"[t]he course of atomic theory from 1895 to 1925 can be construed as the gradual invention of the physics of electrons."[3] Figuring out the physics of electrons was, of course, a wider undertaking than developing an adequate atomic theory. The latter was only a limited part of the former, which was in turn part of a wider investigation concerned with the electron.

The old quantum theory of the atom was an evolving hodgepodge of inconsistent theories and theoretically unjustified empirical rules (the so-called selection rules).[4] It aimed at understanding the behavior and distribution of electrons bound within the atom. By focusing on their putative behavior one can highlight the internal tensions that haunted the theory since its inception and were manifested in the erratic conduct of electrons, which were represented in some respects as obeying and in other respects as disobeying the laws of classical mechanics and electromagnetism. Furthermore, adopting the perspective of the electron qua theoretical entity provides an advantageous standpoint for reconstructing the origins and initial formulation of the old quantum theory of the atom in 1913 by Niels Bohr. It is from that perspective that one can see clearly the most striking aspect of Bohr's initial exposition of the theory: the copresence in that exposition of two incompatible mechanisms for the emission of spectral lines. As we will see, each mechanism was based on a different conception of the electron.[5]

From the above point of view, the major developments in the evolution of the old quantum theory are as follows: First, several quantum constraints, the so-called quantum conditions, were imposed on the putative motion of the electron within the atom. The size, shape, and direction in space of its orbit were quantized; that is, the electron was allowed to move only in those discrete orbits whose size, shape, and direction satisfied certain restrictive

3. J. L. Heilbron, *Historical Studies in the Theory of Atomic Structure* (New York: Arno Press, 1981), p. 11.

4. Imre Lakatos has correctly emphasized the inconsistency of the old quantum theory of the atom in an otherwise seriously flawed discussion of "Bohr's research program." See his "Falsification and the Methodology of Scientific Research Programmes," in I. Lakatos and A. Musgrave (eds.), *Criticism and the Growth of Knowledge* (Cambridge: Cambridge Univ. Press, 1970), pp. 91–196, esp. pp. 140–154. I am also indebted to John Norton's discussion in "The Logical Inconsistency of the Old Quantum Theory of Black Body Radiation," *Philosophy of Science*, 54 (1987): 327–350.

5. This peculiar aspect of Bohr's theory, which was published in three consecutive papers, was first pointed out by John Heilbron and Thomas Kuhn in 1969. See J. Heilbron and T. S. Kuhn, "The Genesis of the Bohr Atom," *Historical Studies in the Physical Sciences*, 1 (1969): 211–290. Kuhn has even said that it was "the nonsense passages [i.e. those involving a classical electron] that gave rise to the project." T. S. Kuhn, "Dubbing and Redubbing: The Vulnerability of Rigid Designation," in C. W. Savage (ed.), *Scientific Theories*, Minnesota Studies in the Philosophy of Science 14 (Minneapolis: Univ. of Minnesota Press, 1990), pp. 298–318, on p. 315.

conditions. These constraints appeared in the form of "quantum numbers" which "characterize the state of the electron in question."[6] By 1916 three of these quantum numbers had been introduced by Bohr and Arnold Sommerfeld. In 1920 another quantum number was added and was initially interpreted as a collective property of the atom. In 1924 Pauli ascribed a fourth quantum number to the valence electron itself. One year later that number was interpreted, without Pauli's approval, as a manifestation of the internal rotation of the electron, the so-called spin.

Second, the so-called selection rules were introduced. According to those rules, an electron can "jump" from one orbit to another only if the initial and the final orbit bear certain relations to each other. Although the two developments noted above represented restrictions on the electron's freedom, in another respect the electron's independence from law-governed processes had been reinforced. When Bohr put forward his theory, Rutherford diagnosed as a major defect of the theory that in a quantum transition the electron has to "*decide* what frequency it is going to vibrate at when it passes from one stationary state to the other."[7] That is, Rutherford had realized that Bohr's theory did not give a causal account of quantum transitions. This marks the beginning of a very long process that eventually led to the construction of an acausal quantum mechanics.

In this chapter and in chapters 6 and 8 I will discuss each of these developments in some detail and explain how they shaped the character of the electron. Throughout these chapters the notion of a problem situation will play a key role. The development of the electron's representation in the context of the old quantum theory took place in response to various problems. Many of those were "ordinary" empirical and conceptual problems, of the kind we discussed in chapter 1. Some other problems, however, were due to the agency of the electron qua theoretical entity. It often happened that when physicists attempted to attribute a new property (or behavior) to the electron, undesirable consequences seemed to follow. The implications of the property (or behavior) in question were in conflict with some of the already accepted features of the electron's representation. If those features had been experimentally determined, the conflict in question amounted to an empirical problem. If, on the other hand, the postulated property clashed with a theoretical prin-

6. "[K]ennzeichnen wir den Zustand des fraglichen Elektrons durch gewisse Quantenzahlen." A. Sommerfeld, *Atombau und Spektrallinien*, 5th ed., vol. 1 (Braunschweig: F. Vieweg, 1931), p. 162.

7. Rutherford to Bohr, 20 Mar. 1913; in N. Bohr, *Collected Works*, vol. 2, ed. U. Hoyer (Amsterdam: North Holland, 1981), p. 583.

ciple that was supposed to govern the electron's behavior, the upshot was a conceptual problem.

Thus, we will see how the electron's previously formed traits (its already established properties and the laws it was supposed to obey) posed limits to the manipulation of its representation by theoreticians and guided their practice. Furthermore, I will highlight the relative independence of the representation in question from the evolving theoretical framework in which it was embedded—an independence that is manifested in two ways. First, its experimentally determined components, most notably the value of the charge and mass of the electron, remained relatively stable. Second, the authenticity of the electron's writings, available in the experimental situations attributed to it, was not challenged, and their qualitative and quantitative features guided the further articulation of its representation. These characteristics of the electron qua theoretical entity rendered it an active participant in the development of the old quantum theory.

## 2. The Electron Migrates to the Quantum World

The electron ceased to be an inhabitant of the classical world toward the end of 1913.[8] The transition process from a classical to a quantum context was effected through the work of Niels Bohr. Bohr's achievement has been reconstructed in meticulous detail by John Heilbron and Thomas Kuhn in their seminal article "The Genesis of the Bohr Atom."[9] In what follows I will discuss briefly, drawing on that analysis and Bohr's papers, the problems that occupied Bohr prior to the publication of the series of three papers that paved the way for the transition. I will then reconstruct the fundamentally incoherent picture of the electron that emerges from these papers.

Niels Bohr (1885–1962) obtained his Ph.D. degree from the University of Copenhagen in May 1912.[10] The subject of his thesis was the electron theory

8. In this section as elsewhere I discuss the behavior of the electron *as represented by physicists*.

9. This paper remains the most thorough treatment of Bohr's revolutionary breakthrough. With few exceptions that will be discussed below, I share their interpretation of the episode in question. For a complementary approach to the same episode that explores the cultural resources available to Bohr see M. N. Wise, "How Do Sums Count? On the Cultural Origins of Statistical Causality," in L. Krüger, L. J. Daston, and M. Heidelberger (eds.), *The Probabilistic Revolution*, vol. 1, *Ideas in History* (Cambridge, Mass.: MIT Press, 1987), pp. 395–425.

10. Cf. A. Pais, *Niels Bohr's Times, in Physics, Philosophy, and Polity* (New York: Oxford Univ. Press, 1991), pp. 107–111. Pais's biography of Bohr, despite its lack of historiographical sophistication, is a valuable and comprehensive account of Bohr's life and work and an important bibliographical resource.

of metals. Its aim was "to attempt to carry out calculations for the different phenomena which are explained by the presence of free electrons in metals [e.g., electrical and thermal conductivity], in the most general manner possible consistent with the principles of Lorentz' theory."[11] Lorentz, who along with Paul Drude had made the most important contributions to this field,[12] had assumed that the electrons in a metal obey the laws of classical mechanics; —that is, their motions are governed by Hamilton's equations. Bohr stressed the limitations of this assumption, since "there are . . . many properties of bodies impossible to explain if one assumes that the forces which act *within the individual molecules (which according to the ordinary view consist of systems containing a large number of 'bound' electrons)* are mechanical also. Several examples of this, for instance calculations of heat capacity and of the radiation law for high frequencies, are well-known; we shall encounter another later, in our discussion of magnetism" (emphasis added).[13] Thus, from the time that he wrote his thesis, Bohr was convinced that, because of "well-known" difficulties, one had to abandon the assumption that "bound" electrons were subject to mechanical forces.

As Bohr mentioned, there was another problem that further revealed the limitations of classical mechanics, a problem related to the magnetic behavior of diamagnetic and paramagnetic substances. Paramagnetic materials are those which, when put under the influence of a magnetic field, experience a force in the direction of the field. Diamagnetic materials, on the other hand, experience a force in the opposite direction. Given that all substances contain free and bound electrons (within atoms and molecules) that perform various motions, the application of a magnetic field will alter these motions. In 1900 J. J. Thomson had used the effect of a magnetic field on the motions of free electrons to explain the phenomenon of diamagnetism, and in 1905 Paul Langevin had used the same effect on the motions of bound electrons to ex-

---

11. Quoted in Heilbron and Kuhn, "The Genesis," p. 214. Bohr's thesis was first published in English in 1972. See N. Bohr, *Studies on the Electron Theory of Metals*, in his *Collected Works*, vol. 1, ed. J. R. Nielsen (Amsterdam: North Holland, 1972), pp. 291–395.

12. For a comprehensive study of the development of the electron theory of metals see W. Kaiser, "Electron Gas Theory in Metals: Free Electrons in Bulk Matter," in J. Z. Buchwald and A. Warwick (eds.), *Histories of the Electron: The Birth of Microphysics* (Cambridge, Mass.: MIT Press, 2001), pp. 255–303.

13. See N. Bohr, *Studier over metallernes elektrontheori* (first publ. 1911), in his *Collected Works*, vol. 1, pp. 167–290, on p. 175; trans. in Heilbron and Kuhn, "The Genesis," p. 215. The paragraph that contains this passage is not included in the English translation of Bohr's thesis that is available in his *Collected Works*. Instead it has been replaced by another one that Bohr inserted in a revised version of his dissertation that he prepared during the summer of 1911. See "Translator's Preface," in *Collected Works*, vol. 1, p. 293.

plain both phenomena. In his thesis Bohr criticized both of these explanatory strategies and showed that "if the ordinary mechanics held, neither the free nor the bound electrons contributed at all to the magnetic properties of matter, or at least not to diamagnetism."[14]

Furthermore, Bohr arrived at the conclusion that classical electromagnetic theory was also of limited use for describing the short-term motions of individual electrons. The route to this conclusion was quite complex. It follows from Maxwell's electrodynamics that the electrons in a metal undergo acceleration whenever they are subjected to radiation, and emit radiation whenever they are forced to slow down or to accelerate. If they are in a state of equilibrium, the energy that they radiate and the energy that they absorb must be equal. Only a fraction of the radiation that they are exposed to gets absorbed, and this fraction is expressed by the coefficient of absorption, which is a function of temperature and of the wavelength of radiation. Similarly, the radiation that they emit is expressed by the coefficient of emission, also a function of temperature and wavelength. Lorentz had calculated the ratio of those coefficients for long waves in 1903 and obtained a result that was empirically adequate. Bohr inferred the limited validity of classical electrodynamics by arguing that a Lorentz-type calculation of the ratio of those coefficients for short waves must lead to failure. "This is presumably due to the circumstance that the electromagnetic theory . . . can only give correct results when applied to a large number of electrons (as are present in ordinary bodies) or to determine the average motion of a single electron over comparatively long intervals of time (such as in the calculation of the motion of cathode rays) but cannot [as in shortwave radiation] be used to examine the motion of a single electron within short intervals of time."[15] Thus, as a result of the above difficulties, Bohr was led to the radical conclusion that individual electrons did not always obey the laws of classical mechanics and electromagnetic theory.

Toward the end of September 1911, a few months after defending his thesis, Bohr moved to Cambridge to study with J. J. Thomson. His stay in Cambridge turned out to be "very interesting . . . but . . . absolutely useless."[16] Bohr's inadequate command of the English language and the indifference

14. Heilbron and Kuhn, "The Genesis," p. 218. A detailed analysis of Bohr's criticism is beyond the scope of this chapter. I refer the interested reader to Heilbron's and Kuhn's reconstruction of Bohr's arguments; ibid., pp. 218–223.

15. Bohr, *Collected Works*, vol. 1, p. 378. For a more detailed exposition of the failure of classical electrodynamics see Heilbron and Kuhn, "The Genesis," pp. 216–218.

16. N. Bohr, interview with T. S. Kuhn, L. Rosenfeld, A. Petersen, and E. Rüdinger, 1 Nov. 1962, quoted in Pais, *Niels Bohr's Times*, p. 121.

of Thomson to his research agenda inhibited any profitable interaction between them.

One aspect of Bohr's visit to Cambridge, however, is pertinent to our story. In a letter to his friend C. W. Oseen, dated 1 December 1911, Bohr wrote that "[a]t the moment I am very enthusiastic about . . . the magneton theory." [17] This theory, which had been proposed by Pierre Weiss a few months before, asserted that the magnetic moments of atoms are quantized; that is, they can take only those values that are integral multiples of a fundamental minimal quantity (the magneton). If an atom is composed of orbiting electrons, then its magnetic moment will be the sum of the magnetic moments of its constituent electrons. Thus, the quantization of the magnetic moments of atoms suggests a corresponding quantization of the magnetic moments of individual electrons. Given that there is a fixed ratio of an orbiting electron's magnetic moment to its angular momentum ($M/L = e/2mc$), a quantization of the former implies a nonmechanical constraint on the electron's motion. Heilbron and Kuhn argue persuasively that Bohr realized the above implications of Weiss's magneton, even though there is no direct evidence to that effect.[18] If they are right, then Bohr's acquaintance with Weiss's theory was one of the most important resources that he acquired during his Cambridge stay.

Bohr's Cambridge sojourn lasted until the middle of March 1912. He then transferred to Manchester "primarily to learn something about radioactivity, the specialty of . . . Rutherford." [19] Two months after his arrival Rutherford instructed him to do experimental work on the absorption of α-particles (helium nuclei) in aluminum.[20] While occupied with that problem he "had a little idea about understanding the absorption of α-rays (it so happened, that a young mathematician from here C.G. Darwin (the grandson of the right Darwin) has just published a theory on this question and it seemed to me that it was not only not quite right mathematically (this was however rather trifling) but quite unsatisfactory in its basic conception), and I have worked out a little theory about it, which, however modest, may perhaps throw some light over a few things concerning the structure of atoms." [21] Darwin's paper was the outcome of an

17. See Bohr, *Collected Works,* vol. 1, p. 431.

18. For their argument I refer the interested reader to Heilbron and Kuhn, "The Genesis," pp. 230–232.

19. Ibid., p. 234.

20. Cf. Pais, *Niels Bohr's Times,* p. 125.

21. Niels Bohr to Harald Bohr (his brother), 12 June 1912, cited in L. Rosenfeld's introduction to N. Bohr, *On the Constitution of Atoms and Molecules* (Copenhagen: Munksgaard, 1963), a reprinting of Bohr's 1913 trilogy, p. xvii.

attempt to solve a problem that Rutherford had posed to him, namely, the energy loss that α-particles undergo when they penetrate matter.[22] The details of Darwin's calculations need not concern us here. What is important is that he had based his calculations on Rutherford's nuclear model of the atom, and his central assumption was that the energy loss of α-particles was predominantly due to their collisions with electrons inside the atom, as opposed to their interaction with the nuclei. To facilitate those calculations, he made another simplifying assumption, namely, that during their collisions with α-particles the electrons can be treated as free. That is, the interactions of electrons with the nucleus can be neglected during their collisions with α-particles.

Darwin's theoretical treatment was confronted with difficulties,[23] and Bohr ascribed these to this assumption. Instead, Bohr proposed that in such collisions the electrons within the atom should be considered as oscillating under the action of an elastic force (as "atomic vibrators").[24] In this manner Bohr's attention became focused on the problem of atomic structure.

While in Manchester, Bohr prepared a draft of his ideas on the constitution of atoms and molecules and gave it to Rutherford for comments.[25] This document, widely known as the Rutherford memorandum, has only recently been published in its entirety.[26] Two theories of atomic structure are discussed in the Rutherford memorandum, J. J. Thomson's "plum pudding" model of the atom and Rutherford's planetary model. In Thomson's model (proposed in 1904) electrons moved in coplanar, circular orbits inside a homogeneously and positively charged sphere.[27] Rutherford's atom consisted of

22. C. G. Darwin, "A Theory of the Absorption and Scattering of the α Rays," *Philosophical Magazine,* 6th series, 23 (1912): 901–920.

23. For details see Heilbron and Kuhn, "The Genesis," pp. 239–240; and Pais, *Niels Bohr's Times,* p. 127.

24. See N. Bohr, "On the Theory of the Decrease of Velocity of Moving Electrified Particles on Passing through Matter," *Philosophical Magazine,* 6th series, 25 (1913): 10–31.

25. Pais (*Niels Bohr's Times,* p. 136) says that Bohr sent that draft to Rutherford on 6 July 1912. Given that Bohr left Manchester for Denmark on 24 July 1912, it is not clear why he had to *send* his draft to Rutherford. Cf. Heilbron and Kuhn, "The Genesis," pp. 244–245.

26. See Bohr, *Collected Works,* vol. 2, pp. 136–143. The Rutherford memorandum had been reproduced "with two unessential omissions" as part of the introduction to Bohr, *On the Constitution,* pp. xxi–xxviii. The quote is from p. xxi.

27. See J. J. Thomson, "On the Structure of the Atom: An Investigation of the Stability and Periods of Oscillation of a Number of Corpuscles Arranged at Equal Intervals around the Circumference of a Circle; with Application of the Results to the Theory of Atomic Structure," *Philosophical Magazine,* 6th series, 7 (1904): 237–265. Heilbron argues that "plum pudding" is not an appropriate name for Thomson's model because it implies a representation of the atom as a static entity, whereas Thomson's atom was a dynamic one. See Heilbron, *Historical Studies in the Theory of Atomic Structure,* pp. 26–27.

a massive, positively charged nucleus ansd a crowd of electrons revolving around the nucleus.[28] Bohr decided to develop Rutherford's theory, not only because it could explain the large-angle scattering of α-particles by matter, but also because it "seems to offer a very strong indication of a possible explanation of the periodic law of the chemical properties of the elements."[29] Thomson's model, on the other hand, failed in both of these respects. It is interesting that Bohr's comparative appraisal of the two atomic theories with respect to their capacity to explain the periodic table "rests on a mistake in calculation."[30]

Rutherford's model, however, generated several conceptual problems, which Bohr addressed in the Rutherford memorandum. First, in contrast with Thomson's model, in a Rutherford type atom "there can be no equilibrium figuration, without motion of the electrons."[31] That is, any static system of particles whose mutual interactions are governed by inverse square forces is unstable.[32] Second, any ring of more than one electron revolving around the nucleus is characterized by mechanical instability: "[I]t can very simply be shown, that a ring as the one in question posesses [sic] no stability in the ordinary mechanical sense."[33] Again, in this respect Thomson's model was preferable to Rutherford's model of the atom. Third, whereas the size of Thomson's atom was fixed (albeit arbitrarily) and was determined by the radius of the sphere of positive electricity in which the electrons are embedded, there was no way to fix the size of a Rutherford-type atom. As Bohr remarked, a ring of electrons "can rotate with an infinitely great number of different times of rotation, according to the assumed different radius of the ring; and there seems to be nothing . . . to allow from mechanical considerations to discriminate between the different radii and times of vibration."[34]

---

28. E. Rutherford, "The Scattering of α and β Particles by Matter and the Structure of the Atom," *Philosophical Magazine*, 6th series, 21 (1911): 669 – 688; repr. in *The Collected Papers of Lord Rutherford of Nelson*, vol. 2 (London: George Allen and Unwin, 1963), pp. 238 –254. The invention of Rutherford's model has been extensively discussed by Heilbron. See J. L. Heilbron, "The Scattering of α and β Particles and Rutherford's Atom," *Archive for History of Exact Sciences*, 4 (1968): 247–307; repr. in his *Historical Studies in the Theory of Atomic Structure*, pp. 85 –145.

29. Bohr, Rutherford memorandum, in *On the Constitution*, p. xxii.

30. For details see Heilbron and Kuhn, "The Genesis," pp. 245 –247. The quote is from p. 246.

31. Bohr, Rutherford memorandum, p. xxii.

32. This follows from a theorem that was proved in 1831 by Samuel Earnshaw. See A. Pais, *Inward Bound* (New York: Oxford Univ. Press, 1988), p. 181.

33. Bohr, Rutherford memorandum, p. xxii.

34. Ibid., p. xxiii.

Bohr solved the problems created by Rutherford's model by postulating a "hypothesis, for which there will be given no attempt of a mechanical foundation (as it seems hopeless . . . )."[35] As I mentioned above, from the time that he wrote his doctoral thesis, Bohr had discerned the limitations of mechanics (with respect to magnetism, Planck's radiation law, and the heat capacity of certain substances), and, thus, it is not surprising that he was willing to introduce a nonmechanical hypothesis. His hypothesis was "that there for any stable ring (any ring occurring in the natural atoms) will be a definite ratio between the kinetic energy of an electron in the ring and the time of rotation" (it is obvious from a subsequent page of the Rutherford memorandum that Bohr meant the frequency).[36] In the rest of the memorandum Bohr outlined the empirical support for this hypothesis and employed it to provide a quantitative analysis of the hydrogen molecule. He provided a detailed analysis of the explanatory power of this hypothesis only after his return to Denmark (24 July 1912). The results of that analysis formed the core of the second and third parts of his trilogy (published in September and November of 1913, respectively).

Thus, in response to the problem situation generated by Bohr's adoption of the Rutherford model, the electron's motion was constrained by a nonmechanical law. Two aspects of that problem situation deserve special emphasis. First, it was the mechanical and not the radiative instability of the Rutherford atom that played an important role in Bohr's initial development and modification of the Rutherford model. Radiative instability occupied his attention only several months after he had chosen to adopt Rutherford's planetary model.[37] Second, spectroscopy was not one of the elements of Bohr's problem situation at the time that he wrote the memorandum. This is extremely interesting in view of Bohr's own later explicit denial that a nonmechanical restriction of the electron's motion could have been possible before he came across the Balmer formula (a mathematical expression of the wavelengths of the hydrogen spectral lines). Bohr's denial was recorded during a series of interviews that he gave to Thomas Kuhn. It is worth reproducing here Kuhn's account of his exchange with Bohr: "My first few interviews with Bohr dealt with the background for his atomic model, and I asked what sorts of connections he had made between the Rutherford atom and the quantum during the period before his attention was directed to the Balmer formula. He replied that he could not have had developed ideas on the subject before turning to

35. Ibid.
36. Ibid.
37. This point was first emphasized by Heilbron and Kuhn, "The Genesis," p. 241.

spectra, and his assistant later reported to me that, after I had left the room, Bohr shook his head and said of our exchange, 'Stupid question. Stupid question.'"[38] Kuhn plausibly attributes the disparity between Bohr's account and the historical record to the fact that "quite usually, scientists will strenuously resist recognizing that their discoveries were the product of beliefs and theories incompatible with those to which the discoveries themselves gave rise."[39]

For several months after his return to Copenhagen Bohr's problem situation remained as we have discussed with respect to the Rutherford memorandum. However, a crucial problem shift took place in early February 1913, when Bohr came across the Balmer formula. A couple of months before, Bohr had discovered two articles by J. W. Nicholson entitled "Constitution of the Solar Corona."[40] In those articles Nicholson proposed a quantized planetary model of the atom in order to account for the frequencies of the spectral lines of the solar corona. A central aspect of his model was that it associated those spectral lines with the vibrations that perturbed intraatomic electrons performed in directions perpendicular to the plane of their orbit. Given that these vibrations depended on the orbital frequency of the electrons, Nicholson could calculate their frequencies, radii, and energies. The values that he obtained, however, were "in striking disagreement with those I [Bohr] have obtained; and I therefore thought at first that the one or the other [theory] necessarily was altogether wrong."[41]

By late December 1912 Bohr had managed to find a way out of this "striking disagreement." In a Christmas card to his brother, Harald, he mentioned that "one of us would like to say that he thinks Nicholson's theory is not incompatible with his own. In fact his calculations would be valid for the final, chemical state of the atoms, whereas Nicholson would deal with the atoms sending out radiation, when the electrons are in the process of losing energy before they have occupied their final positions."[42] And in a letter to Rutherford on 31 January 1913 Bohr elaborated on the apparent incompatibility between Nicholson's theory and his own:

> The state of the systems considered in my calculations is however . . . characterized as the one in which the systems possess the smallest possible amount of

38. T. S. Kuhn, "Revisiting Planck," *Historical Studies in the Physical Sciences*, 14, no. 2 (1984): 231–252, on pp. 247–248.

39. Ibid., p. 248.

40. For Nicholson's theory see R. McCormmach, "The Atomic Theory of John William Nicholson," *Archive for History of Exact Sciences*, 3 (1966): 160–184.

41. Bohr to Rutherford, 31 Jan. 1913, published in Bohr, *On the Constitution*, p. xxxvii.

42. Ibid., p. xxxvi.

energy, i.e. the one by the formation of which the greatest possible amount of energy is radiated away.

It seems therefore to me to be a reasonable hypothesis, to assume that the state of the systems considered in my calculations is to be identified with that of the atoms in their permanent (natural) state . . .

According to the hypothesis in question the states of the systems considered by Nicholson are, contrary, of a less stable character; they are states passed during the formation of the atoms, and are the states *in which* the energy corresponding to the lines in the spectrum characteristic for the element in question is radiated out. From this point of view systems of a state as that considered by Nicholson are only present in sensible amount in places in which atoms are continually broken up and formed again; i.e. in places such as excited vacuum tubes or stellar nebulae. (Emphasis added.)[43]

This passage is revealing of the state of Bohr's theory at the end of January 1913. As Heilbron and Kuhn first pointed out, at that time "Bohr still envisaged a radiation mechanism that was in two respects quasi-classical."[44] First, radiation takes place when "atoms are continually broken up and formed again"; that is, ionization is a necessary condition for radiation, which results from the recapturing of an electron by an ionized atom. Second, when an electron is captured by an atom it proceeds through several states; as soon as it reaches a certain state, a specific amount of "energy . . . is radiated out."

Heilbron and Kuhn interpret Bohr as asserting that "an electron falling into its final orbit in the previously ionized atom causes the electrons in the high energy orbits to vibrate at their resonance frequencies, acting rather like a finger drawn across the strings of a harp."[45] However, I do not think that this interpretation is forced on us. On the contrary, it is problematic in one respect: it does not allow that radiation is emitted during the reformation of hydrogen atoms. Because hydrogen ions do not contain any "electrons in . . . high energy orbits to vibrate at their resonance frequencies," there is no mechanism by which to radiate.[46]

---

43. Ibid., p. xxxvii.

44. Heilbron and Kuhn, "The Genesis," p. 263.

45. Ibid.

46. It is true that some people at the time believed that the hydrogen atom had more than one electron. Nicholson, for instance, suggested that it contained three electrons (see Pais, *Niels Bohr's Times*, p. 145). However, Bohr himself claimed "that the experiments on absorption of α-rays very strongly suggest that a hydrogen atom contains only one electron." See Bohr, "On the Theory of the Decrease of Velocity," p. 24. This paper was communicated by Rutherford in *August 1912*.

Since Bohr did not exclude hydrogen atoms from the domain of application of the radiation mechanism that he envisaged, we should look for an alternative interpretation of the above excerpt from his letter to Rutherford. I propose that Bohr could have meant that radiation is emitted by the captured electron itself, as opposed to the other electrons already present in the atom. As soon as the electron reaches a certain state (i.e., a certain orbit), it emits radiation of the same frequency as the frequency of revolution in that state. After (or during) the emission process it sinks to a lower state in which it radiates again, and so on till it reaches the permanent (or natural) state.

Bohr's radiation mechanism, as I have interpreted it, is based on a quasi-classical conception of the electron and was one of the two incompatible mechanisms that Bohr employed in the first part of his trilogy to derive the Balmer formula. The other mechanism was invented by Bohr in a very short period that followed his acquaintance with this formula in early February 1913.[47] Bohr's new account of radiation portrayed the production of spectral lines as the outcome of an electron's transitions between stationary states and uncoupled for the first time the spectral frequencies from the orbital frequencies of the electron. This mechanism is widely considered "perhaps the greatest and most original of Bohr's breaks with existing tradition."[48] The incompatibility of the two mechanisms was the result of Bohr's struggle with two disparate problem situations.

Unfortunately, there are neither letters nor interviews that give a detailed account of what happened in that short interval.[49] Thus, I will proceed directly to discussing the product of that transition process, the publication of the first part of Bohr's trilogy.[50] In the introduction to that paper Bohr mentioned the conceptual problems that haunted Rutherford's model of the atom (mechanical instability and lack of definite size) and suggested that these difficulties, as well as other problems facing physics at the time (specific heats, photoelectric effect, etc.), could be resolved by an "alteration in the laws of motion of the electrons" (p. 2). Even though one could not specify the exact form of the new laws, "it seems necessary to introduce in the laws in question a quantity foreign to the classical electrodynamics, i.e. Planck's constant, or as it often is called the elementary quantum of action"(p. 2). Thus, as a result of

47. For an account of how Bohr became aware of the Balmer formula see Heilbron and Kuhn, "The Genesis," pp. 264–265; and Pais, *Niels Bohr's Times*, p. 144.

48. Heilbron and Kuhn, "The Genesis," p. 266.

49. For a plausible but a priori reconstruction of the impact of Balmer's formula on Bohr's thought see Heilbron and Kuhn, "The Genesis," pp. 265–266.

50. N. Bohr, "On the Constitution of Atoms and Molecules," pt. 1, *Philosophical Magazine*, 6th series, 26 (1913): 1–25; repr. in his *On the Constitution*, pp. 1–25.

conceptual and empirical pressures the character of the electron was about to undergo a significant transformation.

The rest of Bohr's paper can be seen as a preliminary attempt to specify the new laws governing the motion of the electron and to combine these laws with Rutherford's model of the atom in order to obtain "a basis for a theory of the constitution of atoms" (pp. 2–3). In the beginning of part 1 Bohr assumed "[f]or simplicity . . . that the mass of the electron is negligibly small in comparison with that of the nucleus, and further, that the velocity of the electron is small compared with that of light" (p. 3). An electron moving under the influence of the attractive force of the nucleus will follow a fixed elliptical orbit, provided that it does not radiate. The frequency of revolution ($\omega$) and the major axis of the orbit ($2\alpha$) depend on the energy ($W$) that "must be transferred to the system in order to remove the electron to an infinitely great distance apart from the nucleus" (p. 3), the charge ($e$) and mass ($m$) of the electron, and the charge ($E$) of the nucleus. One can obtain the major axis and the frequency by the following straightforward calculation:[51]

$$(5.1) \qquad \frac{eE}{\alpha^2} = \frac{mv^2}{\alpha},$$

where $v$ is the velocity of the electron (the attractive force between the nucleus and the electron provides the centripetal force that is necessary to keep the electron in a circular orbit).

$$(5.2) \qquad W = -(U + T),$$

where $U$ is the potential and $T$ is the kinetic energy of the electron.

$$(5.3) \qquad U = -\frac{eE}{\alpha}.$$

$$(5.4) \qquad T = \frac{1}{2}mv^2.$$

$$(5.1) + (5.4) \rightarrow (5.5) \qquad T = \frac{eE}{2\alpha}.$$

51. Cf. M. Jammer, *The Conceptual Development of Quantum Mechanics* (New York: McGraw-Hill, 1966), p. 76. The calculation that follows is based on the assumption that the electron moves in a circular orbit. However, the results that one obtains are also valid in the case of an elliptical orbit whose major axis coincides with the diameter of the circular orbit in question. Note that Bohr omitted this calculation from his paper.

$$(5.2) + (5.3) + (5.5) \rightarrow (5.6) \quad W = \frac{eE}{2\alpha}.$$

$$(5.6) \rightarrow (5.7) \quad 2\alpha = \frac{eE}{W} \quad \text{(the major axis)}.$$

$$(5.8) \quad v = 2\pi\omega\alpha.$$

$$(5.4) + (5.5) + (5.6) + (5.8) \rightarrow (5.9) \quad 2m(\pi\omega\alpha)^2 = W.$$

$$(5.7) + (5.9) \rightarrow \quad 2m\left(\frac{\pi\omega\, eE}{2W}\right)^2 = W \rightarrow$$

$$(5.10) \quad \omega = \frac{\sqrt{2}}{\pi} \frac{W^{3/2}}{eE\sqrt{m}} \quad \text{(the frequency)}.$$

These results established that "if the value of $W$ is not given, there will be no value of $\omega$ and $\alpha$ characteristic for the system in question."[52]

So far Bohr had drawn on some aspects of the character previously attributed to the electron (namely, its having a certain charge and mass, and its being subject to the laws of Coulomb and classical mechanics), to calculate the characteristics of its motion. Furthermore, another assumption was necessary for that calculation to get off the ground—that is, that the electron did not radiate. Such a behavior, however, was uncharacteristic of the electron. Because of its acceleration, energy would be radiated away and it would gradually collapse into the nucleus. This paradoxical conclusion, which resulted from the unrestricted application of classical electrodynamics to Rutherford's representation of the behavior of electrons within the atom, was incompatible with the behavior of "actual atoms in their permanent state [which] seem to have absolutely fixed dimensions and frequencies."[53] Thus, Rutherford's (and Bohr's) attempt to enrich the representation of the electron encountered resistance, which was due to the theoretical framework in which that representation was embedded *and* the experimental constraints it had to satisfy (cf. chapter 2, p. 37).

The key to the solution of the above paradox could be found, according to Bohr, in Planck's theory of radiation:

> Now the essential point in Planck's theory of radiation is that the energy radiation from an atomic system does not take place in the continuous way as-

---

52. Bohr, *On the Constitution*, p. 3.
53. Ibid., p. 4.

sumed in the ordinary electrodynamics, but that it, on the contrary, takes place in distinctly separated emissions, the amount of energy radiated out from an atomic vibrator of frequency $\nu$ in a single emission being equal to $\tau h\nu$, where $\tau$ is an entire number, and $h$ is a universal constant . . .

Returning to the simple case of an electron and a positive nucleus considered above, let us assume that the electron at the beginning of the interaction with the nucleus was at a great distance apart from the nucleus, and had no sensible velocity relative to the latter. Let us further assume that the electron after the interaction has taken place has settled down in a stationary orbit around the nucleus . . .

Let us now assume that, during the binding of the electron, a homogeneous radiation is emitted of a frequency $\nu$, equal to half the frequency of revolution of the electron in its final orbit; then, from Planck's theory, we might expect that the amount of energy emitted by the process considered is equal to $\tau h\nu$. . . . If we assume that the radiation emitted is homogeneous, the second assumption concerning the frequency of the radiation suggests itself, since the frequency of revolution of the electron at the beginning of the emission is 0. (Pp. 4–5) [54]

Since the radiated energy equals the energy $W$ which must be expended to remove the electron to an infinite distance from the nucleus, it follows that

$$(5.11) \quad W = \tau h\omega/2.$$

From this equation, together with the above-mentioned equations that relate $W$ to the frequency of revolution and the major axis of the orbit of the electron, one obtains the following formulas:

$$(5.10) + (5.11) \rightarrow (5.12) \quad W = \frac{2\pi^2 m e^2 E^2}{\tau^2 h^2}.$$

$$(5.10) + (5.11) \rightarrow (5.13) \quad \omega = \frac{4\pi^2 m e^2 E^2}{\tau^3 h^3}.$$

54. The reasons that led Bohr to set the frequency of radiation "equal to half the frequency of revolution of the electron in its final orbit" are obscure. Heilbron and Kuhn suggest that the actual reasons for Bohr's choice had nothing to do with "his remark that the frequency [of radiation] . . . is an average of the electron's mechanical frequencies . . . in its initial and final states." But then they are forced to conclude that this remark was "an *ad hoc* rationalization, designed to preserve the parallelism between Bohr's radiator and Planck's." Heilbron and Kuhn, "The Genesis," pp. 271–272. I do not have any serious argument here against their interpretation. However, Bohr's remark that this "assumption . . . suggests itself" does not strike me as an "*ad hoc* rationalization." Cf. E. M. MacKinnon, *Scientific Explanation and Atomic Physics* (Chicago: Univ. of Chicago Press, 1982), pp. 167–168.

$$(5.7) + (5.12) \rightarrow (5.14) \quad 2\alpha = \frac{\tau^2 h^2}{2\pi^2 meE}.$$

By substituting different values of $\tau$, one can obtain a series of values for all of these parameters. Assuming further that these values "correspond to states of the system in which there is no radiation of energy,"[55] Bohr managed to avoid all the difficulties that haunted Rutherford's theory.

Three aspects of Bohr's solution to the problem of radiative instability should be emphasized. First, the process of radiation occurs during the capture of the electron by the nucleus. Second, the frequency of the emitted radiation is still coupled with the mechanical frequency of the electron in its final orbit. Both of these assumptions had been formed before he came across the Balmer formula. Third, during the process of radiation the electron emits $\tau$ quanta of the *same* frequency ($\omega/2$).[56] It is important to note that this assumption precludes the derivation of the Balmer formula. This formula is a mathematical representation of the hydrogen spectrum and reflects the variety of frequencies manifested in that spectrum. Bohr's assumption, on the other hand, requires that the hydrogen spectrum be composed of a single spectral line.[57]

Thus, the electron's writings, its experimental manifestations in the hydrogen spectrum, indicated that Bohr was on the wrong track. Some corrective maneuver was necessary to accommodate those writings in the representation of the electron's behavior. Indeed, his subsequent derivation of the Balmer formula was based on a reinterpretation of the radiation mechanism, according to which the electron emits a *single* quantum of frequency $\tau\omega/2$. Bohr reached this reinterpretation through a critical analysis of Nicholson's explanation of the lines observed in the spectra of the stellar nebulae and the solar corona. As I mentioned above, Nicholson associated those spectral lines with the vibrations of perturbed intraatomic electrons perpendicularly to the plane of their orbit. He further assumed that the ratio between the energy of a ring of electrons around the nucleus and their frequency of rotation was "equal to an entire multiple of Planck's constant."[58] Even though Nicholson's

55. Bohr, *On the Constitution*, p. 5.

56. Even though the above excerpts from Bohr's paper do not necessarily imply that radiation takes place in several distinct quanta, there is direct evidence that this was indeed what he had in mind. As he mentioned a few pages later, one of the assumptions underlying his derivation of the energy levels was "that the different stationary states correspond to the emission of *a different number of Planck's energy-quanta*" (emphasis added). Bohr, *On the Constitution*, p. 8.

57. See Heilbron and Kuhn, "The Genesis," pp. 269–270.

58. Bohr, *On the Constitution*, p. 6.

theory gave an empirically adequate account of the spectral lines in question, "Serious objections . . . may be raised against the theory. These objections are intimately connected with the problem of the homogeneity of the radiation emitted." [59] Bohr's main objection was based on a conceptual problem generated by Nicholson's representation of the electron within the atom. If that representation were accurate, then an electron could not emit monochromatic radiation. In Bohr's words, "As a relation from Planck's theory is used, we might expect that the radiation is sent out in quanta; but systems like those considered, in which the frequency is a function of the energy, cannot emit a finite amount of homogeneous radiation; for, as soon as the emission of radiation is started, the energy and also the frequency of the system are altered." [60] Two points are implicit in Bohr's criticism: First, the connection between the frequency of revolution of the electron within the atom and the frequency of the emitted radiation must be broken. Second, during the radiation process the electron must emit a *single* quantum. If its radiation consisted of several quanta it would not be "homogeneous" (monochromatic).[61]

Both of these points are necessary for his first derivation of the Balmer formula. This derivation goes as follows: For the hydrogen atom ($E = e$), it follows from equation (5.12) that

$$W = \frac{2\pi^2 m e^4}{\tau^2 h^2}.$$

The energy emitted by an electron during a transition from state $\tau_1$ to state $\tau_2$ is

$$W_{\tau_2} - W_{\tau_1} = \frac{2\pi^2 m e^4}{h^2}\left(\frac{1}{\tau_2^2} - \frac{1}{\tau_1^2}\right).$$

By the assumption "that the radiation in question is homogeneous, and that the amount of energy emitted is equal to $h\nu$, where $\nu$ is the frequency of the radiation," [62] it follows that

$$W_{\tau_2} - W_{\tau_1} = h\nu.$$

Therefore,

$$(5.15) \quad \nu = \frac{2\pi^2 m e^4}{h^3}\left(\frac{1}{\tau_2^2} - \frac{1}{\tau_1^2}\right).$$

59. Ibid., pp. 6–7.
60. Ibid., p. 7.
61. See Heilbron and Kuhn, "The Genesis," p. 273.
62. Bohr, *On the Constitution*, p. 8.

From this equation, if we put $\tau_2 = 2$, we obtain the Balmer formula.

Equation (5.15) could also accommodate (for $\tau_2 = 3$) the series of ultrared spectral lines observed by Friedrich Paschen in 1908. The factor

$$\frac{2\pi^2 me^4}{h^3}$$

was the theoretical counterpart of the Rydberg constant and was in close agreement with the empirically determined value of that constant. Further empirical support for Bohr's theory was provided by "the series first observed by Pickering . . . in the spectrum of the star ζ Puppis, and the set of series recently found by Fowler . . . by experiments with vacuum tubes containing a mixture of hydrogen and helium."[63] Bohr attributed both of these series of lines to ionized helium. Finally, his theory could explain why the Rydberg constant was "the same for all substances."[64]

Bohr was now in the following paradoxical situation: On the one hand, he had derived an equation (5.15) which was in impressive agreement with observation; on the other hand, his derivation was based on questionable foundations. One of the assumptions employed in that derivation was "that the different stationary states correspond to an emission of a *different number* of energy-quanta" (emphasis added).[65] However, as Bohr observed, "Considering systems in which the frequency is a function of the energy, this assumption . . . may be regarded as improbable; for as soon as one quantum is sent out the frequency is altered."[66] The problem situation that was confronting Bohr was due to the active nature of the electron qua theoretical entity. In order to deduce the energy levels of an atom, he had attempted to transform the representation of the electron, by putting forward a radiation mechanism that presumably described its behavior during its capture by a positive nucleus. This mechanism, which was constructed on the basis of Planck's theory, prescribed that the radiating electron emitted $\tau$ individual quanta of the same frequency ($\omega/2$). As a result of Bohr's proposal, however, the representation of the electron became incoherent. A radiation mechanism such as he envisaged was theoretically impossible. If, as he assumed, the frequency of the electron's radiation depended on its energy, the emission of a single quan-

63. Ibid., p. 10. Cf. Jammer, *The Conceptual Development of Quantum Mechanics,* pp. 82–84.

64. Bohr, *On the Constitution,* p. 12.

65. Ibid. As I mentioned above, Bohr had already questioned the validity of this assumption. However, he had not, up to that point, explicitly disavowed it.

66. Ibid.

tum would reduce its energy and would, therefore, alter its frequency of radiation. This assumption, built as it were into Bohr's notion of the electron, posed limits on his attempt to specify its behavior in more detail.

To accommodate the resistance of the electron qua theoretical entity, Bohr had to change his interpretation of the mechanism of radiation.[67] He reached the new interpretation through a different derivation of the energy levels of the hydrogen atom. He began this derivation with two observations. First, he noted "that it has not been necessary, in order to account for the law of the spectra by help of the expressions (3) for the stationary states [equations (5.12), (5.13), (5.14)], to assume that in any case a radiation is sent out corresponding to more than a single energy-quantum, $h\nu$."[68] Once the expressions that provided the characteristics of the stationary orbits were in hand, the explanation of spectral regularities did not force one to postulate the emission of more than a single quantum. The second observation was the first hint toward what later became known as the correspondence principle. According to this preliminary formulation, "Further information on the frequency of the radiation may be obtained by comparing calculations of the energy radiation in the region of slow vibrations based on the above assumptions with calculations based on the ordinary mechanics. As is known, calculations on the latter basis are in agreement with experiments on the energy radiation in the named region."[69] Bohr's methodological proposal was to employ the "ordinary mechanics" as a guide for constructing a quantum-theoretical account of the radiation process.[70] In its developed form, it turned out to be a major factor in the construction of the old quantum theory.[71]

The derivation itself went as follows: Bohr's first step was to assume "that the ratio between the total amount of energy emitted and the frequency of

67. Cf. Andrew Pickering's analysis of conceptual practice in terms of the notions of resistance and accommodation (in his *The Mangle of Practice: Time, Agency, and Science* [Chicago: Univ. of Chicago Press, 1995], esp. chapter 4).

68. Ibid.

69. Ibid.

70. It is clear from the derivation that follows that by "ordinary mechanics" Bohr meant classical electrodynamics.

71. For the scientific sources of Bohr's suggestion see Heilbron and Kuhn, "The Genesis," p. 276; and T. S. Kuhn, *Black-Body Theory and the Quantum Discontinuity, 1894–1912*, 2nd ed. (Chicago: Univ. of Chicago Press, 1987), pp. 248–249. The philosophical origins of the correspondence principle are explored in Wise, "How Do Sums Count?" pp. 417–419. The role of Bohr's principle in the construction of the old quantum theory is masterfully analyzed in O. Darrigol, *From c-Numbers to q-Numbers: The Classical Analogy in the History of Quantum Theory* (Berkeley: Univ. of California Press, 1992), pp. 79–259.

revolution of the electron for the different stationary states is given by the equation $W = f(\tau) \cdot h\omega$." [72] From this assumption together with formula (5.10) it follows that

$$W = \frac{\pi^2 m e^2 E^2}{2h^2 f^2(\tau)}$$

and

$$\omega = \frac{\pi^2 m e^2 E^2}{2h^3 f^3(\tau)}.$$

His second step was to assume that "the amount of energy emitted during the passing of the system from a state corresponding to $\tau = \tau_1$ to one for which $\tau = \tau_2$ is equal to $h\nu$." [73] From this assumption, together with the previously derived equation for the energy levels, we get

$$\nu = \frac{\pi^2 m e^2 E^2}{2h^3} \left( \frac{1}{f^2(\tau_2)} - \frac{1}{f^2(\tau_1)} \right).$$

In order to make this expression compatible with the Balmer formula, we need to replace $f(\tau)$ with $c\tau$. The final step is to determine the constant $c$. For $\tau_1 = N$ and $\tau_2 = N - 1$ we get

$$\nu = \frac{\pi^2 m e^2 E^2}{2c^2 h^3} \cdot \frac{2N - 1}{N^2(N - 1)^2}.$$

The frequencies of the electron in the two states are

$$\omega_N = \frac{\pi^2 m e^2 E^2}{2c^3 h^3 N^3} \quad \text{and} \quad \omega_{N-1} = \frac{\pi^2 m e^2 E^2}{2c^3 h^3 (N - 1)^3}.$$

For great values of $N$, that is, "in the region of slow vibrations," these two frequencies will be approximately equal. Furthermore, according to the correspondence principle, in this region classical electrodynamics should give approximately valid results, and thus the frequency of the electron should coincide with the frequency of the emitted radiation. That is,

$$\nu = \frac{\pi^2 m e^2 E^2}{c^2 h^3 N^3} \quad \text{should equal} \quad \omega_N = \frac{\pi^2 m e^2 E^2}{2c^3 h^3 N^3}.$$

72. Bohr, On the Constitution, pp. 12–13. For the source of Bohr's suggestion see Kuhn, Black-Body Theory, p. 249.

73. Bohr, On the Constitution, p. 13.

It follows that $c^2 = 2c^3$ and, therefore, $c = 1/2$. At this point the derivation of the energy levels

$$W = \tau h \frac{\omega}{2}$$

is complete.

Three points are worth stressing in the above derivation. First, the Balmer formula functions as an empirical constraint that cannot be fully derived from Bohr's premises. The electron's writings, as expressed in that formula, were indispensable for constructing an adequate representation of its behavior. Second, the "correspondence" between classical electrodynamics and Bohr's theory in the low-frequency region concerns only their respective *results*. The electron's mechanism of radiation in that region, as described by Bohr's theory, does not correspond to the one dictated by classical electrodynamics. Third, the interpretation of the derived formula

$$W = \tau h \frac{\omega}{2}$$

"is not that the different stationary states correspond to an emission of different numbers of energy-quanta, but that the frequency of the energy emitted during the passing of the system from a state in which no energy is yet radiated out to one of the different stationary states, is equal to different multiples of $\omega/2$, where $\omega$ is the frequency of revolution of the electron in the state considered."[74] The electron, according to this interpretation, radiates during its capture by a positive nucleus a *single* quantum of energy, whose frequency depends on the mechanical frequency of the electron in its final stationary state.

Up to this point of his paper, Bohr had imposed a quantum constraint on the energy emitted by an electron during the reformation of an atom. This constraint, however, could be given an alternative interpretation, as a restriction of the electron's *motion* within the atom. In Bohr's words,

> While there obviously can be no question of a mechanical foundation of the calculations given in this paper, it is, however, possible to give a very simple interpretation of the result of the calculation [of the characteristics of the stationary states] . . . by help of symbols taken from the ordinary mechanics. Denoting the angular momentum of the electron round the nucleus by $M$, we have immediately for a circular orbit $\pi M = T/\omega$, where $\omega$ is the frequency of revolu-

74. Ibid., p. 14.

tion and $T$ the kinetic energy of the electron;[75] for a circular orbit we further have $T = W$[76] . . . and from $\left[ W = \tau h \dfrac{\omega}{2} \right]$ . . . we consequently get

$$M = \tau M_0,$$

where

$$M_0 = h/2\pi \ldots$$

If we therefore assume that the orbit of the electron in the stationary states is circular, the result of the calculation [of the energy levels] . . . can be expressed by a simple condition: that the angular momentum of the electron around the nucleus in a stationary state of the system is equal to an entire multiple of a universal value, independent of the charge on the nucleus. (P. 15)

The quantization of the electron's angular momentum formed the basis, as will be shown below, of Sommerfeld's generalization of Bohr's 1913 theory.

The next section of Bohr's paper is devoted to the absorption of radiation and provides further evidence of his break with classical electromagnetic theory. Again, the electron is the central entity implicated in that break. An electron "can absorb a radiation of a frequency equal to the frequency of the homogeneous radiation emitted during the passing of the system [of a nucleus and an electron] between different stationary states" (pp. 15–16), and not equal to "the frequency of vibration of the electrons calculated in the ordinary way" (p. 16). However, even in this section the representation of the electron is not completely liberated from its classical shackles. Bohr continues to mention "the assumption used in this paper that the emission of line-spectra is due to the re-formation of atoms after one or more of the lightly bound electrons are removed" (p. 18). It was only a few months later that this assumption was abandoned, when Bohr gave a lecture before the Danish Physical Society; but more on this below.

In the same section, Bohr established the continuity between free states and stationary states. Free states are those "in which the electron possesses kinetic energy sufficient to remove to infinite distances from the nucleus" (p. 16). The continuity between these two kinds of states follows from the fact

75. For a circular orbit, $T = \tfrac{1}{2}mv^2 = \tfrac{1}{2}m(2\pi\omega r)^2 \Rightarrow T/\omega = \pi m 2\pi\omega r^2 = \pi M$.

76. Bohr had already asserted that "it can easily be shown that the mean value of the kinetic energy of the electron taken for a whole revolution is equal to $W$." Bohr, *On the Constitution*, p. 3.

that "the difference between frequency and dimensions of the systems in successive stationary states will diminish without limit if $\tau$ increases" (pp. 16–17). Bohr called both kinds of states "mechanical" to emphasize the fact that the electron in any of those states obeyed the laws of classical mechanics.

The final section of Bohr's paper explicates the permanent state of an atomic system in terms of the characteristics of its constitutive electrons. For atoms with more than one electron, "a permanent state is one in which the electrons are arranged in a ring round the nucleus" (p. 20). This ring "is formed in a way analogous to the one assumed for a single electron rotating round a nucleus. It will thus be assumed that the electrons, before the binding by the nucleus, were at a great distance apart from the latter and possessed no sensible velocities, and also that during the binding a homogeneous radiation is emitted" (p. 21). Again, traces of a semiclassical radiation mechanism surface in the above passage. More important, each of the electrons within the atom is subject to the *same* quantum constraint:

the angular momentum of each of the electrons is equal to $\dfrac{h}{2\pi}$.   (P. 22)

However, a ring such as the one considered by Bohr is mechanically unstable. Bohr had been aware of this problem since the time that he wrote the Rutherford memorandum. The problem consisted of two parts: "one concerning the stability for displacements of the electrons in the plane of the ring; one concerning displacements perpendicular to this plane" (p. 23). As Nicholson had shown, only displacements of the former kind threatened the stability of the ring; displacements of the latter kind would gradually dissipate, provided that the number of electrons was small.[77] This problem exhibits, once more, the active nature of the electron qua theoretical entity. As portrayed by Bohr, the electron in a stationary state was supposed to obey the laws of classical mechanics. This assumption, built as it were into Bohr's representation of the electron, had the unintended consequence that any ring of electrons would be mechanically unstable. A configuration of electrons such as the one envisaged by Bohr turned out to be prohibited by classical mechanics, the very theory that was supposed to govern the motion of the electron in a stationary state.

The solution to this conceptual problem, generated by the resistance of

77. As Heilbron and Kuhn point out, Bohr was unaware of this distinction when he wrote the Rutherford memorandum. See Heilbron and Kuhn, "The Genesis," p. 280.

the electron's representation to Bohr's attempt to enrich it,[78] rested on a further relaxation of the validity of classical mechanics. In Bohr's words, "[T]he question of stability for displacements of the electrons in the plane of the ring . . . cannot be treated on the basis of the ordinary dynamics."[79] The stability of the electrons is, instead, "secured through the above condition of the universal constancy of the angular momentum, together with the further condition that the configuration of the particles is the one by the formation of which the greatest amount of energy is emitted."[80] The same condition that originated from Bohr's struggle to construct an adequate account of the electron's mechanism of radiation turned out to be the key to the solution of the stability problem. The proposal of this condition, arguably the central hypothesis of Bohr's trilogy, was meant to discipline the electron qua theoretical entity and ensure that it would not generate any undesirable consequences.

One of the concluding remarks of his paper was a generalized formulation of this condition:

> In any molecular system consisting of positive nuclei and electrons in which the nuclei are at rest relative to each other and the electrons move in circular orbits, the angular momentum of every electron round the centre of its orbit will in the permanent state of the system be equal to $\dfrac{h}{2\pi}$, where h is Planck's constant.[81]

This passage highlights the advantages of reading Bohr's trilogy from the perspective of the electron, as opposed to the atom. It was the electron's behavior, *in both atoms and molecules,* that was subject to quantum constraints. One of those constraints, as formulated in the above excerpt, formed the basis of the subsequent parts of his trilogy.

The second part of that trilogy concerned the structure of atoms and was

78. Here one could argue that the resistance to Bohr's suggestion was due to the theoretical framework in which the representation of the electron was embedded and not to the representation per se. Indeed, theoretical entities are *partly* determined by the theoretical framework in which they are embedded, and, therefore, the resistance they exhibit to manipulation may come from that framework. There are cases, however, where it is more appropriate to attribute the resistance in question to aspects of the theoretical entity that transcend any particular theoretical framework. I will have more to say on this in the concluding section of this chapter.

79. Bohr, *On the Constitution*, p. 23.

80. Ibid. It should be noted that Bohr did not show at this point how these two conditions solve the problem of mechanical stability. A detailed analysis to that effect appeared only in the second part of his trilogy.

81. Ibid., pp. 24–25.

published in the September issue of the *Philosophical Magazine*.[82] It was the outcome of Bohr's struggle with the problems discussed in the Rutherford memorandum.[83] Bohr began his paper with an outline of the "general assumptions" that he employed. Most of those assumptions (concerning the mechanism of formation of atoms, the quantum restriction on the electron's angular momentum, the mechanical stability of electronic configurations) had already been put forward in part 1. There were, however, two novel hypotheses in part 2. The first was an interpretation of an element's position in the periodic table in terms of the number of electrons in an atom of the element in question: "The total experimental evidence supports the hypothesis . . . that the actual number of electrons in a neutral atom with a few exceptions is equal to the number which indicates the position of the corresponding element in the series of elements arranged in order of increasing atomic weight."[84] As Bohr noted, this hypothesis had already been suggested by van den Broek.[85] The second assumption, which was an elaboration of a hypothesis that was found in part 1, asserted "that the electrons are arranged at equal angular intervals in coaxial rings rotating round the nucleus."[86]

Such electronic configurations, however, are "unstable for displacements of the electrons in the plane of the ring"[87]—a characteristic that Bohr had already discussed in his previous paper. In that paper he had assumed that the stability problem is solved by the application of two conditions: the quantum restriction of the angular momentum of the electrons and the demand "that the total energy of the system in the configuration in question [be] less than in any neighbouring configuration."[88] Bohr now showed how the former condition entailed the latter. If a ring of electrons, whose angular momentum was constant, were transformed, "under influence of extraneous forces," into a neigbouring ring, then the resulting configuration would have a greater total energy than the original.[89]

The same two conditions were employed to investigate the stability of a

82. N. Bohr, "On the Constitution of Atoms and Molecules," pt. 2, "Systems Containing Only a Single Nucleus," *Philosophical Magazine*, 6th series, 26 (1913): 476–502; repr. in his *On the Constitution*, pp. 28–54.

83. This has been pointed out by Heilbron and Kuhn. See "The Genesis," p. 283.

84. Bohr, *On the Constitution*, p. 29.

85. See A. van den Broek, "Die Radioelemente, das periodische System und die Konstitution der Atome," *Physikalische Zeitschrift*, 14 (1913): 32–41. Cf. Heilbron, *Historical Studies in the Theory of Atomic Structure*, pp. 35–36.

86. Bohr, *On the Constitution*, p. 29.

87. Ibid., p. 32.

88. Ibid., p. 29.

89. For details ibid., p. 32.

ring "for displacements of the electrons perpendicular to the plane of the ring."[90] Bohr deduced a stability condition which was "identical with the condition . . . deduced by help of ordinary mechanical considerations."[91] His calculations suggested "that a ring of $n$ electrons cannot rotate in a single ring round a nucleus of charge $ne$ unless $n < 8$."[92] Furthermore, if, as he had assumed, the electrons in an atom were distributed in several coaxial rings, then "calculation indicates that only in the case of systems containing a great number of electrons will the planes of the rings separate; in the case of systems containing a moderate number of electrons, all the rings will be situated in a single plane through the nucleus."[93] Thus, the electron's motion within the atom was constrained within a single plane.

After investigating further the stability of multielectronic structures, Bohr proceeded to examine "what configurations of the electrons may be expected to occur in the atoms."[94] These configurations, however, could not be obtained solely from the principles of his theory, which had to be supplemented for that purpose with empirical information about the chemical properties of the elements. Further details of Bohr's attempt to represent several atoms by specific electronic structures need not concern us here, since no new property of the electron emerged in that process. Suffice it to say that in his "scheme . . . the number of electrons in . . . [the outer] ring is arbitrarily put equal to the normal valency of the corresponding element."[95] The roots of this "arbitrary" suggestion can be found in the Rutherford memorandum, where Bohr had assumed that "the chemical properties [of the elements] . . . depend on the stability of the outermost ring, the 'valency electrons.'"[96] Through that assumption the properties of the chemical elements were linked with the number and behavior of the outer electrons in the atoms of each element. Chemistry would, thus, become a source of problems and resources for the subsequent articulation of the representation of the electron (see chapter 7).

90. Ibid., p. 33.

91. Ibid. The classical derivation that Bohr mentioned had been carried out by Nicholson. See J. W. Nicholson, "The Spectrum of Nebulium," *Monthly Notices of the Royal Astronomical Society*, 72 (1911): 49–64, on pp. 50–52. It should be noted that Bohr miscited this paper as 72 (1912): 52.

92. Bohr, *On the Constitution*, p. 34.

93. Ibid., p. 35. The calculation that Bohr mentioned was left unspecified.

94. Ibid., p. 38.

95. Ibid., p. 48.

96. Bohr, Rutherford memorandum, p. xxii. Cf. Pais, *Niels Bohr's Times*, p. 137; and Heilbron and Kuhn, "The Genesis," pp. 246–247.

In the remaining of his paper, Bohr explained the production of characteristic roentgen radiation and analyzed radioactive phenomena in terms of his theory. Again, these explanatory attempts did not alter the conception of the electron that he had employed so far. It is worth noting that even at that stage he continued to espouse the semiclassical radiation mechanism that he had employed in the first part of his trilogy: "[T]he ordinary line-spectrum of an element is emitted during the reformation of an atom when one or more of the electrons in the outer rings are removed."[97] The representation of the electron had not yet been freed from its classical roots.

The final part of Bohr's trilogy developed the implications of his theory for the constitution of molecules and was published in the November issue of the *Philosophical Magazine*.[98] The assumptions that he employed were those that he had already put forward in his previous papers. As he noted, "we have not made use of any new assumption on the dynamics of the electrons."[99] After some preliminary remarks, he investigated the structure and stability of molecular systems and reinterpreted a familiar chemical notion, the chemical bond, in terms of electrons. According to this interpretation, the chemical bond consists of a "few of the outer electrons . . . rotating in a ring round the line connecting the nuclei. The latter ring . . . keeps the system together."[100] The rest of his paper discussed the structure of the hydrogen molecule, the way in which two atoms combine to form a molecule, and "the configuration of the electrons in systems containing a greater number of electrons."[101]

Bohr concluded his article with a summary of the assumptions that formed the backbone of his trilogy, a summary that is worth quoting in detail:

> The main assumptions used in the present paper are :—
>
> 1. That energy radiation is not emitted (or absorbed) in the continuous way assumed in the ordinary electrodynamics, but only during the passing of the systems between different "stationary" states.

97. Bohr, *On the Constitution*, p. 50.

98. N. Bohr, "On the Constitution of Atoms and Molecules," pt. 3, "Systems Containing Several Nuclei," *Philosophical Magazine*, 6th series, 26 (1913): 857–875; repr. in his *On the Constitution*, pp. 55–73.

99. Bohr, *On the Constitution*, p. 60. Bohr's remark referred to his investigation of the stability of "a system consisting of two positive nuclei of equal charges and a ring of electrons rotating round the line connecting them" (ibid., p. 57). However, it applies equally to the rest of his paper.

100. Ibid.

101. Ibid., p. 69.

2. That the dynamical equilibrium of the systems in the stationary states is governed by the ordinary laws of mechanics, while these laws do not hold for the passing of the systems between the different stationary states.

3. That the radiation emitted during the transition of a system between two stationary states is homogeneous, and that the relation between the frequency $\nu$ and the total amount of energy emitted $E$ is given by $E = h\nu$, where $h$ is Planck's constant.

4. That the different stationary states of a simple system consisting of an electron rotating round a positive nucleus are determined by the condition that the ratio between the total energy, emitted during the formation of the configuration, and the frequency of revolution of the electron is an entire multiple of $h/2$. Assuming that the orbit of the electron is circular, this assumption is equivalent with the assumption that the angular momentum of the electron round the nucleus is equal to an entire multiple of $h/2\pi$.

5. That the "permanent" state of any atomic system—i.e., the state in which the energy emitted is maximum—is determined by the condition that the angular momentum of every electron round the centre of its orbit is equal to $h/2\pi$.[102]

These assumptions, with the exception of the last one, continued to govern the behavior of the electron throughout the evolution of the old quantum theory of the atom.[103] The final assumption, on the other hand, was abandoned. Data from x-ray spectroscopy suggested that some electrons in the "permanent" state were endowed with more than a single quantum of angular momentum.[104]

It is worth noting that at various points in Bohr's trilogy the first assumption, which attributed radiation to a transition of the electron between two stationary states, was compromised. As I mentioned above, he repeatedly portrayed the production of a line-spectrum as due to the reformation of a previously ionized atom.[105] In particular, traces of this radiation mechanism

102. Ibid., pp. 72–73.

103. The second assumption, which prescribed that the motion of the electron in a stationary state was governed by classical mechanics, was later qualified as a result of several unsuccessful attempts to obtain, on the basis of that mechanics, an empirically adequate model of the helium atom. Cf. Darrigol, *From c-Numbers to q-Numbers*, pp. 177–178.

104. See Heilbron, *Historical Studies in the Theory of Atomic Structure*, pp. 83–84, 294–295; and H. Kragh, "Niels Bohr's Second Atomic Theory," *Historical Studies in the Physical Sciences*, 10 (1979): 123–186, on p. 126.

105. That was a common view at the time. It had been put forward by Philipp Lenard, and many physicists subscribed to it. See Heilbron, *Historical Studies in the Theory of Atomic Structure*, p. 274.

surfaced in both of his two (imperfect) derivations of the Balmer formula. Toward the end of 1913, in a speech before the Danish Physical Society of Copenhagen, Bohr deduced the energy levels of the hydrogen atom without recourse to this semiclassical picture of radiation.[106] By interpreting the Balmer formula

$$\frac{1}{\lambda} = R\left(\frac{1}{n_1^2} - \frac{1}{n_2^2}\right)$$

in terms of his frequency condition

$$h\nu = E_1 - E_2 \quad \text{or} \quad \nu = \frac{E_1}{h} - \frac{E_2}{h},$$

he obtained an expression for the energy levels:[107]

$$W = \frac{Rhc}{n^2}.$$

Radiation was portrayed as the result of an unfathomable transition process. No explanation was given "about how or why the radiation is emitted."[108] Furthermore, the analogy between the electron's energy in any of the stationary states and the kinetic energy of Planck's resonators, which played a foundational role in the former exposition of his theory, was abandoned. Even though

> this analogy suggests another manner of presenting the theory, . . . [w]hen we consider how differently the equation $\left[ W = \frac{1}{2}nh\omega \right]$ is employed here and in Planck's theory it appears to me misleading to use this analogy as a foundation.[109]

What made the analogy "misleading" was that in Planck's theory the mechanical frequency of the resonator coincided with its frequency of radiation, whereas in Bohr's account of radiation the connection between the frequency of revolution of the electron and the frequency of the emitted radiation was broken.

106. For an English translation of this speech see N. Bohr, *The Theory of Spectra and Atomic Constitution* (Cambridge: Cambridge Univ. Press, 1922), pp. 1–19.

107. Ibid., p. 12.

108. Ibid., pp. 12–13.

109. Ibid., p. 14. Cf. Heilbron and Kuhn, "The Genesis," pp. 276–277.

However, at the end of his address, Bohr referred once more to the semi-classical mechanism of radiation that he had employed in his trilogy. In an attempt to explain the "occurrence of Rydberg's constant in all spectral formulae," he assumed "that the spectra under consideration, like the spectrum of hydrogen, are emitted by a neutral system, and that they are produced by the binding of an electron previously removed from the system."[110] The break of the electron qua theoretical entity with its classical past was not yet complete.

The main aspect of that break consisted of Bohr's proposal that there is no connection between the electron's frequency of radiation and its frequency of revolution within the atom. The radical nature of that proposal was perceived by several physicists at the time.[111] Georg Hevesy, for instance, reported that when Einstein found out that the spectral series observed by Charles Pickering and Alfred Fowler were "due, as Bohr had suggested, to ionized helium, he was extremely astonished and told me: 'Than [sic] the frequency of the light does not depand [sic] at all on the frequency of the electron. . . . And this is an *enormous achiewement* [sic].'"[112] Other physicists, however, were less favorably disposed toward Bohr's revolutionary proposal. Erwin Schrödinger, for example, several years later still thought of Bohr's electron as a "monster."[113] Altogether, the reception of Bohr's new representation of the electron within the atom was mixed, ranging from hostile rejection to enthusiastic acceptance.[114]

## 3. Concluding Remarks

In this chapter I attempted to reconstruct the problem situations out of which the character of the electron was transformed. While the properties attrib-

---

110. Bohr, *The Theory of Spectra*, p. 18. This passage shows that Heilbron's and Kuhn's claim that by the time of Bohr's address "radiation had for him become entirely a transition process, and he no longer referred at all to radiation during the binding of an electron initially at rest," needs qualification. See Heilbron and Kuhn, "The Genesis," p. 276.

111. Cf. L. Rosenfeld's "Introduction" in Bohr, *On the Constitution*, p. xli.

112. G. Hevesy to N. Bohr, 23 Sept. 1913. A part of this letter, from which the quote is taken, is reprinted in Rosenfeld, "Introduction," p. xlii.

113. He characterized Bohr's hypothesis as "monstrous." See Schrödinger to Lorentz, 6 June 1926, in K. Przibram (ed.), *Letters on Wave Mechanics: Schrödinger, Planck, Einstein, Lorentz,* trans. M. J. Klein (New York: Philosophical Library, 1967), p. 61.

114. For some reactions of Bohr's contemporaries see Heilbron, *Historical Studies in the Theory of Atomic Structure*, pp. 46–47; Jammer, *The Conceptual Development of Quantum Mechanics*, pp. 86–87; and J. Mehra and H. Rechenberg, *The Historical Development of Quantum Theory*, vol. 1, *The Quantum Theory of Planck, Einstein, Bohr, and Sommerfeld: Its Foundation and the Rise of Its Difficulties, 1900–1925* (New York: Springer-Verlag, 1982), p. 193. This last work should be read with caution. See Paul Forman's review essay ("A Venture in Writing History," *Science,* 220 (1983): 824–827).

uted to it (being a point particle, with a certain mass and charge) were inherited from its previous, classical representation, the laws governing its behavior were revised, in response to specific empirical and conceptual problems. The electron could now move only in certain discrete orbits, corresponding to the stationary states of the atom, and its energy and angular momentum were accordingly restricted. In those states it was still subject to classical mechanics and obeyed Coulomb's law, but its mechanism of radiation violated classical electrodynamics. The radiation emitted by the electron was now the outcome of a new process (transitions between two stationary states), and a new law specified its frequency, which differed from the electron's frequency of revolution.

In my analysis of these developments I portrayed the representation of the electron as an active agent that participated in the construction of its quantum identity. This participation took two forms, positive and negative. First, the electron's writings played a crucial, positive role in the transformation of its representation. The experimental situations attributed to it provided information that guided the transformation in question. For instance, the discrete structure of the hydrogen spectrum obtained a theoretical counterpart in the discrete series of energy levels occupied by the electron within the hydrogen atom; and the number of spectral lines was linked with the number of transitions between those levels. Once those experimental situations had been linked with the properties and behavior of the electron, their characteristics functioned as empirical constraints on the construction of its representation. Those constraints transcended any particular theory. In that sense, the electron's putative writings were partly independent of the theoretical framework in which its representation was embedded. However, this partial independence should not tempt us to distinguish the electron qua theoretical entity (i.e., the theoretical representation of the electron) from the electron qua experimental entity (i.e., the experimentally determined representation of the electron). There is just one unified representation, *part of which* is determined by experiment. This part, moreover, is not totally independent of theory. In the absence of a theoretical framework, experimentally obtained information would not suffice to construct a representation of the electron (or any other entity, for that matter).[115]

Second, the agency of the representation of the electron was exhibited as resistance to Bohr's attempts to change it. The electron, qua theoretical entity, displayed its resistance by generating undesirable consequences whenever he tried to employ it to further his theoretical aims. These consequences

115. Cf. the discussion of Hacking's entity realism in chapter 9.

were mainly due to the theoretical framework in which the representation of the electron was embedded. The theoretical framework determined, among other things, the laws obeyed by the electron.[116] When a novel theoretical move was attempted, for example, by proposing a model of the electron's behavior in a certain setting, it often turned out that the move in question was prohibited by the very laws that were supposed to govern that behavior.[117] We saw, for instance, that when Bohr suggested that electrons rotate in circular rings inside the atom, it turned out that according to the laws of classical mechanics those rings would have to be unstable.

Unwanted effects like these gave rise, in turn, to problem situations whose resolution produced a new representation of the electron. Thus, the development of the electron's "personality" depended not only on Bohr's attempts to manipulate its representation, but also on its previously established characteristics. Its novel representation was the outcome of a continuous interaction between its former self and Bohr. The emerging representation was not always consistent, and I have tried to highlight the contradictions in Bohr's texts, the locus of the electron's new identity, by reading them from the perspective of the electron. These contradictions can be seen most clearly in his two derivations of Balmer's formula and his subsequent derivation of the energy levels of the hydrogen atom, each of which was based on a different conception of the electron.

The following chapter will reconstruct the next phase of the electron's journey in the quantum world, Sommerfeld's generalization and development of Bohr's theory.

116. It also determined the interpretation of the properties attributed to the electron. As I pointed out above, however, there were aspects of the electron's representation that were not wholly determined by the framework. Besides the experimental situations associated with it, I have in mind the quantitative magnitude of its properties, which was determined by experiment.

117. We will see in the following chapters that there were other cases where the enrichment of the representation of the electron violated constraints from its putative experimental manifestations, or negated its experimentally determined properties. In those cases the origins of resistance transcended the particular theoretical framework in which the representation was embedded.

# Chapter 6 | Between Relativity and Correspondence

## 1. Maturing under the Guidance of the Quantum Technologist

In the next phase of the electron's journey its quantum identity was firmly established and refined. The central aspects of Bohr's theory were empirically confirmed.[1] For my purposes, it suffices to discuss briefly the experiments performed by James Franck and Gustav Hertz, which supported Bohr's ideas about stationary states, the "waiting places" of the electron, and quantum transitions.[2] In 1913 Franck and Hertz investigated the "collisions between gas molecules and slow electrons."[3] In 1914 they published another paper along the same lines, which treated collisions between mer-

1. The experimental confirmation of Bohr's theory is discussed in J. L. Heilbron, *Historical Studies in the Theory of Atomic Structure* (New York: Arno Press, 1981), pp. 48–52; M. Jammer, *The Conceptual Development of Quantum Mechanics* (New York: McGraw-Hill, 1966), pp. 82–85; J. Mehra and H. Rechenberg, *The Historical Development of Quantum Theory*, vol. 1, *The Quantum Theory of Planck, Einstein, Bohr, and Sommerfeld: Its Foundation and the Rise of Its Difficulties, 1900–1925* (New York: Springer-Verlag, 1982), pp. 191–192, 195–200; and A. Pais, *Niels Bohr's Times, in Physics, Philosophy, and Polity* (New York: Oxford Univ. Press, 1991), pp. 181–184.

2. This formulation of stationary states was suggested by Bohr in his lecture to the Danish Physical Society on 20 Dec. 1913. See N. Bohr, *The Theory of Spectra and Atomic Constitution* (Cambridge: Cambridge Univ. Press, 1922), p. 11.

3. See J. Franck and G. Hertz, "Über Zusammenstösse zwischen Gasmolekülen und langsamen Elektronen," *Verhandlungen der Deutschen Physikalischen Gesellschaft,* 15 (1913): 373–390; and Frank and Hertz, "Über Zusammenstösse zwischen langsamen Elektronen und Gasmolekülen," pt. 2, *Verhandlungen der Deutschen Physikalischen Gesellschaft,* 15 (1913): 613–620.

cury molecules and electrons.[4] Their results were as follows: If the kinetic energy of an electron was below a certain value (4.9 eV), then its collision with a mercury molecule was completely elastic; that is, there was no transfer of energy from the electron to the molecule. After such a collision the electron would move in a different direction but with the same (scalar) velocity. If the kinetic energy of the electron reached that value, then after the collision it would come to a stop, all of its kinetic energy having been absorbed by the mercury molecule with which it had collided. If the kinetic energy of the electron was above that value, then after the collision it would continue to move with a velocity corresponding to a kinetic energy lower by 4.9 eV than its original amount. Results of this kind had been already predicted by Bohr in the first part of his trilogy. There he maintained that "an electron of great velocity in passing through an atom and colliding with the electrons bound will loose [sic] energy in distinct finite quanta."[5] From that perspective, what happened in the Franck and Hertz experiments was that an electron with sufficient kinetic energy (more than 4.9 eV) knocked a bound electron out of its original state ($E_1$) to another excited state ($E_2$). The energy difference between these two states ($E_2 - E_1$) was the minimum kinetic energy required to produce an inelastic collision. A further consequence of Bohr's theory was that the excited electron would jump to its original state, emitting in the process radiation with a frequency given by the relation $h\nu = E_2 - E_1$. This consequence of Bohr's theory was also confirmed by Franck and Hertz. They observed that if the kinetic energy of the colliding electrons was greater than 4.9 eV, then mercury vapor would emit ultraviolet radiation with the frequency predicted by Bohr's theory. However, the interpretation of the Franck and Hertz experiments sketched above was not adopted by Franck and Hertz themselves. Bohr's theory had not played any role in the design of those experiments, whose purpose was to determine the ionization potential of the molecules of mercury gas. Thus, in 1914 they suggested that 4.9 eV was the minimum energy required to ionize a mercury molecule (as opposed to exciting it). In 1915 Bohr challenged that interpretation, but only in 1919 did they accept Bohr's view.[6]

4. J. Franck and G. Hertz, "Über Zusammenstöße zwischen Elektronen und den Molekülen des Quecksilberdampfes und die Ionisierungsspannung desselben," *Verhandlungen der Deutschen Physikalischen Gesellschaft,* 16 (1914): 457–467.

5. N. Bohr, *On the Constitution of Atoms and Molecules* (Copenhagen: Munksgaard, 1963), p. 19.

6. For details and references see Heilbron, *Historical Studies in the Theory of Atomic Structure,* pp. 49–51; G. Hon, "Franck and Hertz versus Townsend: A Study of Two Types of Experimental Error," *Historical Studies in the Physical Sciences,* 20, no. 1 (1989): 79–106; and Pais, *Niels Bohr's Times,* pp. 183–184.

In 1914 a new problem confronted Bohr's theory. William Edward Curtis had determined empirically the wavelengths of the hydrogen spectral lines and concluded that "Balmer's formula has been found to be inexact."[7] In response to this anomaly, Bohr came up with an explanation that utilized the relativistic character of the electron. "Assuming that the orbit of the electron is circular . . . but replacing the expressions for the energy and the momentum of the electron by those deduced on the theory of relativity," the Balmer formula would not hold exactly.[8] Furthermore, he suggested that a more accurate representation of the electron's motion within the atom would provide the key for unlocking the fine structure riddle, namely, the fact that most of the hydrogen spectral lines exhibit a doublet structure.[9] Recall that his 1913 theory had been developed on the simplifying assumption that the electrons within the atom move in circular orbits. By taking into account that the electronic orbits are, instead, elliptical, "It might . . . be supposed that we would obtain a doubling of the lines."[10] The doubling of the lines would be a result of the relativistic variation of the electron's mass. If the electron moves in an ellipse, then its velocity will be nonuniform (Kepler's second law) and, therefore, according to relativity theory, its mass will not be the same at every point of its orbit.

A detailed analysis of this relativistic effect and how it explains fine structure will be given below, in connection with Sommerfeld's generalization of Bohr's theory. Suffice it to say here that the relativistic correction of Bohr's theory highlights the active nature of the representation of the electron. In this case its agency unfolded in two stages. First, some of the writings of the electron were to be found, according to its representation, in the hydrogen spectrum. Those writings turned out to have a complexity (doublet structure) that had no counterpart in the representation. So they indicated that such a counterpart had to be found.[11] Second, certain features of the repre-

7. W. E. Curtis, "Wave-Lengths of Hydrogen Lines and Determination of the Series Constant," *Proceedings of the Royal Society of London (A),* 90 (1914): 605–620, on p. 620. Cf. Jammer, *The Conceptual Development of Quantum Mechanics,* pp. 89–90.

8. N. Bohr, "On the Series Spectrum of Hydrogen and the Structure of the Atom," *Philosophical Magazine,* 29 (1915): 332–335, on p. 334.

9. For the history of fine structure see H. Kragh, "The Fine Structure of Hydrogen and the Gross Structure of the Physics Community, 1916–26," *Historical Studies in the Physical Sciences,* 15 (1985): 67–125.

10. Bohr, "On the Series Spectrum of Hydrogen," p. 335.

11. Here, as elsewhere, I assume a threefold distinction between the electron qua entity in nature, its theoretical representation, and its writings. By the term "writings" I mean something more inclusive than the electron's readily detected behavior in specially designed experimental settings, for example, in a cloud chamber. As I explained in chapter 2 (pp. 51–52), the term refers to the purported manifestations of the electron in all the experimental situations

sentation provided heuristic guidance to Bohr on how to improve it. In the context of his 1913 theory several simplifying assumptions had been built into it. He had assumed, for example, that the velocity of the electron was small in comparison with the velocity of light and that its orbit was circular.[12] As soon as empirical problems appeared, a plausible remedy was to eliminate some of those simplifying assumptions and see whether this realistic maneuver provided a solution to the problems.[13] Thus, the electron's writings implied that something was wrong with its representation, which, in turn, suggested what went wrong and how it could be fixed.

Bohr's attempt to construct a relativistic version of his 1913 theory was further developed by Arnold Sommerfeld, who had occupied the chair of theoretical physics at Munich since 1906.[14] There are competing accounts of how he found out about Bohr's work on atomic structure.[15] It is clear, however, from a card that he sent to Bohr on 4 September 1913 that he had studied the first part of Bohr's trilogy very shortly after its publication. Although he "remain[ed] for the present in principle somewhat skeptical toward atomic models," he was greatly impressed by Bohr's calculation of the Rydberg constant.[16]

---

associated with it. The inference from the characteristics of those situations to aspects of its behavior may be either direct, as in the cloud chamber case, or indirect, as in the spectroscopic cases I examine in this chapter.

12. See Bohr, *On the Constitution*, pp. 3–4.

13. For this point I am indebted to Dudley Shapere. See his "Scientific Theories and Their Domains," in F. Suppe (ed.), *The Structure of Scientific Theories*, 2nd ed. (Urbana: Univ. of Illinois Press, 1977), pp. 518–565, esp. pp. 563–564. However, Shapere's account of Sommerfeld's relativistic modification of Bohr's theory is inaccurate. In particular, he argues that this "more realistic treatment . . . [was] given without the problem [of fine structure] ever having come up" (p. 564). In fact, exactly the opposite was the case.

14. See P. Forman and A. Hermann, "Sommerfeld, Arnold," in C. C. Gillispie (ed.), *Dictionary of Scientific Biography*, 16 vols. (New York: Charles Scribner's Sons, 1970–1980), vol. 12, pp. 525–532, on p. 527. The circumstances of Sommerfeld's appointment are detailed in C. Jungnickel and R. McCormmach, *Intellectual Mastery of Nature: Theoretical Physics from Ohm to Einstein*, 2 vols. (Chicago: Univ. of Chicago Press, 1986), vol. 2, pp. 277–278. David Cassidy has referred to Arnold Sommerfeld as the "quantum technologist." See D. C. Cassidy, *Uncertainty: The Life and Science of Werner Heisenberg* (New York: Freeman, 1992), p. 129. Cassidy's apt characterization is drawn from Sommerfeld's remark to Einstein that "I can only promote the technology of quanta; you must make your philosophy" (Ich kann nur die Technik der Quanten fördern, Sie müssen Ihre Philosophie machen). Sommerfeld to Einstein, 11 Jan. 1922, in A. Einstein and A. Sommerfeld, *Briefwechsel*, ed. A. Hermann (Basel and Stuttgart: Schwabe, 1968), p. 97; the translation is from Cassidy, *Uncertainty*, p. 124.

15. See J. L. Heilbron, "The Kossel-Sommerfeld Theory and the Ring Atom," *Isis*, 58 (1967): 451–485, on p. 456.

16. Sommerfeld to Bohr, 14 Sept. 1913. The German original of that letter was published in Rosenfeld's introduction to Bohr's trilogy. See Bohr, *On the Constitution*, p. lii. The translation is from Forman and Hermann, "Sommerfeld," p. 528.

The predictive potential of Bohr's theory may explain why Sommerfeld chose to develop it mathematically. That choice was in tune with his style of doing physics. He was interested in the mathematical articulation of scientific theories and the extraction of testable predictions from them, without worrying about their physical underpinnings.[17]

The unfolding of his thought from the summer of 1913 till December 1915, when he published the first results of his research, is difficult to reconstruct, since neither his surviving correspondence from that period nor the interviews for the Archive for the History of Quantum Physics provide any relevant clues.[18] It is known, however, that he was looking for a generalized formulation of the quantum theory that would extend its range of application to systems with more than one degree of freedom. In its original formulation by Planck, the quantum theory was applicable only to physical systems with one degree of freedom, like the harmonic oscillator. The problem facing physicists was to develop a more general formulation of the theory that would render it applicable to more complex systems, like Bohr's model of the hydrogen atom.[19] The intermediate step toward that generalization would be to impose two quantum constraints on the electron's motion within the atom, as opposed to the single quantum restriction that Bohr had imposed on its angular momentum. Sommerfeld expected that this maneuver would not only point the way to the solution of the more general problem, but would also resolve the problem of fine structure.[20]

Sommerfeld's generalization of Bohr's theory was the subject of a series of lectures that he delivered before his students during the winter of 1914–1915.[21] Several months later he presented the outcome of his research to the Bavarian Academy of Sciences.[22] The definitive version of his theory was pub-

17. For some very interesting remarks on Sommerfeld's style see Heilbron, *Historical Studies in the Theory of Atomic Structure*, pp. 52–53. Cf. also Jungnickel and McCormmach, *Intellectual Mastery of Nature*, vol. 2, pp. 284, 354.

18. See U. Benz, *Arnold Sommerfeld: Lehrer und Forscher an der Schwelle zum Atomzeitalter, 1868–1951* (Stuttgart: Wissenschaftliche Verlagsgesellschaft, 1975), p. 87.

19. This was a widely recognized problem at the time. It had been posed by Poincaré at the Solvay Congress in 1911 and was solved by Planck four years later. See Jammer, *The Conceptual Development of Quantum Mechanics*, p. 90. Cf. also T. S. Kuhn, *Black-Body Theory and the Quantum Discontinuity, 1894–1912*, 2nd ed. (Chicago: Univ. of Chicago Press, 1987), pp. 250–251.

20. See O. Darrigol, *From c-Numbers to q-Numbers: The Classical Analogy in the History of Quantum Theory* (Berkeley: Univ. of California Press, 1992), p. 102; and Jammer, *The Conceptual Development of Quantum Mechanics*, p. 91.

21. See Forman and Hermann, "Sommerfeld," p. 528; and Mehra and Rechenberg, *The Historical Development of Quantum Theory*, vol. 1, p. 213.

22. See A. Sommerfeld, "Zur Theorie der Balmerschen Serie," *Sitzungsberichte der Mathematisch-Physikalischen Klasse der Königlich-Bayerischen Akademie der Wissenschaften zu München*,

lished in the *Annalen der Physik* in 1916.[23] The aim of the first part of that paper ("Theorie der Balmerschen Serie") was to show that, despite the successes of Bohr's theory, "even the theory of the Balmer series is still considerably completed through the consideration of noncircular (thus in the case of hydrogen elliptical) orbits."[24] Before discussing the special case of elliptical orbits, Sommerfeld considered the general case of a system with $f$ degrees of freedom. For the description of such a system $f$ pairs of coordinates ($q_i$, $p_i$) were necessary, where $q_i$ is the position coordinate and $p_i$ the corresponding momentum. Each of this pair was subject to the so-called phase integral,

$$\int p_n dq = nh,$$

where $n$ is a positive integer and the integration ranges over a complete period of the position coordinate.[25] These postulates were regarded by Sommerfeld as the "unproved and . . . perhaps unprovable foundation of the quantum theory."[26] It is worth mentioning that these quantum conditions had been also put forward by William Wilson and Jun Ishiwara, who were grappling with problems very similar to those that occupied Sommerfeld.[27]

Sommerfeld proceeded with an application of this general formulation to the analysis of the Kepler motion, that is, the motion of an object that revolves under the influence of an inverse square force around a fixed cen-

45 (1915): 425–458 (read on 6 Dec. 1915); and Sommerfeld, "Die Feinstruktur der Wasserstoff- und der Wasserstoff-ähnlichen Linien," *Sitzungsberichte der Mathematisch-Physikalischen Klasse der Königlich-Bayerischen Akademie der Wissenschaften zu München*, 45 (1915): 459–500 (read on 8 Jan. 1916).

23. A. Sommerfeld, "Zur Quantentheorie der Spektrallinien," *Annalen der Physik*, 51 (1916): 1–94, 125–167; repr. in his *Gesammelte Schriften*, ed. F. Sauter, 4 vols. (Braunschweig: F. Vieweg, 1968), vol. 3, pp. 172–308. The analysis that follows is based on this paper. For how it compares with his two former papers on this subject see Jungnickel and McCormmach, *Intellectual Mastery of Nature*, vol. 2, p. 352; and S. Nisio, "The Formation of the Sommerfeld Quantum Theory of 1916," *Japanese Studies in History of Science*, 12 (1973): 39–78.

24. "[A]uch die Theorie der Balmerserie noch wesentlich zu vervollständigen ist durch die Betrachtung nichtkreisförmiger (also im Falle des Wasserstoffatoms elliptischer) Bahnen." Sommerfeld, *Gesammelte Schriften*, vol. 3, p. 175.

25. This is the simplest version of the quantum conditions. See Sommerfeld, *Gesammelte Schriften*, vol. 3, p. 180.

26. "[U]nbewiesene und . . . vielleicht unbeweisbare Grundlegung der Quantentheorie." Ibid., p. 177.

27. See Jammer, *The Conceptual Development of Quantum Mechanics*, pp. 91–92; and Kuhn, *Black-Body Theory*, p. 251. Wilson's priority was acknowledged by Sommerfeld. See his *Gesammelte Schriften*, vol. 3, p. 180.

ter.[28] An instance of such motion is the elliptical orbit of the electron within the hydrogen atom. This is a system with two degrees of freedom, since two coordinates are necessary to specify the position of the electron, namely (in polar coordinates), the azimuthal angle $\phi$ and the radius vector $r$. The azimuthal coordinates (angle and angular momentum) are subject to the following quantum condition:

$$(6.1) \qquad \int_0^{2\pi} p\,d\phi = 2\pi p = nh,$$

where $n$ is the azimuthal quantum number. It follows from classical mechanics that the energy of the electron in that orbit is

$$W = T + V = -\frac{me^4}{2p^2}(1 - \varepsilon^2),$$

where $T$ is the kinetic and $V$ the potential energy of the electron, $p$ its angular momentum, and $\varepsilon$ the eccentricity of its orbit.[29] The value of $p$ can be derived from equation (6.1): $p = nh/2\pi$. By substituting this value into the energy equation, we get

$$(6.2) \qquad W = -\frac{2\pi^2 me^4}{h^2}\frac{1 - \varepsilon^2}{n^2}.$$

If the eccentricity varies continuously, then the energy of the electron ceases to be a quantized magnitude. However, the sharpness of the hydrogen spectral lines testifies to the quantization of the electron's energy levels. To preserve that quantization, an additional quantum restriction on the eccentricity of the electron's orbit is necessary.

The electron's representation was, once more, an active agent that guided Sommerfeld's attempts to construct an adequate theory of atomic structure. If one allowed the electron to move in elliptical orbits, then the quantization of its energy was canceled. However, the quantization of the electron's energy was forced by its writings, its observable manifestations in the hydrogen spectrum. Certain features of those writings (the sharpness of the hydrogen

---

28. Sommerfeld regarded this application as the "main object" of his paper: "Es ist der Hauptgegenstand dieser Arbeit, die Anwendung des Ansatzes (I) [ ∫ $p_n\,dq$ = $nh$] auf die Keplersche Bewegung zu studieren." Sommerfeld, *Gesammelte Schriften*, vol. 3, p. 182.

29. This equation was the outcome of a straightforward classical derivation. See Sommerfeld, *Gesammelte Schriften*, vol. 3, pp. 185–187.

spectral lines) were given a theoretical interpretation (the discreteness of energy levels) that was built into the representation of the electron. Thus, to preserve the coherence of that representation a further maneuver was necessary, namely, the quantization of the eccentricity of the electron's orbit. As a result of that interplay between Sommerfeld and the electron's representation, the following quantum condition was introduced:

$$(6.3) \quad \int p_r \, dr = \int m\dot{r} \, dr = \int_0^{2\pi} m\dot{r} \frac{dr}{d\phi} \, d\phi = n'h,$$

where $n'$ is the radial quantum number. By computing this integral, Sommerfeld deduced that

$$2\pi p \left( \frac{1}{\sqrt{1 - \varepsilon^2}} - 1 \right) = n'h.$$

Since $p = nh/2\pi$, it follows that

$$(6.4) \quad \frac{1}{\sqrt{1 - \varepsilon^2}} - 1 = \frac{n'}{n}, \qquad 1 - \varepsilon^2 = \frac{n^2}{(n + n')^2}.$$

What is the physical interpretation of these equations? An ellipse is characterized by two parameters, the major and the minor axis. An ellipse with a given major axis may have infinitely many shapes, the actual shape being determined by the magnitude of the minor axis. According to classical mechanics, an electron under the influence of an inverse square force is allowed to move in any conceivable ellipse. Bohr had introduced a quantum condition that restricted this continuum of orbits to a discrete set of ellipses with specific axes. However, there was nothing in Bohr's scheme that restricted the shape of those orbits or, equivalently, the magnitude of their minor axes. Sommerfeld's additional quantum condition (equation [6.3] above) restricted further the putative freedom of the electron and quantized the possible shapes of its orbits. The minor axis of an ellipse ($2b$) is given by the formula

$$b = \sqrt{1 - \varepsilon^2} \cdot \alpha,$$

where $2\alpha$ is the major axis of the ellipse. By substituting equation (6.4) into this formula we get

$$b = \frac{n}{n + n'} \cdot \alpha;$$

that is, an elliptical orbit with a given major axis may have a limited variety of shapes depending on the ratio

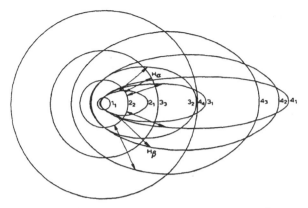

**Fig. 6.1** This diagram "reproduces the relative size and form of the electron orbits." (From N. Bohr, "The Structure of the Atom," Nobel Lecture, 11 Dec. 1922, in *Nobel Lectures: Physics, 1922–1941* [Amsterdam: Elsevier, 1965], pp. 7–43, on p. 26.)

$$\frac{n}{n + n'}.$$

The quantum number $n$ was allowed to take any of the following values: 1, 2, ..., $n + n'$.[30] For $n = n + n'$ Sommerfeld's elliptical orbits reduce to Bohr's circular orbits (see fig. 6.1). Zero was excluded from the possible values of $n$, since it implied that the angular momentum of the electron was also 0. This would, in turn, imply that the electron's orbit was a straight line that passed through the nucleus, a physically impossible result. We see here how the physical model of the electron's motion within the atom constrained its mathematical counterpart. The construction of the latter had to be carried out without violating any of the established features of the former. The mathematical and the physical aspects of the representation of the electron's behavior had to match.

However, the new quantum constraint introduced by Sommerfeld did not proliferate the energy levels of the electron within the hydrogen atom. If one combines formula (6.4) with the energy equation (6.2), the following expression results:

$$W = -\frac{2\pi^2 me^4}{h^2}\frac{1}{(n + n')^2}.$$

30. The term "quantum number" was introduced by Sommerfeld in 1916. See Jammer, *The Conceptual Development of Quantum Mechanics*, p. 93.

This result showed that all electronic orbits with the same major axis, regardless of their shape, were equivalent in energy. It thus explained why Bohr had successfully derived the energy levels of the hydrogen atom, despite the fact that his derivation had been based on the simplifying assumption that the electron's orbits are circular. The validity of that assumption was immaterial to the determination of the energy levels.

The quantum numbers introduced so far had determined the size and the shape of the electron's orbit.[31] Sommerfeld now considered the question "whether the *position* of the orbit [i.e., its direction] can also be 'quantized.'"[32] If one takes into account the orientation of the electron's orbit in space, three coordinates are necessary to specify the position of the electron. For this reason, Sommerfeld chose a system of spatial polar coordinates $r$, $\theta$, $\psi$, whose origin coincided with the nucleus and whose axis was parallel to a certain reference direction (established, say, by an external electric or magnetic field). For each of these coordinates a corresponding quantum condition was introduced:

$$\int_0^{2\pi} p_\psi \, d\psi = n_1 h, \qquad \int p_\theta \, d\theta = n_2 h, \qquad \int p_r \, dr = n'h.$$

In the first of these integrals $p_\psi$ is the component of the total angular momentum in the direction of the axis. Thus, $p_\psi = p \cos \alpha$, where $\alpha$ is the angle between the direction of the electron's orbit and the axis. It follows from the first integral that

(6.5)    $2\pi p \cos \alpha = n_1 h.$

The second integral, after a relatively complicated computation,[33] yields

(6.6)    $\int p_\theta \, d\theta = 2\pi p(1 - \cos \alpha) = n_2 h.$

From equations (6.5) and (6.6) Sommerfeld deduced that

(6.7)    $2\pi p = (n_1 + n_2)h,$

(6.8)    $\cos \alpha = \dfrac{n_1}{n_1 + n_2}.$

---

31. "Von unseren beiden Quantenzahlen $n$ und $n'$ bestimmt $n$ die *Größe* der Bahn . . . , $n'$ die Gestalt der Bahn." Sommerfeld, *Gesammelte Schriften*, vol. 3, pp. 199–200.

32. "[O]b sich auch die *Lage* der Bahn 'quanteln' läßt." Ibid., p. 200.

33. Ibid., pp. 201–202.

At this point Sommerfeld's aim had been accomplished. He had shown that the electron's orbit may assume only certain directions (defined by the normal to the plane of the orbit). If we put $n_1 + n_2 = n$, then the cosine of the angle between the direction of the orbit and the axis can take only $n + 1$ values, including 0 and 1 (for $n_1 = 0$ and $n_2 = 0$, respectively). The angle itself is accordingly quantized. The quantum number $n_1$ is an index, so to speak, of the set of allowed directions of the electron's orbit.

As Sommerfeld observed, the introduction of "'orientation quanta' give[s] the first possibility of a connection between polarization and spatial position of the plain of the orbit." [34] The polarization of spectral lines was one feature of the electron's writings that, so far, had no counterpart in the representation of its behavior within the atom. The task of filling that gap could not be accomplished as long as the only quantized features of the electron's motion were the size and shape of its orbit. None of these features was relevant to the task in question. On the other hand, the polarization of spectral lines could be linked with a further attribute of the electron's motion, the direction of the plane of its orbit. This link was spelled out by Paul Sophus Epstein, who had studied with Sommerfeld, in his quantum-theoretical analysis of the Stark effect—the splitting of spectral lines under the influence of an electric field (observed by Johannes Stark in 1913).[35] Epstein connected the direction of polarization (parallel or perpendicular to the external field) with the change of the quantum number $n_1$ in a quantum transition.[36] A change by an even number gives rise to spectral lines whose direction of polarization is parallel to the external field. A change by an odd number corresponds to spectral lines whose direction of polarization is perpendicular to the field. The polarization of spectral lines had finally found a theoretical counterpart. This experimentally detected property was now appropriately translated and built into the quantum representation of the electron.

Notwithstanding its utility for accommodating polarization, the introduction of an additional quantum number ($n_1$) did not generate any new energy levels. When no external field was present, electronic orbits with the same $n$,

34. "Die hier eingeführten 'Richtungsquanten' geben die erste Möglichkeit einer Beziehung zwischen Polarisation und räumlicher Lage der Bahnebene." Ibid., p. 203.

35. See P. Epstein, "Zur Theorie des Starkeffektes," *Annalen der Physik*, 50 (1916): 489–520. Epstein's analysis is discussed in detail in Mehra and Rechenberg, *The Historical Development of Quantum Theory*, vol. 1, pp. 223–225.

36. I should note that Epstein designated the quantum number corresponding to Sommerfeld's $n_1$ with the term $n_3$. His different terminology reflected the fact that he employed a system of parabolic (as opposed to polar) coordinates. See Sommerfeld, *Gesammelte Schriften*, vol. 3, p. 204.

$n'$ but with different $n_1$'s had exactly the same energy. In the presence of an external field, however, the previously equivalent orbits obtained slightly different energies. Epstein and Karl Schwarzschild capitalized on this multiplication of the electron's energy levels to explain the Stark effect.[37] Furthermore, this proliferation of energy levels was the basis of Sommerfeld's and Peter Debye's explanation of the Zeeman effect.[38] The results of Sommerfeld's and Debye's analyses were identical. The discussion that follows is based on Sommerfeld's paper.[39]

It follows from a theorem proved by Joseph Larmor in 1897 that the effect of a magnetic field on an electron that rotates under the influence of an inverse square force is a precession of the electron's orbit around an axis parallel to the magnetic lines of force.[40] The angular velocity of that precession is

$$\frac{e}{2m}\frac{H}{c}.$$

An observer in a framework that rotates with this angular velocity does not perceive the precession in question. In such a framework, therefore, the motion of the electron can be studied without taking into account the presence of a magnetic field. In a stationary framework with polar coordinates $r$, $\theta$, $\phi$, the momentum associated with the coordinate $\phi$ is $p_\phi = mr^2 \sin^2 \theta \dot{\phi}$. Thus, in the rotating framework the azimuthal coordinate is

$$\chi = \phi - \frac{e}{2m}\frac{H}{c}t,$$

37. Schwarzschild's work is discussed in Jammer, *The Conceptual Development of Quantum Mechanics*, pp. 103–104, 107–108; and Mehra and Rechenberg, *The Historical Development of Quantum Theory*, vol. 1, pp. 225–227.

38. See P. Debye, "Quantenhypothese und Zeeman-Effekt," *Nachrichten von der Gesellschaft der Wissenschaften zu Göttingen, Mathematisch-Physikalische Klasse*, 1916, 142–153; Debye, "Quantenhypothese und Zeeman-Effekt," *Physikalische Zeitschrift*, 17 (1916): 507–512; and A. Sommerfeld, "Zur Theorie des Zeeman-Effekts der Wasserstofflinien, mit einem Anhang über den Stark-Effekt," *Physikalische Zeitschrift*, 17 (1916): 491–507; repr. in his *Gesammelte Schriften*, vol. 3, pp. 309–325.

39. For Debye's interpretation of the Zeeman effect see Jammer, *The Conceptual Development of Quantum Mechanics*, pp. 123–124; and Mehra and Rechenberg, *The Historical Development of Quantum Theory*, vol. 1, pp. 227–228. Sommerfeld's treatment of the normal Zeeman effect is discussed in P. Forman, "Alfred Landé and the Anomalous Zeeman Effect, 1919–1921," *Historical Studies in the Physical Sciences*, 2 (1970): 153–261, on p. 185.

40. See chapter 4, pp. 94–95. Cf. also A. Warwick, "Frequency, Theorem and Formula: Remembering Joseph Larmor in Electromagnetic Theory," *Notes and Records of the Royal Society of London*, 47, no. 1 (1993): 49–60.

and the corresponding momentum $p_\chi = mr^2 \sin^2 \theta \, \dot\chi$, which is constant. The value of that momentum is determined by Sommerfeld's azimuthal quantum condition:

$$(6.9) \qquad \int_0^{2\pi} p_\chi \, d\chi = 2\pi p_\chi = n_1 h.$$

The total energy of the electron in the original (stationary) framework is the sum of its kinetic and its potential energies:

$$W = \frac{m}{2}(\dot r^2 + r^2\dot\theta^2 + r^2 \sin^2 \theta \, \dot\phi^2) - \frac{e^2}{r}.$$

By putting

$$\dot\phi = \dot\chi + \frac{e}{2m}\frac{H}{c}$$

and neglecting the term that depends on

$$\left(\frac{e}{2m}\frac{H}{c}\right)^2,$$

we get

$$W = \frac{m}{2}(\dot r^2 + r^2\dot\theta^2 + r^2 \sin^2 \theta \, \dot\chi^2) - \frac{e^2}{r} + mr^2 \sin^2 \theta \, \dot\chi \frac{e}{2m}\frac{H}{c}.$$

The first two terms on the right side of this equation are equivalent with the total energy of the electron in the rotating framework. That energy is, in turn, identical with the total energy of the electron in the original framework when no external field is present ($W_0$). Therefore,

$$W = W_0 + \Delta W = W_0 + mr^2 \sin^2 \theta \, \dot\chi \frac{e}{2m}\frac{H}{c}$$

$$= W_0 + p_\chi \frac{e}{2m}\frac{H}{c} = W_0 + \frac{eH}{4\pi mc} n_1 h.$$

It follows from Bohr's frequency condition ($h\nu = W_m - W_n$) that the frequency of spectral lines in the presence of a magnetic field is

$$h(\nu_0 + \Delta\nu) = W_{m0} + \Delta W_m - (W_{n0} + \Delta W_n).$$

Thus, the frequency shift due to the magnetic field is

$$\Delta\nu = \frac{\Delta W_m - \Delta W_n}{h} = \frac{eH}{4\pi mc}(m_1 - n_1).$$

For $m_1 - n_1 = 0, \pm1$ this expression coincides with the results that Lorentz had obtained two decades before by means of his theory of "ions." [41] Furthermore, by employing Epstein's rule, which related the polarization of the components of a split spectral line with changes in the quantum number $n_1$ (or $m_1$), Sommerfeld obtained results concerning polarization identical with those predicted by Lorentz. Sommerfeld's theory, however, proved unable to provide a satisfactory explanation of the "anomalous" Zeeman effect—any form of magnetic splitting that did not conform to Lorentz's predictions. As late as 1923, Sommerfeld remarked that "in its present state the quantum treatment of the Zeeman effect achieves just as much as Lorentz's theory, but no more. It can account for the normal triplet, including the conditions of polarisation, but hitherto it has not been able to explain the complicated Zeeman types." [42] The subsequent development of the electron's representation was intertwined with the physicists' attempts to construct an adequate explanation of the "anomalous" phenomena. The taming of these phenomena became possible only with the proposal of the spin concept by Samuel Goudsmit and George Uhlenbeck in 1925; but more on this in chapter 8.

As I mentioned above, Sommerfeld developed Bohr's suggestion that relativistic effects should be taken into account in analyzing the electron's motion within the atom. [43] That route led him to a proliferation of the electron's energy levels, which in turn provided an explanation of the fine structure of the hydrogen spectral lines. By removing the simplifying assumptions from the electron's representation one could account for the complexities of the observable phenomena. To put it another way, in Sommerfeld's more realistic representation of the electron those complexities were linked with its complex (relativistic) behavior, which now became the unobservable counterpart of the fine structure of the hydrogen spectrum.

Sommerfeld first figured out the shape of the electron's orbit within the

41. For details about Lorentz's explanation of the Zeeman effect see chapter 4.

42. A. Sommerfeld, *Atomic Structure and Spectral Lines,* trans. from the 3rd German ed. by H. L. Brose (London: Methuen, 1923), p. 304.

43. See the second part of Sommerfeld's *Annalen der Physik* paper, entitled "Die Feinstruktur der Wasserstoff- und der wasserstoff-ähnlichen Linien," in Sommerfeld, *Gesammelte Schriften,* vol. 3, pp. 215–265.

hydrogen atom, by taking into account the velocity dependence of its mass. The angular momentum of the electron is (in polar coordinates)

$$p = mr^2\dot{\phi}, \quad \text{where} \quad m = \frac{m_0}{\sqrt{1 - \beta^2}} \quad \text{and} \quad \beta = \frac{v}{c}.$$

The equation of motion is[44]

$$\frac{d^2\sigma}{d\phi^2} + \sigma = \frac{eEm}{p^2} = \frac{eEm_0}{p^2}\frac{1}{\sqrt{1 - \beta^2}},$$

where $\sigma = 1/r$. Since the energy of the electron is

$$W = m_0 c^2 \left( \frac{1}{\sqrt{1 - \beta^2}} - 1 \right) - \frac{eE}{r},$$

it follows that

$$\frac{d^2\sigma}{d\phi^2} + \sigma \left[ 1 - \left( \frac{eE}{pc} \right)^2 \right] = \frac{eEm_0}{p^2} \left( 1 + \frac{W}{m_0 c^2} \right).$$

By integrating this equation, Sommerfeld obtained

(6.10)  $\dfrac{1}{r} = \sigma = c(1 + \varepsilon \cos \gamma\phi),$

where

$$\gamma^2 = 1 - \left( \frac{eE}{pc} \right)^2 \quad \text{and} \quad c = \frac{eEm_0}{\gamma^2 p^2} \left( 1 + \frac{W}{m_0 c^2} \right).$$

Equation (6.10) is a mathematical representation of a precessing elliptical orbit. During a revolution of the electron round the nucleus, the perihelion of its orbit describes an angle with value $2\pi/\gamma - 2\pi$. The precession of the perihelion takes place "in the same sense as that of the orbit."[45]

Sommerfeld employed equation (6.10) along with his radial and azimuthal quantum conditions to calculate the energy of the electron in a stationary orbit. The value he obtained was

(6.11)  $W = m_0 c^2 \left[ \left\{ 1 + \dfrac{\left( \alpha \dfrac{E}{e} \right)^2}{\left[ n' + n\sqrt{1 - \left( \dfrac{\alpha E}{ne} \right)^2} \right]^2} \right\}^{-1/2} - 1 \right],$

44. Ibid., p. 217.
45. "[I]m Sinne des Umlaufs." Ibid., p. 218.

where $\alpha = 2\pi e^2/hc = 7 \cdot 10^{-3}$ is the so-called fine structure constant. This expression for the energy levels implied that electronic orbits with the same quantum sum $n' + n$ but different $n$'s had slightly different energies. Owing to the variable mass of the electron, orbits that had been previously identical (in energy) were now differentiated. This differentiation provided the key to the solution of the problem of the fine structure of the hydrogen spectrum.

Sommerfeld's next step was to expand the energy function (6.11) in a series of powers of $\alpha$. By omitting powers of $\alpha$ greater than four, he obtained the following simplified expression:

$$W = \frac{Nh}{(n + n')^2}\left(\frac{E}{e}\right)^2\left[1 + \frac{\alpha^2}{(n + n')^2}\left(\frac{E}{e}\right)^2\left(\frac{1}{4} + \frac{n'}{n}\right)\right],$$

where $N$ was Sommerfield's symbol for the Rydberg constant. Thus, the supposition that the electron obeyed special relativity had a twofold effect on the energy levels that corresponded to its orbits: First, it led to a *"general increase of the absolute value of the energy"* that depended only on the quantum sum $n + n'$ (the major axis) of an orbit and not on its shape. Sommerfeld named this effect the "[r]elativity correction of the energy for circular orbits." [46] This consequence of relativity had been already deduced by Bohr. [47] Second, it produced a differential increase of the energy of an orbit depending on the ratio $n'/n$. This ratio, in turn, depended on the eccentricity of the orbit (see equation [6.4] above). Thus, orbits with the same major axis but different eccentricities (shapes) were associated with different energy levels. Since $n + n'$ orbits with the same major axis were possible, it followed that an electron with a given quantum sum $n + n'$ could occupy $n + n'$ different energy levels. The electron's quantum transitions were accordingly multiplied. The desired "energy splitting" had been achieved. [48] The fine structure of spectral lines became the observable manifestation of the relativistic behavior of the electron.

Sommerfeld's relativistic calculations were in agreement with Friedrich Paschen's very precise measurements of the spectral lines due to ionized helium. [49] To bring about the agreement in question, Sommerfeld introduced the so-called selection rules, which arbitrarily limited the allowed quantum

46. Ibid., p. 226.

47. More precisely, it had been implicit in Bohr's relativistic derivation of the frequency of the hydrogen spectral lines. See Bohr, "On the Series Spectrum of Hydrogen," p. 334.

48. "[W]ir nennen diesen Teil die *'Aufspaltung der Energie.'*" Sommerfeld, *Gesammelte Schriften*, vol. 3, p. 226.

49. F. Paschen, "Bohrs Heliumlinien," *Annalen der Physik*, 50 (1916): 901–940.

transitions.[50] This agreement suggested that the various orbits corresponding to a given quantum sum $n + n'$ had "real existence."[51] Furthermore, it constituted a "spectroscopic confirmation of the theory of relativity."[52] In particular, Paschen's measurements confirmed Einstein's relativistic representation of the electron. The special theory of relativity implied that a moving electron undergoes a contraction in the direction of its motion and specified how its mass depends on its velocity. However, Einstein's was not the only quantitative account of that dependence. Max Abraham had developed a theory that portrayed the electron as "rigid," that is, denied a velocity dependence of its dimensions, and led to a different prediction with respect to the velocity dependence of its mass.[53] Karl Glitscher employed that prediction to calculate the fine splitting of the helium spectral lines, but he obtained results that were incompatible with Paschen's measurements.[54] We see how a quantitative aspect of the electron's representation was constructed so as to reflect the quantitative features of the corresponding experimental situation. As soon as that situation was interpreted as the observable manifestation of the electron's variable mass, its quantitative features could be employed to quantify that aspect of the electron's behavior.

In this section I have followed the refinement of the electron's quantum identity. I have highlighted the active nature of its representation by focusing on the heuristic guidance it provided to Bohr and Sommerfeld in their attempts to explain the fine structure of spectral lines. The successful explanation of that phenomenon amounted to a coordination between the electron's writings and its representation, where the structure of the former obtained a

50. Cf. Heilbron, *Historical Studies in the Theory of Atomic Structure*, p. 55. The introduction of selection rules will be discussed below. It should be mentioned that the accuracy of Paschen's results as well as the validity of their interpretation by Sommerfeld was a subject of considerable debate that lasted well into the 1920s. This controversy, all but neglected in the historiography of the period, has been reconstructed by Helge Kragh. See his article "The Fine Structure of Hydrogen."

51. Sommerfeld, *Gesammelte Schriften*, vol. 3, p. 215.

52. Sommerfeld, *Atomic Structure and Spectral Lines* (1923), p. 525. Several opponents of relativity theory contested that confirmation by challenging the validity of Paschen's results. See Kragh, "The Fine Structure of Hydrogen," esp. pp. 81–102.

53. See M. Abraham, "Prinzipien der Dynamik des Elektrons," *Annalen der Physik*, 10 (1903): 105–179. Abraham's position was still favored by a minority of physicists who opposed Einstein's theory. See Kragh, "The Fine Structure of Hydrogen," pp. 82–83.

54. K. Glitscher, "Spektroskopischer Vergleich zwischen den Theorien des starren und des deformierbaren Elektrons," *Annalen der Physik*, 52 (1917): 608–630. Cf. Jammer, *The Conceptual Development of Quantum Mechanics*, p. 95; and Sommerfeld, *Atomic Structure and Spectral Lines* (1923), pp. 525ff.

counterpart in the latter. In the attainment of this coordination both the writings and the representation of the electron played a heuristic role. The former indicated that the electron could occupy more energy levels than previously thought and, thus, suggested how its representation should be transformed. The latter, in turn, pointed the way toward its own transformation. Its previous version had been deliberately inaccurate, since effects due to the high-speed motion of the electron had been ignored. A plausible way to achieve the desired multiplication of the electron's energy levels was to take into account those effects and, thereby, enhance the accuracy of its representation.[55]

The following section will examine the reaction of the electron qua theoretical entity to Sommerfeld's attempt to attribute to it a relativistic-quantum character, and will show how that reaction led to the introduction and theoretical justification of selection rules.

## 2. Being Disciplined by the "Magic Wand"

As we have seen, by taking into account the relativistic change of the electron's mass when it moves within the atom, orbits which had been, from the point of view of classical mechanics, indistinguishable in energy were now assigned slightly different energies. The consequence of that was a multiplication of the energy levels of the electron within the atom, and therefore of its possible quantum transitions between levels. The question whether all such transitions were permissible then arose. To put it another way, did an electron in a certain orbit have unrestricted freedom to "jump" to any other orbit? The answer turned out, on experimental grounds, to be negative. The possibility of investigating this question by experimental means was predicated on a central feature of the representation of the electron, namely, the electron's postulated mechanism of radiation. Since every quantum jump was supposed to give rise to a spectral line, the number and intensity of spectral lines provided information for the possibility and frequency of the corresponding quantum transitions.

As I mentioned above, Sommerfeld's calculations indicated that $n + n'$ different energy levels were assigned the same quantum sum $n + n'$. This meant that an electron whose original and final states were characterized by

---

55. By emphasizing the heuristic role of the electron's representation I do not mean to underplay the significance of the theoretical framework in which it was embedded. Part of that framework, that is, the theory of relativity, not only revealed the limited accuracy of the representation, but also provided the resources for its transformation.

$M = m + m'$ and $N = n + n'$, respectively, could, in principle, undergo $N \cdot M$ transitions. However, the experiments undertaken by Friedrich Paschen on the spectrum of ionized helium indicated that not all of those transitions were possible. Paschen's research concerned the electron's transitions between energy levels corresponding to $M = 4$ and $N = 3$, respectively. If no restrictions were imposed on the variation of the azimuthal quantum number, then one would have transitions from four different energy levels ($m = 1, 2, 3, 4$) to three lower energy levels ($n = 1, 2, 3$). Since each transition gives rise to a spectral line, one should observe twelve different spectral lines. However, fewer spectral lines were actually observed by Paschen.[56] There were two options: either to suggest that certain transitions were not permissible, or to assume that some transitions occurred so infrequently that the low intensity of the corresponding spectral lines did not allow their experimental detection. In fact, both options were exploited.

Before examining those options, it is worth reading the situation from my biographical perspective. Sommerfeld subjected the electron qua theoretical entity to a relativistic ordeal, hoping that in this way he would solve the problem of fine structure. However, as soon as he did that, the representation of the electron reacted back, so to speak, by multiplying the possible transitions of the electron beyond the amount dictated by empirical considerations. The writings of the electron, as manifested in the spectrum of ionized helium, ruled out the number of quantum transitions calculated by Sommerfeld. This unintended consequence of his attempt to enrich the electron's representation reveals, once more, the representation's active nature. Even though it had been the product of scientific construction, it became largely independent of its makers.[57] The electron's representation embodied the features of various experimental situations that were attributed to it and was, therefore, constrained by those features. Any attempt to enrich it would have been subject to those constraints. If, as it turned out, those constraints were violated, some further adjustment was necessary to restore the coherence of the representation.

This adjustment was provided by Sommerfeld, whose next move was

---

56. Paschen's experiments involved two different kinds of spectra, spark and direct current. In the former case all twelve lines were observed. Only in the latter case were fewer lines detected. For details about this disparity see Kragh, "The Fine Structure of Hydrogen," pp. 73–79.

57. Cf. K. R. Popper, *Unended Quest: An Intellectual Autobiography* (La Salle, Ill.: Open Court, 1976), p. 196.

to discipline the electron by arbitrary "selection rules," which limited its allowed quantum jumps. He proposed that only those quantum transitions are possible in which neither the azimuthal nor the radial quantum numbers increases:

$$m \geq n, \qquad m' \geq n'.$$

He considered the first "inequality" "exceptionless" "under normal circumstances," and the second only "roughly correct."[58] Furthermore, he suggested that the intensity of a spectral line was determined by the frequency of occurrence of the corresponding transition. The latter was, in turn, determined by the probabilities of the initial and the final orbit of the radiating electron. Thus, the intensity of spectral lines obtained for the first time a counterpart in the quantum representation of the electron.

In Sommerfeld's view, circular orbits were the most probable. In general, the probability of an orbit was proportional to the ratio $m/(m + m')$. In other words, the more elliptical an orbit the less probable it was.[59] The intensity of the line that was produced by the most probable transition (i.e., the transition whose starting and ending points were circular orbits) was assigned the value 1. The intensity of any other spectral line relative to that, most intense, line was determined by the following product:

$$(6.12) \quad J = \frac{n}{n + n'} \cdot \frac{m}{m + m'}.$$

This equation was valid only when the selection rules did not come into play.

Sommerfeld emphasized the provisional and approximate character of his conjectures concerning the intensity of spectral lines and pointed out that the selection rules as well as the equation of intensity (6.12) were context dependent, that is, they "depend on the particular circumstances of the stimulation of light."[60] The "particular circumstances" that he had in mind were either discharges of large quantities of electricity (spark spectra) or weak discharges (direct current spectra). In the former case, as Paschen's experiments indicated, the quantum inequalities were not operative and, thus, the intensity of spectral lines could be calculated by equation (6.12).

58. Sommerfeld, *Gesammelte Schriften*, vol. 3, p. 195.

59. This assignment of probabilities was suggested by the fact that the most eccentric orbit ($n = 0$, $\varepsilon = 1$) was physically impossible and was, therefore, assigned probability zero (see p. 153 above).

60. "[V]on den näheren Umständen der Anregung des Leuchtens abhängen." Sommerfeld, *Gesammelte Schriften*, vol. 3, p. 199.

Sommerfeld elaborated on these preliminary remarks in a subsequent paper that he delivered before the Bavarian Academy of Sciences on 3 March 1917.[61] There he attempted to give a quantum-theoretical justification of his "intensity rule." He assumed that the relative frequency ($W$) of a state with energy $E$ was given by the well-known Boltzmann distribution:

$$W = Ae^{-E/kT},$$

where $A$ was "the number of the various ways of production" of the state.[62] To put it another way, $A$ was the number of all the orbits that had the same energy ($E$). As we have seen, the energy of an orbit was determined by the sum of the radial ($n'$) and the azimuthal ($n$) quantum numbers, provided that one ignored relativistic effects (see pp. 153–154 above). Since all orbits with the same quantum sum ($n + n'$) were equivalent in energy, it followed that, in determining the frequency of occurrence of any of those orbits relative to any other, the exponential term in the above equation could be disregarded.

Thus, the frequency of occurrence of a state corresponding to a pair of quantum numbers $n$, $n'$ ($W_{n,\,n'} = A$) was given by the number of possible realizations of the state in question. As noted earlier,[63] for each value ($n$) of the azimuthal quantum number $n + 1$ orientations of the corresponding orbit were possible and, therefore, the corresponding state could be produced in $n + 1$ different ways ($W_{n,\,n'} = n + 1$). Consequently, the probabilities of the various fine structure orbits would be related as follows:

$$s + 1 : s : s - 1 : \cdots : 3 : 2 : 1,$$

where $s = n + n'$. The leftmost term in this formula corresponds to the most probable, circular orbit ($n' = 0$, $n = s$). As we move to the right the probability of the orbits diminishes, while their shape becomes more elliptical. The rightmost term, corresponding to $n' = s$, $n = 0$, was excluded on dynamical grounds (see on p. 153 above). Thus, the intensity of a fine structure component was determined by the following product:

$$J = W_{n,\,n'} \cdot W_{m,\,m'},$$

provided that the initial ($m$, $m'$) and the final ($n$, $n'$) orbits were statistically independent.

---

61. A. Sommerfeld, "Zur Quantentheorie der Spektrallinien. Intensitätsfragen," *Sitzungsberichte der Mathematisch-Physikalischen Klasse der Königlich-Bayerischen Akademie der Wissenschaften zu München* 47 (1917): 83–109; repr. in his *Gesammelte Schriften,* vol. 3, pp. 432–458.

62. Sommerfeld, *Gesammelte Schriften,* vol. 3, p. 437.

63. See p. 155.

Sommerfeld then assigned intensity 1 to the group of transitions whose beginning and ending points were circular orbits:

$$W_{s,0} \cdot W_{r,0} = 1,$$

where $s = n + n'$ and $r = m + m'$. This choice enabled him to derive the relative intensity of the remaining transitions:

$$(6.13) \quad J = W_{n,n'} \cdot W_{m,m'} = \frac{n + 1}{s + 1} \cdot \frac{m + 1}{r + 1} = \frac{n + 1}{n + n' + 1} \cdot \frac{m + 1}{m + m' + 1}.$$

Recall, however, that his aim was to justify the intensity rule (6.12), which differed slightly from the above equation (6.13). Rule (6.12) could be derived by reducing the probabilities of all those orbits, one of whose quantum numbers was zero, to half of their formerly assumed values, and otherwise following the same procedure that led to equation (6.13).[64] Sommerfeld recognized the "artificial" (künstlich) and "forced" (gezwungen) character of this reduction but suggested that it was supported by Paschen's observations of the direct current spectrum of helium.[65]

Sommerfeld's statistical computations of the intensity of spectral lines were based on two assumptions: first, that the initial and final orbits of a quantum transition were statistically independent; and second, that all orbits, except those whose azimuthal quantum number was zero, were dynamically possible. The first assumption could be upheld only for spark spectra. Direct current spectra, on the other hand, pointed to the existence of selection rules, which were the manifestation of *unknown* dynamical laws that limited the transitions between orbits.[66] Epstein's investigations concerning the number and intensity of spectral lines in the Stark effect supported further this conclusion.[67] Again, the electron's writings showed the way toward the articulation of its representation. Once spectral lines were assigned unobservable counterparts (transitions of electrons between energy levels), the number, frequencies, and intensities of the former prescribed the number, magnitude, and probability of the latter. In this way they constrained and guided the practice of theoretical physicists.

The second assumption was also problematic. Bohr had already suggested in 1913 that in gases with not very low density the electron could not move in orbits corresponding to very high quantum numbers, because the

64. For details see Sommerfeld, *Gesammelte Schriften*, vol. 3, pp. 439–440.
65. See ibid., pp. 454–455.
66. Ibid., p. 457.
67. Ibid., p. 458.

radii of those orbits would exceed the mean distance between the gas molecules. Only in very rarefied gases (as can be found, for example, in stars) were such orbits possible.[68] Sommerfeld developed Bohr's proposal in an attempt to come up with selection rules. By the exclusion of certain orbits on the ground that they were dynamically impossible he derived certain selection rules, which, however, were incompatible with the "quantum inequalities" that he had formerly introduced (see p. 164 above).[69] Moreover, Paschen's experiments on the direct current spectrum of helium did not indicate that any orbits were excluded.[70] Thus, Sommerfeld's attempt to provide a dynamical derivation of selection rules and thereby justify his arbitrary restriction of the electron's behavior failed. One more time, his struggle to discipline the electron in a principled way ran into difficulties with its writings.[71]

The theoretical justification of selection rules was made possible by Bohr's "magic wand," the correspondence principle,[72] which, in one of its versions, asserted that the predictions of quantum theory and those of classical electrodynamics should coincide in the domain of large quantum numbers.[73]

---

68. See Bohr, *On the Constitution,* pp. 9–10; and Sommerfeld, *Gesammelte Schriften,* vol. 3, p. 444.

69. Sommerfeld, *Gesammelte Schriften,* vol. 3, pp. 444–447.

70. Ibid., pp. 455–456.

71. Here, as elsewhere, I am talking about the electron qua theoretical entity. The electron qua real entity was, obviously, totally indifferent to Sommerfeld's theoretical maneuvers. Cf. Ian Hacking's useful discussion of "indifferent kinds" in his *The Social Construction of What?* (Cambridge, Mass.: Harvard Univ. Press, 1999), esp. pp. 104–107.

72. The "magic wand" is Sommerfeld's expression for the correspondence principle. In Sommerfeld's words, "Bohr has discovered in his *principle of correspondence* a magic wand (which he himself calls a 'formal' principle), which allows us immediately to make use of the results of the classical wave theory in the quantum theory." Sommerfeld, *Atomic Structure and Spectral Lines* (1923), p. 275.

73. Bohr offered the first full-blown definition of the correspondence principle in 1920. See H. Kragh, "Niels Bohr's Second Atomic Theory," *Historical Studies in the Physical Sciences,* 10 (1979): 123–186, on p. 155. However, he had employed earlier versions of this principle as early as 1913. Cf. Darrigol, *From c-Numbers to q-Numbers,* p. 90; and J. Hendry, *The Creation of Quantum Mechanics and the Bohr-Pauli Dialogue* (Dordrecht: Reidel, 1984), p. 140. Darrigol has argued that "the relevant correspondence is not between classical and quantum theory, but between atomic motion and radiation" (see his "Classical Concepts in Bohr's Atomic Theory [1913–1925], *Physis,* 34 [1997]: 545–567, on p. 550). While this reading of the correspondence principle is valid, Bohr's use of the term "correspondence" suggests that it had a dual significance, which also concerned the relation between quantum theory and classical electrodynamics. For instance, in the section of his 1920 paper where he explicates the principle, he says: "The frequencies of the spectral lines calculated according to both methods [classical and quantum] agree completely in the region where the stationary states deviate only little from one another. . . . *This correspondence between the frequencies determined by the two methods* must have a deeper significance." N. Bohr, "On the Series Spectra of the Elements," speech to the German Physical Society on 27 Apr. 1920, trans. in his *The Theory of Spectra,* pp. 20–60, on p. 27 (emphasis added).

Bohr moved beyond Sommerfeld's statistical treatment of the intensity of spectral lines and managed to restrict the electron's choices on dynamical grounds. In a paper that he published in 1918 he employed the correspondence principle to calculate transition probabilities, deduce selection rules, and represent the polarization of spectral lines.[74]

Bohr's paper built on two contributions to quantum theory, one by Paul Ehrenfest and the other by Einstein. In 1913 Ehrenfest had introduced the so-called adiabatic principle, in an attempt to "trace the boundary between the 'classical region' and the 'region of quanta.'"[75] This principle asserted that if one changed very slowly the parameters of a dynamical system (e.g., by imposing an external electric or magnetic field) *"allowed motions are transformed into allowed motions."*[76] Thus, there were certain magnitudes, the "adiabatic invariants," which remained unaffected by changes of this kind. It turned out that Sommerfeld's phase integrals were among those magnitudes that remained invariant during adiabatic transformations.[77] So they obtained a dynamical justification and ceased to be the "unprovable" foundation of the quantum theory. In other words, the adiabatic principle helped to make mechanical sense of Sommerfeld's phase integrals.[78] Furthermore, it extended the applicability of classical mechanics to the description of the electron's motion in the presence of slowly varying external fields. Finally, it solved an important conceptual problem that was generated by the simultaneous presence in Bohr's theory of two incompatible concepts: on the one hand, quantum transitions were supposed to be nonmechanical processes; on the other hand, mechanics was essential for explicating the energy difference between stationary states. In Bohr's words, "[W]e have assumed that the direct transition between two . . . [stationary] states cannot be described by ordinary mechan-

---

74. N. Bohr, "On the Quantum Theory of Line-Spectra," pt. 1, *Kongelige Danske Videnskabernes Selskabs Skrifter Naturvidenskabelig og mathematisk afdeling,* series 8, IV (1918), 1, 1–36; repr. in B. L. van der Waerden (ed.), *Sources of Quantum Mechanics* (New York: Dover, 1967), pp. 95–137.

75. P. Ehrenfest, "Adiabatic Invariants and the Theory of Quanta," *Philosophical Magazine,* 33 (1917): 500–513; repr. in van der Waerden (ed.), *Sources of Quantum Mechanics,* pp. 79–93, on p. 79. For historical information on Ehrenfest's principle see M. J. Klein, *Paul Ehrenfest,* vol. 1, *The Making of a Theoretical Physicist* (New York: Elsevier, 1970), pp. 264–292.

76. Ehrenfest, "Adiabatic Invariants and the Theory of Quanta," in van der Waerden (ed.), *Sources of Quantum Mechanics,* p. 80. Bohr renamed Ehrenfest's hypothesis the principle of "mechanical transformativity."

77. See Bohr, "On the Quantum Theory of Line-Spectra," pt. 1, in van der Waerden (ed.), *Sources of Quantum Mechanics,* pp. 106–107, 112–113; and Ehrenfest, "Adiabatic Invariants and the Theory of Quanta," in van der Waerden (ed.), *Sources of Quantum Mechanics,* pp. 87–88.

78. Cf. Jammer, *The Conceptual Development of Quantum Mechanics,* pp. 100–101.

ics, while on the other hand we possess no means of defining an energy difference between two states if there exists no possibility for a continuous mechanical connection between them. It is clear, however, that such a connection is just afforded by Ehrenfest's principle." [79]

The other point of departure of Bohr's treatise was Einstein's introduction of transition probabilities. In 1916 Einstein had managed to provide a probabilistic derivation of Planck's radiation law that dispensed with every classical element.[80] Einstein associated with each transition between stationary states a certain probability, which, for the time being, could not be given a causal interpretation. This probability was not determined, pace Sommerfeld, by the "a priori probabilities" of the initial and final states of the transition, but was an intrinsic attribute of each transition.[81]

However, Einstein did not provide any means of calculating these probabilities. Bohr's correspondence arguments aimed at providing such a means by connecting "the probability of a transition between any two stationary states and the motion of the system in these states." [82] In particular, transition probabilities were related to the harmonic components of the motion of the electrons in stationary states. Every complex periodic motion can be analyzed, à la Fourier, in a series of simple periodic motions. For a nondegenerate system with $s$ degrees of freedom the motion of the particles can be represented as follows:

$$(6.14) \quad \xi = \sum C_{\tau_1,\ldots,\tau_s} \cos 2\pi \left[ (\tau_1 \omega_1 + \cdots + \tau_s \omega_s) t + c_{\tau_1,\ldots,\tau_s} \right],$$

where the summation extends from $-\infty$ to $+\infty$, "and where the $\omega$'s are the . . . mean frequencies of oscillation for the different [coordinates]." [83] According to classical electrodynamics, the coefficients $C_{\tau_1,\ldots,\tau_s}$ in the above for-

79. Bohr, "On the Quantum Theory of Line-Spectra," pt. 1, in van der Waerden (ed.), *Sources of Quantum Mechanics*, p. 102. Cf. Darrigol, *From c-Numbers to q-Numbers*, pp. 132–137; and M. N. Wise, "How Do Sums Count? On the Cultural Origins of Statistical Causality," in L. Krüger, L. J. Daston, and M. Heidelberger (eds.), *The Probabilistic Revolution*, vol. 1, *Ideas in History* (Cambridge, Mass.: MIT Press, 1987), pp. 416–417.

80. See A. Einstein, "Zur Quantentheorie der Strahlung," *Physikalische Zeitschrift*, 18 (1917): 121–128; first publ. in *Mitteilungen der Physikalischen Gesellschaft, Zürich*, 18 (1916): 47–62; trans. in van der Waerden (ed.), *Sources of Quantum Mechanics*, pp. 63–77. For a concise discussion of this paper see M. J. Klein, "The First Phase of the Bohr-Einstein Dialogue," *Historical Studies in the Physical Sciences*, 2 (1970): 1–39, on pp. 7–8.

81. This was true only for "spontaneous" emissions. In the presence of a radiation field the transition probabilities also depended on the radiation density $\rho$. The expression "a-priori probability" is from Bohr, "On the Quantum Theory of Line-Spectra," pt. 1, in van der Waerden (ed.), *Sources of Quantum Mechanics*, p. 100.

82. Ibid., p. 101.

83. Ibid., p. 129.

mula "determine the intensity . . . of the emitted radiation of the corresponding frequency $\tau_1\omega_1 + \cdots \tau_s\omega_s$."[84] In quantum theory, on the other hand, that intensity is given by the probability of the corresponding transition between stationary states. Bohr suggested that in the region of high quantum numbers there is a direct connection between transition probabilities and classical coefficients. In particular, the coefficient of the harmonic component with frequency $\tau_1\omega_1 + \cdots + \tau_s\omega_s$ determines the probability of the transition from a state characterized by the quantum numbers $n_1', \ldots, n_s'$ to a state with quantum numbers $n_1'', \ldots, n_s''$, where $\tau_1 = n_1' - n_1'', \ldots, \tau_s = n_s' - n_s''$. If this coefficient were zero, the corresponding probability would also be zero and, therefore, the transition in question would be forbidden (selection rule).

Bohr's analysis was general and concerned many-particle systems with multiple periodicities. In these systems, "we find much more complicated motions with correspondingly complicated harmonic components."[85] To handle the complexity of these motions, Bohr formulated the correspondence principle in terms of generalized coordinates, the so-called action-angle variables.[86] By means of these variables, he could express the periodic character of the motions and the quantum constraints imposed on them, without describing them in full detail. For simplicity, in what follows I will focus on the implications of Bohr's analysis for one-electron systems, where the correspondence between the electron's motion and the emitted radiation is more straightforward.

In the domain of high quantum numbers the radiation frequency and the frequency of oscillation of the electron approximately coincide. Therefore, in that region the harmonic components of the electron's motion should coincide with the frequencies of the emitted spectral lines. Furthermore, the coefficients of those components should determine the intensities of the corresponding lines. Since in quantum theory the intensities of spectral lines are specified by the probabilities of the corresponding transitions, it follows that the characteristics of the electron's motion in a stationary state determine the probabilities of its transition to other stationary states.

As I mentioned above, the intensities of spectral lines could be legitimately inferred from the periodic properties of the motion of the electrons in stationary states only for large quantum numbers. Bohr, however, extended

84. Ibid., p. 130.

85. Bohr, "On the Series Spectra of the Elements," p. 30.

86. See Darrigol, *From c-Numbers to q-Numbers*, pp. 113–114; and Jammer, *The Conceptual Development of Quantum Mechanics*, pp. 102ff.

the validity of that inference by assuming that "also for values of the $n$'s which are not large, there must exist an intimate connection between the probability of a given transition and the values of the corresponding Fourier coefficient in the expressions for the displacements of the particles in the two stationary states." [87] Since $\tau$ ranged over both positive and negative values, this bold conjecture enabled him to conclude that, pace Sommerfeld, "transitions will be possible for which some of the $n$'s increase while others decrease." [88]

It should be emphasized that, despite the quantitative correspondence between the classical and the quantum descriptions of a radiating electron, "the mechanism of emission in both cases is different. The different frequencies corresponding to the various harmonic components of the motion are emitted simultaneously according to the ordinary theory of radiation. . . . But according to the quantum theory the various spectral lines are emitted by entirely distinct processes, consisting of transitions from one stationary state to various adjacent states." [89] Thus, the correspondence that Bohr had in mind was between the effects of a single radiating electron that obeyed the laws of classical electrodynamics and the effects of several electrons that performed quantum jumps. The periodic characteristics of the electron's motion determined its propensity to undergo each of those transitions. Rutherford's complaint that the electron, as portrayed by Bohr, had to "*decide* what frequency it is going to vibrate when it passes from one stationary state to the other" was finally addressed.[90] The electron's "decisions" inhered in its behavior in its "waiting places," the stationary states. The periodic properties of its initial state determined, in a probabilistic fashion, its final destination.[91]

The correspondence principle also enabled Bohr to represent the polarization of spectral lines in terms of the periodicities in the motion of the electrons in stationary states. If, for instance, the coefficient of a harmonic component of that motion in a given direction were zero, then the corresponding quantum transition would "be accompanied by a radiation which is polarised

87. Ibid., pp. 130–131.

88. Ibid., p. 131. Sommerfeld later admitted that his quantum inequalities (see p. 164 above) were not correct and should be replaced by the "theoretically justified selection principle" obtained by Bohr. See W. Kossel and A. Sommerfeld, "Auswahlprinzip und Verschiebung bei Serienspektren," *Verhandlungen der Deutschen Physikalischen Gesellschaft,* 21 (1919): 240–259; repr. in Sommerfeld, *Gesammelte Schriften,* vol. 3, pp. 476–495, on p. 485.

89. Bohr, "On the Series Spectra of the Elements," p. 27.

90. Rutherford to Bohr, 20 Mar. 1913, in N. Bohr, *Collected Works,* vol. 2, ed. U. Hoyer (Amsterdam: North Holland, 1981), p. 583.

91. My analysis here has profited from M. Norton Wise, "Forman Reformed," unpublished manuscript, pp. 14–16.

in a plane perpendicular to this direction."[92] By applying this reasoning Bohr derived the empirical polarization rule that Epstein had introduced in his analysis of the Stark effect (see p. 155 above).[93]

In 1919 the quantitative aspects of Bohr's analogy were spelled out by Hendrik Kramers, Bohr's student and collaborator, in his investigation of the fine structure and electrical splitting of the hydrogen spectral lines. Kramers's calculations led to conclusions, regarding the intensities and polarizations of the fine structure and Stark components, which were supported by the available spectroscopic evidence.[94] Thus, the application of Bohr's "magic wand" made possible the construction of an empirically adequate and theoretically grounded representation of the electron. By linking various properties of spectral lines with characteristics of the motion of electrons inside the atom, the correspondence principle provided a satisfactory "translation" of the former to the latter.

However, one cannot exclude the possibility that several incompatible "translations" of the same experimental data can be produced. Indeed, Bohr's attempt to come up with theoretically justified selection rules was not the only one. The Polish physicist Adalbert Rubinowicz, who at that time was Sommerfeld's assistant, deduced similar selection rules from the conservation of the angular momentum of an atomic system (atom and emitted radiation) during a quantum transition. In particular, he concluded that only those transitions were allowed in which the azimuthal quantum number changed, at most, by one unit ($\Delta n = 0, \pm 1$).[95] Bohr's corresponding selection rule, on the other hand, required that $\Delta n = \pm 1$, a demand that was supported by observations on the fine structure components of the hydrogen spectral lines.[96] Sommerfeld, who was more favorably disposed to Rubinowicz's method, managed to exclude the transitions for which $\Delta n = 0$ without invoking the correspondence principle but recognized the superiority of Bohr's procedure: "Whereas we, in the case $\Delta n = 0 \ldots$, attained our object only as a result of plausible reflections, the principle of correspondence comes to its decisions

92. Bohr, "On the Quantum Theory of Line-Spectra," pt. 1, in van der Waerden (ed.), *Sources of Quantum Mechanics*, p. 131.

93. See Bohr, "On the Series Spectra of the Elements," pp. 41–42.

94. See Darrigol, *From c-Numbers to q-Numbers*, pp. 127–128; Heilbron, *A History of the Problem of Atomic Structure*, pp. 354–355; and Jammer, *The Conceptual Development of Quantum Mechanics*, pp. 115–116.

95. See A. Rubinowicz, "Bohrsche Frequenzbedingung und Erhaltung des Impulsmomentes," *Physikalische Zeitschrift*, 19 (1918): 441–445, 465–474. For a concise presentation of Rubinowicz's procedure see Sommerfeld, *Atomic Structure and Spectral Lines* (1923), pp. 264–266; and Darrigol, *From c-Numbers to q-Numbers*, pp. 139–140.

96. Cf. Kragh, "The Fine Structure of Hydrogen," p. 78.

by unambiguous analytical criteria."[97] Furthermore, Rubinowicz's approach could not yield transition probabilities.[98] Sommerfeld, however, had an ambivalent attitude toward the correspondence principle, despite its advantages, because he considered it a classical remnant "which is added to the quantum theory as something foreign to its nature."[99]

This perspective was not shared by Bohr, who regarded the correspondence principle as an integral part of the quantum theory.[100] In fact, this principle formed the main heuristic element of Bohr's subsequent research program, which culminated in Heisenberg's formulation of matrix mechanics. Within that program it played a dual role. On the one hand, to the extent that orbits of electrons within the atom could be calculated by classical mechanics and quantum conditions, it was a predictive tool. On the other hand, it allowed the construction of a representation of those orbits out of experimental data concerning the characteristics of spectral lines.[101]

## 3. Concluding Remarks

In this chapter we saw that the representation of an unobservable entity has to be rich enough so as to be able to embody all the relevant qualitative and quantitative features of the experimental situations that are taken to be the observable manifestations of the entity in question. For example, the production of spectral lines was interpreted as the observable manifestation of an underlying electronic reality. That underlying reality had to be such as to account for all the relevant observable features of spectral lines (their frequency, intensity, polarization, magnetic and electric splitting, fine structure, etc.). In other words, every observable feature had to have its counterpart in the properties and behavior of electrons inside the atom. When a quantum representation of the electron was introduced, only the frequency of spectral lines could be accounted for. Their intensity, polarization, and so on, were left unexplained. The task faced by Bohr and other physicists was to enrich the representation of the electron's behavior, so that every relevant observable feature would have its unobservable counterpart. In that process of translation from the observable to the unobservable domain Bohr's correspondence principle played a central role.

97. Sommerfeld, *Atomic Structure and Spectral Lines* (1923), p. 276.
98. Cf. Bohr, "On the Series Spectra of the Elements," p. 52.
99. Sommerfeld, *Atomic Structure and Spectral Lines* (1923), p. 274.
100. See Darrigol, *From c-Numbers to q-Numbers*, p. 138.
101. Cf. ibid., pp. 83, 151, 159, 171.

The physicists' attempts to attribute additional properties to the electron were sometimes facilitated and sometimes met with resistance from the electron qua theoretical entity. On the positive side, the electron's writings guided, time and again, the enrichment of its representation, by indicating what was missing or problematic in it. On the negative side, the newly ascribed properties gave rise to unintended and undesirable consequences, which rendered incoherent the new representation of the electron. The representation's resistance to manipulation was a theme that continued to be of importance for the further development of the quantum theory of atomic structure and manifested itself most clearly in the proposal of spin by Goudsmit and Uhlenbeck. The road to spin passed through other developments, like the introduction of new quantum numbers, Bohr's "unmechanische Zwang," and Pauli's exclusion principle. It was also anticipated in the chemistry literature, where an intrinsic magnetic disposition had been attributed to the electron well before the physicists did something similar. Thus, before continuing the story of the physicists' representation of the electron I will discuss the chemists' perspective on it and the problem situations that shaped it.

## Chapter 7 | "How the Electrons Spend Their Leisure Time": The Chemists' Perspective

### 1. Introduction

The aim of Millikan's Faraday lecture before the members of the Chemical Society on 12 June 1924 was to convince his audience that the electron "is a definitely and directly observable fact," attempting throughout his presentation "to differentiate . . . between . . . fact and . . . fancy."[1] Even though no one, according to Millikan, could doubt the atomicity of electric charge, there was considerable disagreement between chemists and physicists about the behavior of atomic charges inside of the atom. In particular, the controversy concerned "how the electrons spend their leisure time, the portions of their lives within the atom when they are not radiating."[2] The electron qua theoretical entity was leading a double life, one in chemistry and one in physics, exhibiting symptoms of a split personality:[3] "The chemist has in general been content with what I will call the 'loafer' electron theory. He has imagined these electrons

1. R. A. Millikan, "Atomism in Modern Physics," *Journal of the Chemical Society,* 125, pt. 2 (1924): 1405–1417, on pp. 1406, 1405.

2. Ibid., p.1411. A few months earlier, on 22 Apr. 1924, Millikan had presented the disagreement, in virtually the same terms, to the American Chemical Society. See R. A. Millikan, "The Physicist's Present Conception of an Atom," *Science,* 59, no. 1535 (30 May 1924): 473–476, on p. 474.

3. These metaphors presuppose the biographical approach I have adopted. Only a single entity can have a split personality or a double life. Furthermore, this approach captures the pervasive feeling among many physicists and chemists that both were dealing with the *same* entity, despite their divergent representations of it.

sitting around on dry goods boxes at every corner ready to shake hands with, or hold on to, similar electrons in other atoms. The physicist, on the other hand, has preferred to think of them as leading more active lives, playing ring-around-the-rosy, crack-the-whip and other interesting games. In other words, he has pictured them as rotating with enormous speeds in *orbits,* and as occasionally flying out of these orbits for one reason or another."[4]

In the historiography of the period the controversy has been portrayed as a conflict between two different conceptions of the atom, the physicists' dynamic atom and the chemists' static atom.[5] While this account of the conflict has its merits, there are certain advantages in portraying the controversy as a clash between two different conceptions of the electron. Even though the clash dates from before the "discovery of the electron,"[6] after that discovery it was transformed into a conflict between two incompatible conceptions of the electron. The incompatibility between the physicists' dynamic atom and the chemists' static atom became a mere epiphenomenon of an underlying conflict between the physicists' and the chemists' electrons.[7] Given what I said in chapters 3 and 4, I obviously do not mean to suggest that the discovery of the electron was an unproblematic and straightforward event. Despite its complexity, however, the process that has been seen retrospectively as the discovery of the electron rendered the incompatibility between the two atoms an epiphenomenon of the incompatibility between their subatomic constituents. A dynamic atom resulted from a nonmagnetic electron which obeyed Coulomb's law (except in quantum transitions), whereas a static atom resulted from a magnetic electron which (at small distances) violated Coulomb's law. Furthermore, chemists were not very interested in detailed pictures of the structure of the atom; they were, rather, concerned with "an atom's outermost electrons," the only electrons which entered in the process of chemical combination.[8] As J. J. Thomson remarked in 1923, "The electron

---

4. Millikan, "The Physicist's Present Conception of an Atom," p. 474.

5. See, e.g., R. E. Kohler, "The Lewis-Langmuir Theory of Valence and the Chemical Community, 1920–1928," *Historical Studies in the Physical Sciences,* 6 (1975): 431–468.

6. A dynamic model of the atom was deemed necessary to account for the data of spectroscopy, whereas a relatively inert atom was required for the explanation of chemical bonding.

7. Cf. Kostas Gavroglu, "The Physicists' Electron and Its Appropriation by the Chemists," in J. Z. Buchwald and A. Warwick (eds.), *Histories of the Electron: The Birth of Microphysics* (Cambridge, Mass.: MIT Press, 2001), pp. 363–400, esp. p. 367.

8. See A. N. Stranges, *Electrons and Valence: Development of the Theory, 1900–1925* (College Station: Texas A and M Univ. Press, 1982), p. 75. The same view is endorsed by Kragh. See H. Kragh, "Bohr's Atomic Theory and the Chemists," *Rivista di storia della scienza,* 2, no. 3 (1985): 463–486, on p. 467. Note, however, that some chemists did not accept this view. See J. D. Main Smith, *Chemistry and Atomic Structure* (London: Ernest Benn, 1924), pp. 7, 194–195.

is the dominating factor in these problems which are just those with which the chemist is most concerned."[9]

This early-twentieth-century conflict between chemistry and physics gives us the opportunity to address and explore in a concrete fashion the following set of interrelated questions: When a theory is put forward in a certain discipline (e.g., chemistry), are the criteria employed for its evaluation specific to the specific discipline? In other words, how local is the experimental evidence employed for theory appraisal? What happens when, say, a chemical theory successfully accounts for the relevant chemical evidence but fails to account for the relevant physical evidence and vice versa? In general, what criteria determine the relevance of a specific piece of evidence to the evaluation of a theory? What is the role of the theoretical entity or entities employed by the theory in the specification of these criteria? To put it another way, what is the significance of theoretical entities for the coordination between theory and evidence? When a theoretical entity that "belongs" to a discipline (physics) is appropriated by another discipline (chemistry), how does this affect the appropriating discipline? In particular, which of its notions and regularities have to be reinterpreted in terms of the appropriated entity? In general, how does the disciplinary context affect its "personality"?

While addressing these questions, the aim of this chapter is to describe the origins and development of that conflict and to account for that episode in terms of the different methodological outlooks and problem situations of two distinct scientific cultures, the physicists and the chemists.

## 2. The Emergence of the Conflict: G. N. Lewis's "Loafer" Electron

Gilbert N. Lewis (1875–1946) commenced his graduate studies at Harvard in 1897 and was trained by the physical chemist T. W. Richards.[10] In 1900, a year after he had completed his dissertation, he went to Germany and worked with Wilhelm Ostwald and Walther Nernst. The instrumentalism that pervades his mature writings can be traced, perhaps, to his early interaction with Ostwald. Lewis's radical instrumentalism was displayed in his *Valence and the Structure of Atoms and Molecules.* There he argued that the empirical success of Bohr's model of the hydrogen atom did not provide adequate grounds for be-

---

9. J. J. Thomson, *The Electron in Chemistry* (Philadelphia: Franklin Institute, 1923), on the second page of the preface.

10. For a short biographical essay on G. N. Lewis see Robert Kohler, "Lewis, Gilbert Newton," in C. C. Gillispie (ed.), *Dictionary of Scientific Biography,* 16 vols. (New York: Charles Scribner's Sons, 1970–1980), vol. 8, pp. 289–294.

lieving that it was "something more than a mere working hypothesis, and may represent an ultimate reality." [11] Furthermore, he emphasized that "we must . . . focus our attention upon our actual experimental facts, and give less heed to those conventional abstractions of the mind, such as force and fields of force, energy and the conservation of energy, or even space and time. Some of these abstractions may have to be abandoned as the conventional ether was abandoned after the acceptance of relativity." [12]

While in Europe he formed the habit, as a result of his interaction with Jacobus van't Hoff, "of shocking the chemical establishment with his novel ideas." [13] One of those novel ideas was his cubic model of the atom. Lewis's cubic model was originally conceived in 1902: "In the year 1902 (while I was attempting to explain to an elementary class in chemistry some of the ideas involved in the periodic law) becoming interested in the new theory of the electron, and combining this idea with those which are implied in the periodic classification, I formed an idea of the inner structure of the atom which, although it contained certain crudities, I have ever since regarded as representing essentially the arrangement of electrons in the atom." [14] As Lewis mentioned in the above excerpt, the new ideas on the electronic constitution of matter along with the periodic properties of the chemical elements were the central elements of his problem situation. The two main desiderata were to portray the atom as composed primarily of electrons [15] and to provide a physical interpretation, based on the new model of the atom, of the periodic table. These desiderata indicate that the importation of the electron in chemistry had to be accompanied by a reconceptualization of preexisting chemical notions (e.g., the atom) in terms of electrons. Furthermore, chemical regularities (e.g., the periodic properties of the elements) had to be reinterpreted in terms of the behavior of electrons; the former had to obtain a counterpart in the representation of the latter.

Both of these desiderata were fulfilled with Lewis's 1902 proposal of a cubic atom. According to that proposal, the atom had an elaborate cubic struc-

11. G. N. Lewis, *Valence and the Structure of Atoms and Molecules* (New York: Chemical Catalog Co., 1923), p. 50.

12. Ibid., p. 165.

13. R. E. Kohler, "The Origin of G. N. Lewis's Theory of the Shared Pair Bond," *Historical Studies in the Physical Sciences, 3* (1971): 343–376, on p. 348. Kohler's remark was drawn from a letter from Lewis to his former adviser, T. W. Richards (13 Jan. 1901, T. W. Richards Archive, Harvard Univ.).

14. Lewis, *Valence,* p. 29.

15. Lewis had also to include in his model a "positive charge which balanced the electrons in the neutral atom" and about which his "ideas were very vague . . . at that time." Ibid., p. 30.

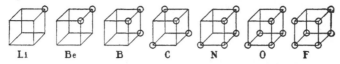

**Fig. 7.1** These pictures represent the outer shell of electrons in various atoms, from lithium to fluorine, and "are taken from a memorandum dated March 28, 1902." (From G. N. Lewis, "The Atom and the Molecule," *Journal of the American Chemical Society,* 38 [1916]: 762–785, on p. 767.)

ture and was composed of electrons which occupied the vertices of a series of concentric cubes (see fig. 7.1). The most important aspect of that theory, for my purposes, was that it portrayed the electron within the atom as a static particle to suit the chemical static atom. Despite the fact that "this theory of structure seemed to offer a remarkably simple and satisfactory explanation" of the formation of polar compounds, it failed to explain "chemical combinations of a less polar type" such as take place in the formation of the majority of organic compounds.[16] This is why Lewis's theory, while it was "discussed freely with my colleagues and in my classes, . . . [was] given no further publicity."[17]

The next stage in his development of the theory began in 1916. In the meantime several developments took place in both chemistry and physics that were decisive for the further elaboration of the 1902 theory. In 1904 Richard Abegg (1869–1910) proposed a new interpretation of valence based on the recently "discovered" electron.[18] Before the advent of the electron "the valence of an atom . . . [was] defined as the number of bonds which attach it to other atoms."[19] In Abegg's theory the "normal" valence of an atom was reinterpreted as the definite number of electrons that the atom would give (positive valence) or take (negative valence) upon its combination with another atom to form a molecule or a compound. Besides its normal valence, each atom was characterized by a "contra-valence," an alternative number of electrons that the atom would take or give during the process of chemical com-

16. Lewis, *Valence,* p. 30. Polar compounds were those that were composed of two oppositely charged ions, whereas compounds "of a less polar type" were those that were composed of neutral parts. The former were mainly inorganic substances, whereas the latter were predominantly organic substances. Lewis proposed the replacement of the organic-inorganic distinction with the polar-nonpolar distinction in 1913. See G. N. Lewis, "Valence and Tautomerism," *Journal of the American Chemical Society,* 35 (1913): 1448–1455.

17. Lewis, *Valence,* p. 30.

18. See R. Abegg, "Die Valenz und das periodische System: Versuch einer Theorie der Molekularverbindungen," *Zeitschrift für Anorganische Chemie,* 39 (1904): 330–380. For a brief exposition of Abegg's theory see Kohler, "The Origin," pp. 349–350.

19. Lewis, *Valence,* p. 104.

bination. The normal and the contra-valence were always numbers of opposite sign, and the sum of their absolute values was always eight. Thus, the introduction of the electron in chemistry led to a reconceptualization and refinement of the notion of valence.

It should be emphasized that Abegg attached, as Lewis had done before him, special significance to the number eight, a "number which for all atoms represents the points of attack of electrons,"[20] that is, the total number of points on the atom which were potentially occupied by valence electrons. The number eight corresponds to the eight groups of the periodic table; that is why it was a privileged number in Abegg's and Lewis's theories. However, Abegg did not explain why that number was always eight, whereas for Lewis the value of that number followed from the cubic structure of his atom. Further details of Abegg's scheme need not concern us here. Suffice it to say that for him the affinity between two atoms was a purely electrostatic phenomenon due to the transfer of one or more electrons from one atom to another and the concomitant electrostatic attraction between the oppositely charged atoms.

In the same year in which Abegg published his theory of electrovalence J. J. Thomson proposed the "plum pudding" model of the atom, in which corpuscles (electrons) moved in coplanar, circular orbits inside a homogeneously and positively charged sphere.[21] Thomson's model provided a qualitative understanding of the periodic table of Mendeleev and of chemical combination, among other phenomena. The demand for a mechanically stable atom, formulated in terms of Thomson's model, provided the key for the interpretation of the periodic properties of the elements. Only those rings of corpuscles which contained a number of corpuscles below a certain threshold would be stable. Above that threshold a ring would break into two distinct rings, the additional ring being formed from the extra electrons which threatened the stability of the previous ring. Since the chemical properties of the elements were determined by those extra electrons, a periodic occurrence of the same number of electrons in the "incomplete" ring of different elements would result in the periodicity of their chemical properties.

Furthermore, stability considerations were crucial for understanding the formation of chemical compounds. Atoms with an "incomplete" ring, which

20. "[D]er Zahl, die für alle Atome die Angriffsstellen der Elektronen darstellt." Abegg, "Die Valenz," p. 380.

21. J. J. Thomson, "On the Structure of the Atom: An Investigation of the Stability and Periods of Oscillation of a Number of Corpuscles Arranged at Equal Intervals around the Circumference of a Circle; with Application of the Results to the Theory of Atomic Structure," *Philosophical Magazine*, 6th series, 7 (1904): 237–265.

contained a number of electrons close to the stability threshold, would tend to capture additional electrons ("electronegative elements") and thus increase their stability. Otherwise, atoms would tend to increase their stability by shedding the electrons of their incomplete ring ("electropositive elements"). The classification of the elements in these two categories led to a very simple interpretation of chemical combination: "When atoms like the electronegative ones . . . are mixed with atoms like the electropositive ones . . . the detachment of corpuscles from the electropositive atoms and their transference to the electronegative [will probably take place]. The electronegative atoms will thus get a charge of negative electricity, the electropositive atoms one of positive, the oppositely charged atoms will attract each other, and a chemical compound of the electropositive and electronegative atoms will be formed." [22] It should be noted that for Thomson, as for Abegg, the chemical bond between two atoms resulted from the transfer of electrons from one atom to the other and the concomitant attraction between the oppositely charged atoms.

In 1907 Thomson developed further the chemical applications of his atomic theory in a highly influential book, *The Corpuscular Theory of Matter*.[23] Again, despite the theoretical possibility of nonpolar bonds (i.e., bonds "binding the two systems together without a resultant charge on either system"),[24] Thomson admitted the existence only of polar bonds "established between two atoms [by] the transference of one corpuscle from the one atom to the other." [25] In the the second decade of the twentieth century, because of the popularity of Thomson's theory among American chemists, the exclusive reality of polar bonds became part of the orthodoxy of the day.

It is ironic that Thomson became one of the first to revolt against the orthodoxy that his 1907 theory had helped to consolidate. In 1914 he published a paper in which he admitted the existence of two kinds of bonds, polar and nonpolar.[26] In nonpolar bonds there was no electron transfer between the atoms forming the chemical compound. Instead, such bonds consisted of two tubes of force, one between an electron of the first atom and the nucleus of the second and another between an electron of the second atom and the nu-

22. Ibid., pp. 262–263.

23. J. J. Thomson, *The Corpuscular Theory of Matter* (New York: Charles Scribner's Sons, 1907).

24. Ibid., p. 121. Cf. C. A. Russell, *The History of Valency,* (New York: Humanities Press, 1971), p. 281.

25. Thomson, *The Corpuscular Theory of Matter,* p. 138. For a brief discussion of Thomson's reasons for denying the existence of nonpolar bonds see Kohler, "The Origin," p. 354.

26. J. J. Thomson, "The Forces between Atoms and Chemical Affinity," *Philosophical Magazine,* 6th series, 27 (1914): 757–789.

cleus of the first.[27] The important innovation of Thomson's paper that proved significant for Lewis's further elaboration of his cubic atom was the idea that a pair of electrons was involved in a nonpolar bond.[28]

Another crucial development that formed part of Lewis's problem situation was Alfred Parson's proposal in 1915 of an electron endowed with magnetic properties.[29] Parson portrayed the electron as a revolving circular band of negative electricity, which, as a result of its rotation, had a magnetic moment and was, thus, called "magneton." Parson had not been the first to associate magnetic properties with the electron. Ten years before Parson's proposal and in a different context the French physicist Paul Langevin (1872–1946) had considered the electron the ultimate magnetic particle.[30] The important innovation of Parson, however, was the utilization of his magnetic electron for understanding the process of chemical bonding. Furthermore, Parson's magnetons were part of a larger picture of the structure of the atom, which had considerable affinities with both Lewis's 1902 hypothesis of a cubic atom and Thomson's 1904 theory of a "plum pudding" atom. An atom with a cubic structure was supposed to be the result of the magnetic interactions between the magnetons, which were embedded, as in Thomson's model, in a sphere of positive electricity.[31]

Parson had shown his paper, prior to its publication, to Lewis and, thus, stimulated Lewis to revive his 1902 theory. The most important aspect of Parson's paper for the development of Lewis's theory was a new mechanism of chemical binding based on the magnetic properties of the electron. As I mentioned above, the then dominant conception of chemical bonding portrayed the formation of chemical compounds as due exclusively to the transfer of electrons from one atom to another and the concomitant electrostatic attraction between the oppositely charged atoms. That theory of chemical bonding, however, could not explain the formation of molecules consisting of neutral parts, like the hydrogen molecule ($H_2$) and the majority of organic

27. At that point Thomson had accepted Rutherford's "planetary" model of the atom.

28. For a detailed discussion of the origins and characteristics of Thomson's two-electron bond see Kohler, "The Origin," pp. 358–361.

29. See A. L. Parson, "A Magneton Theory of the Structure of the Atom," *Smithsonian Miscellaneous Collections*, 65, no. 11 (1915): 1–80. See also A. Baracca, "Early Proposal of an Intrinsic Magnetic Moment of the Electron in Chemistry and Magnetism (1915–1921) before the Papers of Goudsmit and Uhlenbeck," *Rivista di storia della scienza*, 3, no. 3 (1986): 353–374.

30. P. Langevin, "Magnétisme et Théorie des Électrons," *Annales de Chimie et de Physique*, 5 (1905): 70–127.

31. I added the qualifier "supposed to be" because Parson was unable to provide a strict mathematical demonstration of the cubic arrangement of his magnetons. Cf. Kohler, "The Origin," p. 365.

compounds. Because of these difficulties the view that the only mechanism of chemical bonding was a process of electron transfer had been already challenged. A group of chemists at Berkeley, under the leadership of Lewis, "suggested that there are two distinct types of union between atoms: polar, in which an electron has passed from one atom to the other, and nonpolar, in which there is no motion of an electron." [32] No physical mechanism was proposed, however, for the formation of nonpolar bonds. It was this elusive mechanism that Parson supplied in his 1915 paper.

Like J. J. Thomson, Lewis, and his colleagues, Parson admitted the existence of two kinds of bonds, polar and nonpolar. Unlike polar bonds, which were formed by magneton transfer, nonpolar bonds consisted of a pair of magnetons which was formed as a result of the magnetic attraction between two magnetic electrons, an attraction strong enough to overcome the electrostatic repulsion between the negatively charged particles. It was, thus, the magnetic character of Parson's electron that enabled the formation of nonpolar bonds. Parson's idea of a magnetic electron was a significant resource for the subsequent development of Lewis's theory of atomic structure.

The most innovative aspect of that development was Lewis's idea that the chemical bond between two atoms in a molecule consisted of a *pair* of electrons that did not belong exclusively to either of the combining atoms but were, instead, shared by both of them. [33] The origin of Lewis's conception of the shared pair bond is a question that has not been definitively answered. One thing that "seems fairly clear [is] that Lewis derived the shared pair bond in some way from the rich and suggestive speculations of Thomson . . . and Parson on the nonpolar bond." [34] Lewis had already introduced the distinction between polar and nonpolar bonds in 1913, but he had not offered any physical mechanism that could account for the formation of nonpolar bonds. Such a mechanism was provided by Thomson in 1914 and by Parson in 1915. The problem then facing Lewis was to translate Thomson's and Parson's conceptions into the terms of his 1902 cubic model of the atom and to retain only those aspects of their theories that could fit into his own theory of atomic structure. I will discuss Lewis's seminal paper "The Atom and the Molecule" (1916) and the problems that it tried to address, to clarify his debt to Thomson's and Parson's ideas, while at the same time, exhibiting the features of

32. W. C. Bray and G. Branch, "Valence and Tautomerism," *Journal of the American Chemical Society*, 35 (1913): 1440–1447, on p. 1443; quoted in Kohler, "The Origin," p. 357.

33. This idea turned out to be Lewis's "lasting contribution to the explanation of valency." W. G. Palmer, *A History of the Concept of Valency to 1930* (Cambridge: Cambridge Univ. Press, 1965), p. 135.

34. Kohler, "The Origin," p. 371.

Lewis's representation of the electron that conflicted with those ascribed to it by the physicists.

In the beginning of that paper Lewis acknowledged the similarity between his ideas and those found in J. J. Thomson's "extremely interesting paper on the 'Forces between Atoms and Chemical Affinity' [1914] in which he reached conclusions in striking accord with my own."[35] Furthermore, he offered as a reason for "presenting this theory [of atomic structure] briefly in the present paper . . . [that], while it bears much resemblance to some current theories of the atom, it shows some radical points of departure from them" (p. 764). Lewis's cubic model of the atom was supposedly developed a "number of years ago, to account for the striking fact which has become known as Abegg's law of valence and countervalence, and according to which the total difference between the maximum negative and positive valences or polar numbers of an element is frequently eight and is in no case more than eight" (p. 767). This statement on the origin of the theory of the cubic atom is puzzling in view of the fact that "Abegg's law" was proposed by Abegg in 1904, and Lewis's theory in 1902. This puzzle, to the best of my knowledge, has not been addressed in the historical literature on the origins of Lewis's theory, and his original motivation remains a mystery.

Whatever its origins, the cubic model of the atom seemed to Lewis "more probable intrinsically than some of the other theories of atomic structure which have been proposed" (p. 767). Lewis's atom consisted of a core of negative and positive charges that did not partake in chemical phenomena and an outer shell of electrons that governed chemical changes. The net positive charge of the inner core was balanced by the negative charge of the outer shell so that the atom as a whole was electrically neutral. The most important aspect of that model, for my purposes, was that it portrayed the electrons as static particles, which, moreover, did not strictly obey Coulomb's law. As Lewis remarked, "Electric forces between particles which are very close together do not obey the simple law of inverse squares which holds at greater distances" (p. 768).

It is important to reconstruct the reasoning that led Lewis to abandon the unrestricted validity of Coulomb's law in order to see in what respects the physicists' electron, invariably subject to Coulomb's inverse square formula, was an obstacle for the understanding of chemical phenomena. There were two ways, according to Lewis, "in which one body can be held by another. It may, owing to a force of attraction, be drawn toward the second body until

---

35. G. N. Lewis, "The Atom and the Molecule," *Journal of the American Chemical Society*, 38 (1916): 762–785, on p. 763.

this force is gradually offset by a more rapidly increasing force of repulsion. In this case it comes to rest at a point where the net attraction or repulsion is zero." (p. 772).[36] This was exactly the way in which electrons were held together in the atom, presumably under the attractive-repulsive action of the positive charges that were also in the atom and under their own mutual repulsion-attraction.[37] It should be emphasized that Lewis kept silent about the nature and arrangement of the positive charges inside an atom, and he did not offer any mathematically formulated alternative to Coulomb's law. As a result he was unable to show that the cubic structure and the stability of his atom followed from the collective interaction between electrons and positive charges. The qualitative nature of Lewis's model of the atom explains, to some degree, why it was not taken very seriously by the physicists.[38]

The second way in which the electrons could be held together in the atom was the physicists' way, most notably Rutherford's. According to that model, electrons were revolving around a massive, positively charged nucleus under the action of an inverse square attractive force. Rutherford did not make any attempt to employ his model in order to explain the chemical properties of the elements, and, not surprisingly, his theory did not appeal to chemists. As Lewis remarked, the so-called planetary model of the atom "seems inadequate to explain even the simplest chemical properties of the atom, and I imagine it has been introduced only for the sake of maintaining the laws of electromagnetics which are known to be valid at large distances."[39] Furthermore, the advantage of "maintaining the laws of electromagnetics" was not

36. This was not a novel idea. It had been suggested by R. J. Boscovich as early as 1763. See L. P. Williams, *Michael Faraday: A Biography* (1965; repr., New York: Da Capo Press, 1987), pp. 73–75.

37. Lewis said explicitly only that Coulomb's law failed at small distances vis-à-vis the interaction between oppositely charged particles. However, we can infer that he held the same to be true for similarly charged particles, since he argued that in atoms of large atomic volume electrons are at a sufficient distance from one another to mutually repel each other.

38. See Kragh, "Bohr's Atomic Theory and the Chemists," pp. 474–477. In 1917 Lewis suggested a mathematical alternative to Coulomb's law, without, however, deducing the cubic arrangement of electrons from it. See G. N. Lewis, "The Static Atom," *Science,* 46 (28 Sept. 1917): 297–302, on p. 299.

39. Lewis, "The Atom and the Molecule," p. 772. Lewis did not elaborate further the inadequacy of Rutherford's model vis-à-vis chemical phenomena. Several years later he was more specific: "If the electrons are to be regarded as taking an essential part in the process of binding atom to atom in the molecule, it seemed impossible that they could be actuated by the simple laws of force, and travelling in the orbits, required by the planetary theory. The permanence of atomic arrangements . . . is one of the most striking of chemical phenomena. . . . It appears inconceivable that these permanent . . . configurations could result from the simple law of force embodied in Coulomb's law." Lewis, *Valence,* pp. 55–56.

preserved in the most sophisticated among the planetary models, namely, Bohr's semiclassical model. As we saw in chapter 5, Bohr's dynamic electron violated classical electromagnetic theory in two respects: First, it was influenced by Coulomb forces only when "the angular momentum of the electron round the nucleus in a stationary state of the system is equal to an entire multiple of a universal value [$h/2\pi$], independent of the charge on the nucleus."[40] Second, in these "stationary" states the electron violated classical electrodynamics, since it did not emit radiation.

Lewis diagnosed another respect in which Bohr's dynamic electron was incompatible with contemporary electromagnetic theory. The motion of Bohr's electrons "produces no effect upon external charges. Now this is not only inconsistent with the accepted laws of electromagnetics but, I may add, is logically objectionable, for that state of motion which produces no physical effect whatsoever may better be called a state of rest."[41] It is not clear from the above passage what, exactly, Lewis's objection to Bohr's theory was. Lewis, however, developed that objection in a paper that was published the following year.[42] In that paper he exposed a hidden consequence of Bohr's model of the atom, which can be seen clearly by means of the following thought experiment:[43]

> Let us . . . represent a hydrogen atom according to Bohr with an electron in the first orbit [by a circle whose diameter is XX′; see fig. 7.2], that is to say in the most stable state, and let us represent by AA′ a small wire which may be brought near to the hydrogen atom. Now if the electron in the orbit exerts any sort of electrical force at a distance, when the electron is in position X there will be a slight flow of positive electricity in the wire toward A, and when the electron is at X′ there will be a slight flow toward A′. Indeed at any finite distance of the wire from the atom there should be set up in the wire a finite alternating current which would continue indefinitely. Such a current should generate heat, but since the atom is supposed to be in the state of lowest possible energy there appears to be no source from which the heat could originate. In other words, we must conclude either that such an alternating current is not produced or that it is produced but meets with no ohmic resistance.

40. N. Bohr, *On the Constitution of Atoms and Molecules* (Copenhagen: Munksgaard, 1963), p. 15.

41. Lewis, "The Atom and the Molecule," p. 773.

42. Lewis, "The Static Atom," on p. 299.

43. The description that follows is from Lewis, *Valence*, pp. 50–51.

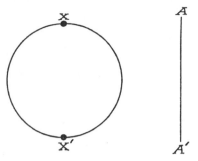

**Fig. 7.2** "Illustrating a doubtful point concerning the Bohr atom." (From G. N. Lewis, *Valence and the Structure of Atoms and Molecules* [New York: Chemical Catalog Co., 1923], p. 51.)

Lewis eliminated the second alternative on experimental grounds and concluded that "an electron, in a Bohr orbit, exerts upon other electrons no force which depends upon its position in the orbit." [44] That conclusion was an essential step toward the reconciliation of Bohr's dynamic and Lewis's static electron and will be further discussed below.

Thus, in Lewis's view, the physicists' dynamic electron was not only incompatible with chemical data but also objectionable on "logical" grounds. Moreover, he claimed that the predictive success of Bohr's theory vis-à-vis the lines of the hydrogen spectrum could be duplicated by his own model if one assumed "that an electron may be held in the atom in stable equilibrium in a series of different positions, each of which having definite constraints, corresponds to a definite frequency of the electron." [45] It is evident that Lewis proposed a totally classical picture of the electron that dispensed with quantum conditions. As we will see, this was one of the sources of resistance to Lewis's proposal by those physicists that were engaged with the development of quantum theory. [46]

So far we have identified three characteristics of Lewis's model that might have led to its negative reception by the physics community: first, its predom-

44. Ibid., p. 51.

45. Lewis, "The Atom and the Molecule," p. 773. He repeated that claim, which was essentially a promissory note, in his "The Static Atom," pp. 297, 299.

46. Bohr, for instance, dismissed rather hastily the Lewis atom and its elaboration by Langmuir. See N. Bohr, "The Structure of the Atom," Nobel Lecture, 11 Dec. 1922, in *Nobel Lectures: Physics, 1922–1941* (Amsterdam: Elsevier, 1965), pp. 7–43, on p. 35. Cf. Mary Jo Nye, "Remodeling a Classic: The Electron in Organic Chemistry, 1900–1940," in Buchwald and Warwick (eds.), *Histories of the Electron*, pp. 339–361, on p. 344.

**Fig. 7.3** A representation of the chemical bond as a pair of electrons. (From Lewis, "The Atom and the Molecule," p. 775.)

inantly qualitative, nonmathematical formulation;[47] second, its disregard for quantum conditions; and third, its violation of the strict validity of Coulomb's law. That final feature of Lewis's theory was closely related with the magnetic properties of the electron. If the electron was a tiny magnet, as Parson had suggested in 1915, the configuration of electrons inside the atom would be the product of the operation of magnetic forces, in addition to the operation of Coulomb-like electric forces. The repulsive, inverse square force between two electrons would be counteracted by an attractive, magnetic force. Assuming the existence of such forces, "Parson was led entirely independently to the conclusion which I have stated above, namely that the most stable condition for the atomic shell is the one in which eight electrons are held at the corners of a cube."[48] Thus, Parson's attribution of magnetic properties to the electron explained why electrons did not strictly obey Coulomb's law and thus justified on physical grounds a crucial aspect of Lewis's theory. It should be noted, however, that, contrary to Lewis's claims, Parson had not been able to demonstrate mathematically that a cubic arrangement of electrons inside the atom followed from the combined action of electric and magnetic forces.[49]

The idea of a magnetic electron also enabled another crucial innovation that Lewis introduced in 1916, namely, the conception of the chemical bond as consisting of a pair of electrons shared between the combined atoms.[50] The two electrons occupied the vertices of an edge that was common to both cubic atoms (see fig. 7.3). A chemical bond of this kind would not be possible if the only force acting between two electrons were electrostatic repulsion.

47. It is interesting that Bohr's theory of the periodic system, which was also qualitative, was not received very positively by quantum physicists. See H. Kragh, "Niels Bohr's Second Atomic Theory," *Historical Studies in the Physical Sciences,* 10 (1979): 123–186, esp. pp. 161–162.

48. Lewis, "The Atom and the Molecule," p. 774.

49. See Parson, "A Magneton," pp. 18–19.

50. Lewis's theory of the shared pair bond could explain the formation of both polar and nonpolar molecules through the same single mechanism and solved problems such as the structure of ammonia and the ammonium ion, "which has proved extremely embarrassing to a number of theories of valence." Lewis, "The Atom and the Molecule," p. 777.

**Fig. 7.4** A transformation of a cubic atom into a tetrahedral one. (From Lewis, "The Atom and the Molecule," p. 780.)

Some other force was necessary to overcome that repulsion and draw the two particles together. In 1916 Lewis suggested that the force in question would be magnetic "if the magneton theory is correct."[51] In the following year, however, he admitted that "we do not at present understand" the nature of the forces that hold together the electrons in a bond.[52]

Finally, the magnetic electron was essential for the representation of the triple bond, that is, the bond between two atoms that shared three pairs of electrons. From a geometrical point of view two cubic atoms could never share three pairs of electrons. The pictorial representation of the triple bond requires that the combined atoms possess a tetrahedral structure. A cubic structure can be transformed into a tetrahedral one if "each pair of electrons [that occupy the vertices of the cube] has a tendency to be drawn together, perhaps by magnetic force," a tendency that can be reasonably assumed "in such very small atoms as that of carbon" (see fig. 7.4).[53] The tetrahedral structure, moreover, "explain[s] the phenomenon of free mobility about a single bond which must always be assumed in stereochemistry."[54]

We have seen how Lewis's representation of the electron emerged out of his problem situation, which consisted of a series of chemical problems along with a set of methodological constraints that determined what counted as an acceptable solution. The difference between these constraints and those that were part of the physicists' problem situation helps to explain the divergence between the physicists' and the chemists' representations of the electron. It was clear, however, at least to Lewis, that both physicists and chemists were

51. Ibid., p. 780.

52. Lewis, "The Static Atom," p. 302. Some years later, having conceded the reality of electronic orbits, he suggested that the pairing of electrons is due to the magnetic coupling of their orbits. See G. N. Lewis, "Introductory Address: Valence and the Electron," *Transactions of the Faraday Society*, 19 (1923–1924): 452–458, on p. 453. Cf. J. W. Servos, *Physical Chemistry from Ostwald to Pauling: The Making of a Science in America* (Princeton: Princeton Univ. Press, 1990), p. 282.

53. Lewis, "The Atom and the Molecule," p. 780.

54. Ibid.

dealing with the same entities.[55] Given that belief, the divergence in the representations of the electron was not acceptable and their reconciliation was imperative. The required synthesis had to take into account evidence from all the experimental situations involving the electron, regardless of their (physical or chemical) origin. Experimental results from physics became relevant to the evaluation of chemical theory and vice versa. The electron qua theoretical entity drew together two prima facie disparate domains.

Before discussing the further development and eventual resolution of the conflict between the chemists' and the physicists' representations of the electron it would be helpful to recapitulate their contrasting characteristics.

## 3. A Recapitulation of the Conflict

Chemists, on the one hand, conceived the electron as a classical, static particle endowed with magnetic properties and not always subject to Coulomb forces. Physicists, on the other hand, regarded the electron as a dynamic non-magnetic particle, endowed with quantum properties, and constantly in very high-speed elliptical motion around a positively charged nucleus under the influence of Coulomb-like attractive forces.[56] We have already discussed the problem situation out of which the chemists' picture of the electron emerged. Central to that problem situation was the requirement of a chemical bond consisting of a pair of electrons that was shared by the atoms that constituted a molecule. From a geometrical point of view, a static electron could easily fulfill that requirement. A dynamic electron, on the other hand, that revolved around the atomic nucleus could not perform the role that was required for the interpretation of chemical phenomena; that is, it could not belong simultaneously to more than one atoms.[57]

55. See the preface of his *Valence*. Even though he referred to the atom and the molecule, his remarks apply equally to the electron. Cf. K. Gavroglu and A. Simoes, "The Americans, the Germans, and the Beginnings of Quantum Chemistry: The Confluence of Diverging Traditions," *Historical Studies in the Physical Sciences,* 25, no. 1 (1994): 47–110, on p. 50.

56. Some qualifications must be added to this generalization. First, there were some physicists prior to Goudsmit and Uhlenbeck, most notably Langevin (1904) and A. H. Compton (circa 1917), who regarded the electron as the ultimate magnetic particle. (See chapter 8 and Baracca, "Early Proposal," pp. 359ff.) Second, the quantum conception of the electron was not accepted by every physicist. J. J. Thomson and several other British physicists continued to advocate classical models of the electron well into the 1920s. Furthermore, there was considerable German opposition to quantum conceptions. My discussion of the physicists' electron is very schematic and presupposes material that was examined in previous chapters.

57. Millikan, however, argued that a dynamic electron was compatible with the data of chemistry if one assumed that the electronic "orbits are properly distributed in space." Millikan, "Atomism," p. 1411. Cf. Millikan, "The Physicist's Present Conception of an Atom," p. 474.

There were many arguments, however, that supported the notion of a dynamic electron.[58] First, Bohr had derived from the postulates of his atomic theory in 1913 a formula for the Rydberg constant that predicted very accurately the corresponding experimental value.

Second, Bohr's theory had predicted that the spectral lines due to ionized helium ($He^+$) should be in a position slightly different from that of the corresponding hydrogen lines. The energy of an electron in an orbit, as calculated by Bohr's theory, depended on the reduced mass $mM/(M + m)$ (where $m$ is the mass of the electron and $M$ is the mass of the nucleus). Given that the nuclear mass of helium is four times that of hydrogen, the energy of an electron in a hydrogen atom and the corresponding energy of an electron in a helium atom should be slightly different. That difference would give rise to a small difference between the positions of the hydrogen and the helium spectral lines. Again, the predicted difference agreed exactly with the one obtained experimentally.[59]

Third, Sommerfeld had shown in 1916 that "the consideration of the relativistic change of mass of the electron leads to slightly different energies for orbits of equal major axis but different eccentricity, so that each energy level will show an $n$-fold fine structure."[60] An observable consequence of this relativistic phenomenon was that the hydrogen spectral line corresponding to a transition from the second energy level to the first would exhibit a doublet structure. This prediction, obtained by Sommerfeld's relativistic correction of Bohr's model of the atom, was experimentally confirmed. Fourth, Bohr's model of the atom was successfully employed by Epstein to account for the Stark effect—the splitting of spectral lines when their source is placed under the influence of an electric field.

As a result of these and other successes predicated on the notion of a dynamic electron Millikan declared in 1924 that "the theory of non-radiating electronic orbits is one of the established truths of modern physics. . . . [a truth that is] as much hypothetical to-day as is the theory of the rotation of the earth

---

58. For a detailed list see Millikan, "Atomism," pp. 1411–1414. Cf. his "Radiation and Atomic Structure," *Science*, 45, no. 1162 (6 Apr. 1917): 321–330, on pp. 324, 327.

59. Cf. A. Pais, *Inward Bound* (New York: Oxford Univ. Press, 1988), pp. 201–202. Millikan did not mention that the prediction of Bohr's theory concerned *ionized helium* and gave the impression that it referred instead to the helium atom. (See Millikan, "Atomism," p. 1412.) However, the spectrum of the helium *atom* remained a puzzle throughout the development of the old quantum theory and was explained only after the advent of quantum mechanics.

60. L. Pauling and S. Goudsmit, *The Structure of Line Spectra* (New York: McGraw-Hill, 1930), p. 17.

upon its axis or of the planets around the sun."[61] Those successes were too dramatic to be ignored by the chemists. Some of them came around to the view that "the atom of the physicist and the chemist is a dynamic atom, and theories based on static electrons must give place to it, no matter how difficult the conception of the dynamic atom may be for the mechanism of chemical combination."[62] Lewis himself, several years after his proposal of a static model of the atom, commented on the "remarkable quantitative success" of "the brilliant theory proposed by Bohr (1913)."[63] There were two available options: either to give a mathematical formulation of Lewis's theory of the static atom and try to emulate the quantitative achievements of Bohr's theory or to attempt to reconcile the two theories through an appropriate reinterpretation of both. The former option was followed by Irving Langmuir; the latter by Lewis himself.

## 4. Irving Langmuir's Elaboration of Lewis's Ideas

In 1903 Irving Langmuir (1881–1957) obtained his degree in metallurgical engineering from the School of Mines at Columbia University.[64] It seems doubtful that he was interested in metallurgy per se. Rather, he entered the School of Mines because, as he remarked later, "[t]he course was strong in chemistry" and, moreover, "[it] had more physics than the chemical course, and more mathematics than the course in physics—and I wanted all three."[65] After the completion of his undergraduate studies, he went to Germany and worked with Walther Nernst "on the dissociation of various gases by a glowing platinum wire, research that became the topic of his 1906 doctoral dissertation."[66] In 1906 he returned to the United States and began his academic

61. Millikan, "Atomism," p. 1414. Millikan expounded the same view before various audiences. See his "Radiation and Atomic Structure"; and his lecture to the American Chemical Society on 22 Apr. 1924 ("The Physicist's Present Conception of an Atom"). Cf. also his *The Electron: Its Isolation and Measurement and the Determination of Some of Its Properties,* 2nd ed. (Chicago: Univ. of Chicago Press, 1924; 1st ed., 1917), p. 221.

62. Smith, *Chemistry and Atomic Structure,* p. 163. Later in the book, however, Main Smith admitted that the issue of "the type of atom, dynamic or static, to be utilised in chemistry" remained open among chemists. See ibid., p. 181.

63. Lewis, *Valence,* pp. 52 and 43, respectively.

64. For biographical information on Langmuir see A. Rosenfeld, "The Quintessence of Irving Langmuir," in *The Collected Works of Irving Langmuir,* ed. C. G. Suits, 12 vols. (Oxford: Pergamon Press, 1962), vol. 12, pp. 3–229. See also Charles Süsskind's short biographical essay, "Langmuir, Irving," in Gillispie (ed.), *Dictionary of Scientific Biography,* vol. 8, pp. 22–25.

65. Quoted in Rosenfeld, "The Quintessence," p. 42.

66. Süsskind, "Langmuir," p. 23.

career at the Stevens Institute of Technology, in Hoboken, New Jersey. His teaching load at the Stevens Institute was very heavy and left him little time for research. So, in 1909, after a brief and unhappy period at Stevens Tech, he accepted an offer from the new research laboratory of General Electric in Schenectady, New York. The laboratory in Schenectady provided an ideal research environment, and Langmuir stayed there until his retirement in 1950. It is indicative of the research climate in that laboratory that its director, Willis R. Whitney, instructed Langmuir to "follow any line of inquiry you like."[67] As one would expect, Langmuir followed Whitney's advice, and, thus, it should occasion no surprise that the study of atomic structure (a subject in the frontiers of "basic" research without immediate technological applications) became Langmuir's central research interest in the period from 1919 to 1921.

According to Robert Kohler, who has reconstructed in detail the origins of Langmuir's concern with atomic structure, Langmuir had been interested in atomic models since 1916.[68] As his notebooks from that period reveal, he "thought that there were better theories than Bohr's, such as Alfred Parson's magneton theory which he had just been studying."[69] Furthermore, he agreed with G. N. Lewis that Bohr's model of the atom "ignored the facts of stereochemistry . . . [and] was designed simply to 'save Coulomb's Law' at short distances."[70] His negative appraisal of Bohr's atomic theory contrasted with his favorable view of Lewis's static model of the atom. As he mentioned a few years later in a letter to Lewis, "When I read your paper on the 'Atom and the Molecule' in 1916, I was immediately struck by the very fundamental nature of the ideas you presented, and of their splendid agreement with the general facts of chemistry, so I very soon began to look upon all chemical phenomena from the viewpoint that you presented."[71] Thus, Lewis's atomic theory and Parson's proposal of an electron endowed with magnetic properties were important resources upon which Langmuir drew for the development of his own theory of atomic structure. A further element of his problem situation was his hope to create "a new chemistry, a deductive chemistry, one

67. Quoted in Rosenfeld, "The Quintessence," p. 77.

68. R. E. Kohler, "Irving Langmuir and the 'Octet' Theory of Valence," *Historical Studies in the Physical Sciences,* 4 (1974): 39–87.

69. Ibid., p. 47.

70. Ibid.

71. I. Langmuir to G. N. Lewis, 22 Apr. 1919; published in Kohler, "Irving Langmuir and the 'Octet' Theory of Valence," pp. 68–70, on p. 68. Cf. also Langmuir's remark that "[t]his work of Lewis [1916] has been the basis and the inspiration of my work on valence and atomic structure." I. Langmuir, "The Structure of Atoms and Its Bearing on Chemical Valence," *Journal of Industrial and Engineering Chemistry,* 12, no. 4 (1920): 386; repr. in his *Collected Works,* vol. 6, pp. 93–100, on p. 98.

in which we can reason out chemical relationships without falling back on chemical intuitions."[72]

Langmuir, however, had to suspend his interest in the riddle of atomic structure because of his engagement with war-related research. His involvement with war-related projects lasted from February 1917 (two months before the entry of the United States into the war) till the end of the war (November 1918).[73] In January 1919 Saul Dushmann, one of his colleagues at General Electric, organized a colloquium that revived Langmuir's interest in Lewis's 1916 theory and the structure of the atom.[74] In his notebook from that period "Langmuir's theory of the atom emerges, differing in only some refinements from the form in which it was published."[75] Thus, since the differences between the published and the unpublished record are minor in this case, I will discuss the published version of Langmuir's theory.

A preliminary version of his theory appeared in the 1919 issue of the *Proceedings of the National Academy of Sciences*.[76] In the beginning of his article Langmuir acknowledged that "[t]his theory, which assumes an atom of the Rutherford type, . . . is essentially an extension of Lewis' theory of the 'cubical atom.'"[77] He continued by laying out his theory in terms of eight postulates, a form of exposition that betrays his ideal of a "deductive chemistry." The first postulate asserted that "[t]he electrons in atoms are *either* stationary *or* rotate, revolve *or* oscillate about definite positions in the atom" (emphasis added).[78] It should be emphasized that Langmuir's model of the atom, in contrast with Lewis's, was not based on the assumption of a completely stationary electron. Rather, it was compatible with the supposition of a dynamic electron. Langmuir modified Lewis's theory in order to reconcile it with Bohr's model of the atom, which was based on the assumption of a rapidly moving electron. As he explained in a subsequent paper, "Bohr, Sommerfeld, and others have developed an extensive and very successful theory of spectra upon the hypothesis that the electrons in atoms are in rapid rotation in plane orbits about the nucleus. . . . Stark, Parson, and G. N. Lewis on the other hand,

72. Quoted in Rosenfeld, "The Quintessence," p. 109.

73. See Rosenfeld, "The Quintessence," pp. 102–108.

74. See Kohler, "Irving Langmuir and the 'Octet' Theory of Valence," p. 51.

75. Ibid., p. 52.

76. I. Langmuir, "The Structure of Atoms and the Octet Theory of Valence," *Proceedings of the National Academy of Sciences,* 5 (1919): 252–259; repr. in his *Collected Works,* vol. 6, pp. 1–8.

77. Ibid. (in *Collected Works*), p. 1.

78. Ibid.

starting from chemical evidence, have assumed that the electrons are stationary in position. . . . The two theories can, however, be reconciled if we consider that the electrons . . . rotate about certain definite positions of the atom which are distributed symmetrically in three dimensions." [79] The first postulate also included a symmetry principle, according to which "[t]he electrons in the . . . atoms . . . of the inert gases, have positions symmetrical with respect to a plane called the equatorial plane, passing through the nucleus at the center of the atom." [80]

The second postulate stated that the electrons in atoms are distributed in successive "concentric (nearly) spherical shells, all of equal thickness." Moreover, the "mean radii of the shells form an arithmetic series 1, 2, 3, 4, and the effective areas are in the ratios $1 : 2^2 : 3^2 : 4^2$" (p. 1). Each of those shells was divided, according to the third postulate, "into cellular spaces or cells occupying equal areas in their respective shells and distributed over the surface of the shells according to the symmetry required by postulate 1" (p. 1). Thus, the first shell should contain two cells (one above and one below the equatorial plane), the second, having an area four times larger than the first, should contain eight cells, and so on.

The cells in the first shell are occupied, according to the fourth postulate, by a single electron, while cells in successive shells can contain, at most, two electrons. The latter are allowed to have two electrons only if every other cell in the same shell is occupied by, at least, one electron. Furthermore, electrons start to populate a shell only after all previous shells get "their full quotas of electrons" (p. 1).

The fifth postulate capitalizes on the magnetic properties of the electron to determine the arrangement of electrons in the outermost shell. These electrons are subject to a magnetic attraction from the "underlying" electrons. When the number of the outside electrons is small, this magnetic attraction predominates over the electrostatic repulsion (also due to the underlying electrons) and determines their arrangement. When, on the other hand, there are many outside electrons, the electrostatic repulsion becomes dominant and pushes them "as far as possible from the underlying ones" (p. 2).

Postulates 6 and 7 concern the stability of electronic arrangements. The

79. Langmuir, "The Structure of Atoms and Its Bearing on Chemical Valence." Cf. also his "Theories of Atomic Structure," *Nature,* 105 (1920): 261; repr. in his *Collected Works,* vol. 6, pp. 104–105.

80. Langmuir, "The Structure of Atoms and the Octet Theory of Valence" (in *Collected Works*), p. 1.

former asserts that a pair of electrons (e.g., in the helium atom) is the most stable grouping of electrons, while, according to the latter, "The next most stable arrangement of electrons is the octet [a group of eight electrons]" (p. 2). Finally, the eighth postulate introduces the idea that an electron (or a pair of electrons) can be shared by, at most, two octets.

By means of those postulates Langmuir could explain the chemical properties of all the elements. Whereas Lewis's theory "could be used to explain the *physical* and chemical properties of the first twenty elements . . . I have been able to extend the theory to cover all the elements and to give a theory of the mechanism of chemical combination which gives a rational explanation of most of the very varied and extensive data underlying the Periodic Table and the different theories of valence."[81] The breadth and empirical success of Langmuir's theory, as well as the fact that he was equally at home in academic and in industrial chemistry, facilitated the reception of Lewis's ideas by the community of chemists, which had a significant industrial component.[82]

Langmuir continued to develop the notion of a static electron. In 1921 he suggested that the reason that the electrons did not fall into the nucleus was a repulsive "quantum force" which balanced the attractive force between them and the nucleus. By choosing ad hoc the magnitude of that force, he obtained a value of the distance between the electrons and the nucleus that was "identical with that for the radius of the orbit in Bohr's theory."[83] Thus, Langmuir's theory reproduced the predictions of Bohr's theory "without assuming electrons moving about the nucleus."[84] However, this turned out to be his last paper on atomic structure. The reason might have been the proposal of Bohr's second atomic theory (1921), which was also very successful vis-à-vis the periodic table without relying on ideas that were alien to the spirit of the quantum theory.[85]

81. I. Langmuir to E. Rutherford, 18 May 1919, Cambridge Univ. Library, Add 7653/L16.

82. Cf. Langmuir's remark that "Lewis' treatment of this subject was rather brief and perhaps for this reason it does not seem to have met with the general acceptance which it deserves." I. Langmuir, "The Arrangement of Electrons in Atoms and Molecules," *Journal of the American Chemical Society,* 41, no. 6 (1919): 868–934; repr. in his *Collected Works,* vol. 6, pp. 9–73, on p. 27. See also Kohler, "Irving Langmuir and the 'Octet' Theory of Valence," p. 57; and Russell, *The History of Valency,* p. 285.

83. I. Langmuir, "The Structure of the Static Atom," *Science,* 53, no. 1369 (1921): 290–293; in his *Collected Works,* vol. 6, pp. 124–127, on p. 125. Cf. Kohler, "Irving Langmuir and the 'Octet' Theory of Valence," p. 65; and Kragh, "Bohr's Atomic Theory and the Chemists," pp. 478–479.

84. Langmuir, *Collected Works,* vol. 6, p. 125.

85. Cf. Kohler, "Irving Langmuir and the 'Octet' Theory of Valence," pp. 65–66.

## 5. Concluding Remarks

This chapter capitalized on "problem situations" to account for the conflicting representations of the electron that were put forward by physicists and chemists. The different problems that occupied the two groups and their different methodological perspectives gave rise to the conflict in question. Some attempts were made to resolve that conflict. Norman Campbell, for instance, suggested that the notion of electronic orbits was not essential to Bohr's theory.[86] Lewis himself came to think of the two representations of atomic structure as "completely reconcilable," provided that "an orbit as a whole and not the electron in some one position within the orbit . . . is the building stone of atomic and molecular structure."[87] However, the conflict would not be fully resolved before the advent of the exclusion principle, spin, and eventually quantum mechanics. The exclusion principle gave a satisfactory explanation of the facts of the periodic table that superseded all previous accounts. Spin, in its original classical interpretation, substantiated the chemists' relatively vague conception of the electron as the ultimate magnetic particle. And quantum mechanics, at least in its dominant interpretation, dispensed with the notion of electronic orbits. Thus, aspects of both conceptions were incorporated into the quantum-mechanical representation of the electron.[88]

We can now address the questions that were raised in the introductory section of this chapter. We have seen that both chemists and physicists felt obliged to account for experimental evidence that was not produced within their discipline. This suggests that a scientific theory has to account for all the data that are deemed part of its explanatory domain. The latter may encompass experimental results from more than one discipline. The relevance of a piece of experimental evidence to the evaluation of a theory is, in turn, determined by the entities that the theory employs. If, for instance, a chemical theory appropriates an entity that is part of the ontology of physics, then two possibilities arise. Either the original representation of the entity remains un-

---

86. See N. R. Campbell, "Atomic Structure," *Nature,* 106 (1920): 408–409. Bohr, however, thought otherwise. See N. Bohr, "Atomic Structure," *Nature,* 107 (1921): 104–107. Cf. E. M. MacKinnon, *Scientific Explanation and Atomic Physics* (Chicago: Univ. of Chicago Press, 1982), pp. 177–179.

87. Lewis, "Introductory Address: Valence and the Electron," p. 452. Cf. Servos, *Physical Chemistry from Ostwald to Pauling,* p. 281.

88. The exclusion principle and spin will be discussed in the following chapter. For the impact of quantum mechanics on the resolution of the conflict see T. Arabatzis and K. Gavroglu, "The Chemists' Electron," *European Journal of Physics,* 18 (1997): 150–163; and Gavroglu, "The Physicists' Electron and Its Appropriation by the Chemists."

affected by the appropriation in question, or it is transformed so as to be adequate to its new explanatory role. In the latter case, the new chemical representation has also to reproduce the explanatory successes of its physical counterpart. The entity's previous "writings" have to be accommodated by its novel representation. Thus, experimental evidence is deemed relevant to the evaluation of a theory, even if the evidence in question comes from a different discipline (physics) than the discipline where the theory belongs (chemistry), when it is attributed to the entity that the theory talks about (the electron). So it is the purported entity, presumed to be the same across disciplinary boundaries, that coordinates the evidence and the theory.

This function of theoretical entities indicates their relative independence from the disciplinary context in which their putative counterparts in nature are probed. In the case in hand, the chemists who employed a representation of the electron for theoretical purposes did not question the electron's previous experimental manifestations, that is, those attributed to it by the physicists. Chemists like Lewis and Langmuir even felt obliged to account, by means of the alternative representations of the electron they put forward, for its physical manifestations. Furthermore, physicists and chemists did not dispute some of the electron's key properties. In particular, the chemists did not challenge the properties of the electron that had been experimentally determined by the physicists, namely, its charge and mass.[89] Finally, for both physicists and chemists the explanatory role of the electron was the same. Both groups agreed on the regularities (from spectroscopy, the facts of chemical combination, and the periodic table) that had to be interpreted according to the properties and behavior of electrons. They all believed, for instance, that the periodicity in the chemical properties of the elements had a counterpart in "the number and arrangement of electrons."[90] There was also a consensus about the preelectronic notions (e.g., the chemical bond, or chemical valence) that had to be reconceptualized in terms of electrons. In all these respects, the electron qua theoretical entity exhibited a transdisciplinary character; it had taken on a life of its own.[91]

89. See, e.g., G. N. Lewis and E. Q. Adams, "Notes on Quantum Theory. A Theory of Ultimate Rational Units; Numerical Relations between Elementary Charge, Wirkungsquantum, Constant of Stefan's Law," *Physical Review*, 3, no. 2 (1914): 92–102, esp. p. 97.

90. I. Langmuir, "Modern Concepts in Physics and Their Relation to Chemistry," *Science*, 70, no. 1817 (1929): 385–396, on p. 393.

91. Cf. the interesting discussion of "boundary objects" in S. L. Star and J. R. Griesemer, "Institutional Ecology, 'Translations' and Boundary Objects: Amateurs and Professionals in Berkeley's Museum of Vertebrate Zoology, 1907–39," *Social Studies of Science*, 19, no. 3 (1989): 387–420.

As we have seen, however, in other respects the electron's chemical personality and its physical personality diverged considerably. The laws it was supposed to obey (Coulomb versus non-Coulomb), its putative behavior inside the atom (dynamic versus static), and some of the properties attributed to it (nonmagnetic versus magnetic, classical versus quantum) differed across disciplines. It was because of these differences that the British physicist Edward Andrade exclaimed (somewhat desperately) that "[t]he electrons in the Langmuir atom have, in fact, so few of the known properties of electrons that it is not immediately clear why they are called electrons at all."[92]

In view of the divergent representations of the electron in early-twentieth-century physics and chemistry, could one still maintain that both physicists and chemists referred to the same thing when they used the term "electron"? From what I have said so far, the reader may guess my answer. Nevertheless, I will postpone taking an explicit stand on this issue, since its adequate treatment would require an examination of the meaning and reference of scientific terms. This will be the subject of chapter 9. Before that, however, we should round off our biographical analysis of the physicists' electron.

92. E. N. da C. Andrade, *The Structure of the Atom,* 3rd rev. ed. (London: G. Bell and Sons, 1927; 1st ed., 1923), p. 642.

# Chapter 8 | Forced to Spin by Uhlenbeck and Goudsmit

## 1. The Setting

The development of the representation of the electron after 1920 was closely tied with the frustrations of multiplet lines in the optical region of the spectra of multielectronic atoms. Those lines were supposed to arise from quantum jumps of the valence electron(s). Whereas in the hydrogen atom orbits with the same main axis but different shapes corresponded to the same energy level (provided that one ignored relativistic effects), in a multielectron atom the shape of the outer electron's orbit was one of the factors that determined its energy. The dependence of the electron's energy on its azimuthal quantum number was due to the deviation of the attractive force between the valence electron and the rest of the atom from an inverse square law. The multiplicity of the spectral lines originating from transitions between orbits characterized by different azimuthal quantum numbers revealed, in turn, a corresponding multiplicity in the energy levels associated with each of those numbers. That multiplicity could not be explained in terms of a physical model. Furthermore, multiplet spectral lines exhibited anomalous Zeeman effects. That is, under the influence of a weak magnetic field those lines showed splitting patterns that departed from the triplet anticipated by Lorentz and Larmor. Those anomalous splittings also defied explanation. Thus, the electron's writings had a complexity that was not reflected in its existing representation. The multiplicity of its observable manifestations had no counterpart in its putative behavior.

Until 1919 the dominant approach in the analysis of spectral observations was a deductive one. One employed classical mechanics to calculate the electron's possible motions within the atom and then selected out of this continuum of motions those that were compatible with quantum constraints. In this way one determined the allowed energy states of an atom. The final step in this process was to examine whether the energy differences between those states (divided by Planck's constant) matched the observed spectral frequencies.

This approach, however, encountered severe difficulties in explaining the complex spectra of multielectronic atoms. The lack of a physical explanation for the multiplet spectral lines and the anomalous Zeeman patterns observed in those spectra rendered impossible a deductive determination of the energy levels of those atoms. Thus, Sommerfeld initiated a different approach, whose aim was to link in an empirical fashion every spectral frequency with two energy levels. Each of those levels was specified by several quantum numbers, and various selection rules were introduced so as to fit the spectral data.[1] To obtain the observed splitting patterns of doublet and triplet spectral lines, he had to characterize each energy level by three quantum numbers. Besides the familiar radial $(n')$ and azimuthal $(n)$ quantum numbers, the so-called inner quantum number $(i)$ was introduced. Its values were $n$, $n - 1$ for doublets and $n$, $n - 1$, $n - 2$ for triplets. By subjecting it to the selection rule $\Delta i = \pm 1, 0$, he was able to reproduce the observed multiplicities.[2]

However, the physical meaning of the new quantum number was obscure. In analogy with his interpretation of the azimuthal quantum number as an index of angular momentum, Sommerfeld associated the inner quantum number with a "hidden [verborgenen] rotation." As for its "geometric significance we are quite as ignorant as we are of those differences in the orbits which underline the multiplicity of the series terms."[3]

The first step toward its interpretation was made by Alfred Landé, who identified it with the quantum number corresponding to the total angular momentum of the atom: "The different 'inner' quantum numbers of a term

1. Cf. P. Forman, "Alfred Landé and the Anomalous Zeeman Effect, 1919–1921," *Historical Studies in the Physical Sciences*, 2 (1970), p. 186; and P. Forman and A. Hermann, "Sommerfeld, Arnold," in C. C. Gillispie (ed.), *Dictionary of Scientific Biography*, 16 vols. (New York: Charles Scribner's Sons, 1970–1980), vol. 12, pp. 525–532, on p. 529.

2. See A. Sommerfeld, "Allgemeine spektroskopische Gesetze, insbesondere ein magnetooptischer Zerlegungssatz," *Annalen der Physik*, 63 (1920): 221–263; repr. in his *Gesammelte Schriften*, ed. F. Sauter, 4 vols. (Braunschweig: F. Vieweg, 1968), vol. 3, pp. 523–565, esp. pp. 555–556. Cf. Forman, "Alfred Landé," pp. 190–191.

3. "Ihre geometrische Bedeutung kennen wir ebensowenig, wie wir die Bahnunterschiede kennen, die den Muliplizitäten der Serienterme zugrunde liegen." Sommerfeld, *Gesammelte Schriften*, vol. 3, p. 533. The translation is from Forman, "Alfred Landé," p. 191.

will probably signify simply the total quantum numbers of the atom about its invariable axis for different spatial orientations of the valence electrons around the atomic core."[4] Landé's work on the anomalous Zeeman effect built on Sommerfeld's and Debye's quantum-theoretical analysis of the normal Zeeman effect (see chapter 6). He considered each component of the anomalous Zeeman patterns as arising from a transition between two stationary states. In the presence of a magnetic field an additional quantum number ($m$), which corresponded to the projection of the total angular momentum vector on the direction of the field, was necessary to specify each of those states. In an attempt to accommodate the anomalous effects, he suggested that the displacement of the energy levels due to the magnetic field was slightly different from the one obtained by Sommerfeld and Debye. The difference was encapsulated in an empirical constant, the so-called $g$-factor (the ratio of the empirically determined shift of the atom's energy in the presence of a magnetic field to the value of that shift obtained by Sommerfeld's and Debye's analysis). The expression giving the energy of the magnetically perturbed state was $E = E_i + gmh(eH/4\pi mc)$. For $g = 1$ one recovers the formula previously derived by Sommerfeld.[5]

The physical model underlying Landé's analysis was linked with his interpretation of the inner quantum number as an index of the total angular momentum of the atom. Since that momentum depended, in part, on the orientation of the orbit of the valence electron, different orientations would result in different values of the atom's angular momentum. The multiplicity of those values would, in turn, give rise to the complexity of the corresponding spectral term. In other words, each value of the total angular momentum would correspond to a value of the atom's energy. For a given value of $i$, $m$ could take any of the following ($2i + 1$) values: $-i, -(i - 1), \ldots, 0, \ldots,$ $i - 1, i$. The total number of $m$'s was, therefore, an odd number. For the case of even multiplicities, however, the anomalous splitting pattern consisted of an even number of components. Thus, with each inner quantum number ($i$) one had to associate an even number of distinct levels. To save the phe-

4. A. Landé to N. Bohr, 16 Feb. 1921, published in Forman, "Alfred Landé," pp. 242–245; the translation is from ibid., p. 202. This interpretation of the inner quantum number also appeared in print. See A. Landé, "Über den anomalen Zeemaneffekt," pt. 1, *Zeitschrift für Physik,* 5 (1921): 231–241, on p. 234. Paul Forman, in the above-mentioned paper, has provided a very thorough analysis of Landé's work on multiplet spectral lines and their intricate magnetic splittings.

5. Landé, "Über den anomalen Zeemaneffekt," pt. 1, p. 233. Cf. D. C. Cassidy, "Heisenberg's First Core Model of the Atom: The Formation of a Professional Style," *Historical Studies in the Physical Sciences,* 10 (1979): 187–224, on pp. 194–195.

nomena, Landé conjectured, without any physical justification, that in the case of doublets $m$ could take any of the following $(2i)$ values: $-(i - 1/2)$, $-(i - 3/2), \ldots, (i - 3/2), (i - 1/2)$. Since $i$ could be either $n$ or $n - 1$, the number of magnetic energy levels should be $2n + 2(n - 1) = 2(2n - 1)$. This number was in agreement with spectral data. Notwithstanding the incomprehensibility of Landé's maneuver, it had the effect of reproducing the empirical regularities associated with the anomalous Zeeman effect.[6] However, the assumption of half-integral quantum numbers had no counterpart in the physical model that Landé had developed. Thus, the model was left implicit in his paper, and the complex structure of spectral lines was left unexplained.[7]

The development of a model that yielded doublet and triplet spectral terms and their associated magnetic decompositions was carried out by Sommerfeld's precocious student Werner Heisenberg.[8] The starting point of Heisenberg's work was Sommerfeld's translation of Woldemar Voigt's empirically successful classical analysis of the anomalous Zeeman effect in the language of quantum theory. In Voigt's model the electrons' interactions with each other and with the nucleus were governed by elastic forces. From this model he had derived the frequencies of the anomalous Zeeman components. By expressing those frequencies as differences between energy levels divided by Planck's constant, Sommerfeld managed to represent them in terms of stationary states that were characterized by familiar quantum numbers. One of the aims of Sommerfeld's project was to derive Landé's results without recourse to half-integral quantum numbers, which did not fit in the framework of the old quantum theory. In this respect he was successful, since he managed to reproduce the empirical regularities by employing only integral quantum numbers. However, another one of his aims, namely, the development of a physical model for the anomalous Zeeman effect, was not fulfilled.[9]

This task was carried out by Heisenberg. Taking off from Sommerfeld's reinterpretation of Voigt's theory, Heisenberg, who did not hesitate to violate well-established principles and introduce theoretically unjustified assump-

6. Cf. Cassidy, "Heisenberg's First Core Model," p. 195; and Cassidy, *Uncertainty: The Life and Science of Werner Heisenberg* (New York: Freeman, 1992), pp. 119–120.

7. Cf. Forman, "Alfred Landé," pp. 203, 234–235.

8. See W. Heisenberg, "Zur Quantentheorie der Linienstruktur und der anomalen Zeemaneffekte," *Zeitschrift für Physik,* 8 (1921–1922): 273–297 (received on 17 Dec. 1921).

9. See A. Sommerfeld, "Quantentheoretische Umdeutung der Voigtschen Theorie des anomalen Zeemaneffektes vom D-Linientypus," *Zeitschrift für Physik,* 8 (1922): 257–272; repr. in his *Gesammelte Schriften,* vol. 3, pp. 609–624. Cf. Cassidy, "Heisenberg's First Core Model," p. 202; and Cassidy, *Uncertainty,* p. 120.

tions, rewrote Sommerfeld's equation for the energy of a magnetically perturbed state in terms of half-integral quantum numbers.[10] His next step was to interpret those numbers as denoting half-integral angular momenta and to incorporate this notion into a physical model that led to the equation in question. In his scheme, valence electrons rotated around the "core" of the atom (i.e., the nucleus and the inner electrons). The core was supposed to have an angular momentum and, therefore, a magnetic moment. Because of its magnetic disposition, its direction was influenced by the magnetic field produced by the motion of the valence electrons. The result of that influence was a precessional motion of the axis of the core about the direction of the (internal) magnetic field. This led to a shift of the atom's energy level equal to

$$\Delta E = \mu H_i = \frac{1}{2} H_i \frac{eH}{4\pi mc} \cos \theta,$$

where $\mu$ is the magnetic moment of the core, $H_i$ the magnetic field of the outer electron, and $\theta$ is the angle between the angular momentum vectors of the electron and the core. By assuming that the core's direction was either parallel or antiparallel to the direction of the magnetic field, the shift of the atom's energy could be either positive or negative. Thus, as a result of the internal magnetic field, the original energy of the atom would split into a doublet.[11]

Half-integral angular momenta were supposed to arise by a mysterious interaction between the core and the valence electrons, as a result of which the core appropriated part of the angular momentum of the valence electron. The valence electron was left with an angular momentum equal to $n - 1/2$, and the core obtained an angular momentum equal to $n \pm 1/2$ (corresponding to its two orientations in relation to the internal magnetic field). The only grounds for this assumption were that it "is justified by its success."[12] Sommerfeld's inner quantum number was interpreted, à la Landé, as the total angular momentum of the atom. The multiplicity of that number followed naturally. If the direction of the core were parallel to the direction of the outer

10. For Heisenberg's bold way of doing physics see Cassidy, "Heisenberg's First Core Model," esp. pp. 189, 212–214, 223.

11. Cf. Heisenberg to Pauli, 19 Nov. 1921, in W. Pauli, *Wissenschaftlicher Briefwechsel mit Bohr, Einstein, Heisenberg, u.a.,* band 1, *1919–1929,* ed. A. Hermann, K. von Meyenn, and V. F. Weisskopf (New York: Springer-Verlag, 1979), p. 38.

12. A. Sommerfeld, *Atomic Structure and Spectral Lines,* trans. from the 3rd German ed. by H. L. Brose (London: Methuen, 1923), p. 406. Cf. Heisenberg's comment that "[s]uccess sanctifies the means" (Der Erfolg heiligt die Mittel). Heisenberg to Pauli, 19 Nov. 1921, in Pauli, *Wissenschaftlicher Briefwechsel,* p. 38.

electron's orbit, then the total angular momentum of the atom would be the sum of the angular momentum of the core (1/2) and the angular momentum of the valence electron $(n - 1/2)$: $n$. If the momenta of the core and the valence electron were antiparallel, the total momentum would be equal to $n - 1/2 - 1/2 = n - 1$.[13]

In the presence of a medium external magnetic field the core oriented itself in the direction of the vector sum of the external field and the magnetic field created by the valence electron. Since the direction of the external field could vary continuously, the direction of the core's axis could also change continuously. The core's lack of space quantization implied that it did not obey Larmor's theorem. According to this theorem, the core should perform a precessional motion about the direction of the internal magnetic field. In Heisenberg's model, however, no such motion was taking place. The core's axis was fixed in the plane defined by the directions of the external and the internal fields.

To explain the splitting of the atom's energy levels due to the external magnetic field Heisenberg capitalized on the space quantization of the valence electron. As he suggested in a letter to Pauli, "The stationary states of the atoms are always given through the space quantization of the *outer* electron."[14] In the presence of an external field all those states were distinguished in energy. The total shift in the energy of the atom due to the combined presence of the external and internal fields was

$$(8.1) \quad \Delta E = \mu_1 H \cos \theta_1 + \mu_2 H \cos \theta_2 + \mu_2 H_i \cos \theta_{12}$$

$$= \left( n - \frac{1}{2} \right) \frac{h}{2\pi} \cdot \frac{e}{2mc} \cdot H \cos \theta_1$$

$$+ \frac{1}{2} \frac{h}{2\pi} \cdot \frac{e}{2mc} \cdot H \cos \theta_2 + \frac{1}{2} \frac{h}{2\pi} \cdot \frac{e}{2mc} H_i \cos \theta_{12}$$

$$= \Delta \nu_n \cdot h \left[ \left( n - \frac{1}{2} \right) \cos \theta_1 + \frac{1}{2} \cos \theta_2 + \frac{1}{2} \frac{H_i}{H} \cos \theta_{12} \right],$$

where $\mu_1$, $\mu_2$ are the magnetic moments of the valence electron and the core, respectively, $\theta_1$ is the angle between the direction of the electron's orbit and

13. See Heisenberg to Pauli, 19 Nov. 1921, in Pauli, *Wissenschaftlicher Briefwechsel,* p. 38. Cf. Cassidy, "Heisenberg's First Core Model," pp. 203–204; and Cassidy, *Uncertainty,* pp. 121–123.

14. "Die stationären Zustände des Atoms sind stets durch räumliche Quantelung des *äußeren* Elektrons gegeben." Heisenberg to Pauli, 17 Dec. 1921, in Pauli, *Wissenschaftlicher Briefwechsel,* p. 48.

the magnetic field, $\theta_2$ is the angle between the axis of the core and the magnetic field, $\theta_{12}$ is the angle between the direction of the electron's orbit and the axis of the core, and $\Delta\nu_n = eH/4\pi mc$ is the normal magnetic splitting. The first term in this expression represents the interaction between the valence electron and the magnetic field, the second term is due to the interaction between the core and the magnetic field, and the third term reflects the coupling between the core and the outer electron.

In a weak external field the first two terms are negligible and the core's direction coincides with the direction of the valence electron's orbit ($\cos \theta_{12} = \pm 1$). In this case, as in the absence of an external field, the energy levels split into doublets. For large values of $H$ the third term is negligible, the core takes the direction of the external field ($\theta_2 = 0$), and the splitting of the energy levels becomes

$$\Delta E = \Delta\nu_n \cdot h\left[\left(n - \frac{1}{2}\right)\cos \theta_1 + \frac{1}{2}\right] = \Delta\nu_n \cdot h\left(m - \frac{1}{2} + \frac{1}{2}\right) = \Delta\nu_n \cdot hm,$$

where $(m - 1/2)$ is a quantum number corresponding to the projection of the angular momentum of the valence electron on the direction of the external field. Thus, in the case of strong external fields the splitting pattern becomes a normal Zeeman pattern (Paschen-Back effect). Moreover, from the assumption that the core's direction is stable (i.e., from the extremum condition $\partial\Delta E/\partial\theta_2 = 0$) Heisenberg could derive Landé's g-factors.[15]

The empirical success of Heisenberg's model was remarkable. It explained (quantitatively) the multiplicities of spectral terms and led to the g-factors that Landé had introduced in his analysis of the anomalous Zeeman effect. However, Heisenberg's proposal was plagued with conceptual problems. First, it did not explain the mysterious sharing of angular momentum between the valence electron and the core and could not account for the half-integral value of the shared momentum. Second, the assumption of half-integral angular momenta did not fit within the framework of the Bohr-Sommerfeld theory. Such momenta were incompatible with Sommerfeld's quantum conditions and undermined the past successes of the quantum theory, which had been obtained by the use of integral quantum numbers. Third, in order to explain the Paschen-Back effect (the transition from an anomalous to a normal Zeeman splitting when one increased the intensity of the magnetic field) Heisen-

15. See Heisenberg to Pauli, 19 Nov. 1921, in Pauli, *Wissenschaftlicher Briefwechsel*, pp. 39–40. Cf. Cassidy, "Heisenberg's First Core Model," pp. 205–206; and Sommerfeld, *Atomic Structure and Spectral Lines* (1923), pp. 408–409.

berg assumed that the direction of the core's angular momentum coincided with the direction of the total magnetic field (internal and external), which could vary continuously. Pace Sommerfeld, the direction in question was not quantized and the core did not obey Larmor's theorem.[16]

Another conceptual problem was the so-called riddle of the statistical weights, created by a clash between the core model and a tenet of Bohr's atomic theory.[17] In his attempt to devise an adequate interpretation of the periodic table in terms of a theory of atomic structure, Bohr had put forward the so-called building-up principle. According to that principle, atoms were "formed by the successive capture and binding of the electrons one by one in the field of force surrounding the nucleus."[18] In that process the quantum numbers of the already bound electrons (and, thus, the quantum numbers corresponding to the old atom as a whole) were supposed to remain unaffected by the capture of additional electrons. Furthermore, the "statistical weight" of the old atom (the number of possible orientations of its total angular momentum) was also supposed to remain invariant. Heisenberg's model violated both of those aspects of Bohr's principle. Whereas, according to the latter the angular momentum of the core of an alkali atom, because of its noble-gas structure, should retain the value that it had before the outer electron's capture, namely, zero, according to the former it was endowed with a half unit of angular momentum. Moreover, whereas the statistical weight of a noble-gas atom was $2 \cdot 0 + 1 = 1$, the core of an alkali atom, according to Heisenberg, had a statistical weight $2 \cdot 1/2 + 1 = 2$.[19]

Despite the conceptual problems facing Heisenberg's model, one could not overlook its empirical success. The task facing the physicists that espoused Bohr's theoretical agenda was to reproduce its successes and dispense with its problematic assumptions. In this context, Bohr proposed two hypotheses that confronted the problems faced by the core model of the atom. First, he attacked the problem of spectral terms with even multiplicities. That prob-

16. Cf. Cassidy, "Heisenberg's First Core Model," p. 207; and Cassidy, *Uncertainty,* p. 123.

17. The phrase "riddle of the statistical weights" comes from D. Serwer, *"Unmechanischer Zwang:* Pauli, Heisenberg, and the Rejection of the Mechanical Atom, 1923–1925," *Historical Studies in the Physical Sciences,* 8 (1977): 189–256, on p. 201.

18. See Bohr's lecture to the Physical and Chemical Societies of Copenhagen (18 Oct. 1921), titled "The Structure of the Atom and the Physical and Chemical Properties of the Elements"; trans. in his *The Theory of Spectra and Atomic Constitution* (Cambridge: Cambridge Univ. Press, 1922), pp. 61–126; the quote is from p. 75.

19. Cf. Cassidy, "Heisenberg's First Core Model," pp. 207–208, 218; and Serwer, *"Unmechanischer Zwang,"* pp. 204, 206.

lem could be solved either with half-integral quantum numbers or by the exclusion of one of the relative orientations between core and valence electron. Bohr followed the second alternative and eliminated by fiat the outer electron's orbit that was perpendicular to the direction of the core. Thus, the total number of the orientations of that orbit decreased from $2k + 1$ to $2k$, where $k$ was Bohr's symbol for the azimuthal quantum number, and the need for half-integral values for that number disappeared. Second, he introduced a nonmechanical force (*Zwang*) governing the interaction between the outer electron and the atomic core. As a result of Zwang, one of the $2k$ orientation possibilities of the outer electron's orbit was eliminated and an additional orientation of the core with respect to an external magnetic field was added. It followed that the total number of magnetic states was $2(2k - 1)$, the number indicated by experimental data on the spectra of alkali atoms.[20]

Thus, Bohr's proposal reproduced the empirically adequate results of the core model for the total number of magnetic energy levels of an alkali atom. Furthermore, Bohr avoided the riddle of statistical weights, by restricting the applicability of the principle of the permanence of quantum numbers to those numbers which he had employed in his quantum-theoretical construction of the periodic table and which denoted properties of individual electrons. The numbers in question ($n$ and $k$) fixed the size and shape of an electron's orbit.[21] Sommerfeld's inner quantum number, on the other hand, denoted a collective property of the atom and was not involved in Bohr's construction of the periodic table. Its half-integral values and its variation in the transition from an atom to its adjacent in the periodic table could be explained by recourse to the mysterious Zwang that governed the interaction between the core and the valence electron.[22]

As we have seen, there were two alternative ways to come to terms with the complex structure of spectral lines and their anomalous Zeeman effects: on the one hand, the introduction of half-integral quantum numbers, half-integral angular momenta, and a strangely behaving core; on the other hand, the proposal of a nonmechanical Zwang. A few years later those puzzling features of Heisenberg's model would receive an explanation and Bohr's Zwang

20. N. Bohr, "Linienspektren und Atombau," *Annalen der Physik,* 71 (1923): 228–288, esp. pp. 274–276. Cf. Cassidy, "Heisenberg's First Core Model," p. 222; and J. L. Heilbron, "The Origins of the Exclusion Principle," *Historical Studies in the Physical Sciences,* 13 (1983): 261–310, on pp. 275–277.

21. The letter $n$ was Bohr's symbol for Sommerfeld's "quantum sum" (see chapter 6).

22. See Bohr, "Linienspektren und Atombau," pp. 277–278. Cf. Serwer, "*Unmechanischer Zwang,*" p. 208.

would be eliminated with the proposal of a spinning electron. Two crucial elements of the problem situation that led to the hypothesis of spin were Pauli's introduction of a fourth quantum number and his formulation of the exclusion principle. The next section focuses on those developments.

## 2. Becoming Antisocial in the Land of the Formalism-Philistines

Early in 1923 Landé formulated an expression for his *g*-factor in terms of *R* (angular momentum of the core), *K* (angular momentum of the valence electron), and *J* (total angular momentum of the atom):[23]

$$g = 1 + \frac{J^2 - \frac{1}{4} + R^2 - K^2}{2\left(J^2 - \frac{1}{4}\right)}.$$

Landé professed that the above formula led to an empirically correct representation of the splitting pattern in the anomalous Zeeman effect.

Several weeks later Pauli published an "unfortunate" (unglückseligen) paper on the phenomenology of the anomalous Zeeman effect.[24] By examining the regularities in the anomalous Zeeman patterns, he came up with a formula equivalent with the one obtained by Landé.[25] However, he abstained from interpreting those regularities in terms of a model. Such an interpretation "will probably not be possible . . . within the framework of the quantum theory of multiply periodic systems and will require conceptual innovations."[26]

In his attempt to represent the splitting pattern in the presence of strong magnetic fields (the Paschen-Back effect), he introduced two new quantum

23. See A. Landé, "Termstruktur und Zeemaneffekt der Multipletts," *Zeitschrift für Physik,* 15 (1923): 189–205 (received on 5 Mar. 1923), on p. 193. Cf. P. Forman, "The Doublet Riddle and Atomic Physics *circa* 1924," *Isis,* 59 (1968): 156–174, on p. 162.

24. The derogatory characterization comes from Pauli himself. See Pauli to Sommerfeld, 19 July 1923, in Pauli, *Wissenschaftlicher Briefwechsel,* pp. 105–107, on p. 105.

25. W. Pauli, "Über die Gesetzmässigkeiten des anomalen Zeemaneffektes," *Zeitschrift für Physik,* 16 (1923): 155–164 (received on 26 Apr. 1923); repr. in his *Collected Scientific Papers,* ed. R. Kronig and V. F. Weisskopf, 2 vols. (New York: Interscience, 1964), vol. 2, pp. 151–160.

26. "[D]ürfte nicht . . . innerhalb des Rahmens der Quantentheorie der mehrfach periodischen Systeme möglich sein und begriffliche Neuerungen erfordern." Ibid. (in *Collected Scientific Papers*), p. 151.

numbers $(m_1, \mu)$ whose sum was equivalent with Landé's magnetic quantum number $m$. The perturbed energy levels were given by the formula

$$E = (m_1 + 2\mu)h\frac{eH}{4\pi mc}.$$

Even though in his published paper Pauli did not attach any physical meaning to those quantum numbers, in letters to several physicists he interpreted them in terms of the core model of the atom. The quantum number $m_1$ corresponded to the component of the angular momentum of the valence electron in the direction of the external field, and $\mu$ signified the component of the angular momentum of the core in the same direction. Furthermore, he traced the origin of the anomalous effect to the anomalous (double) gyromagnetic ratio of the core.[27] The problem with this interpretation was that it required the existence of two kinds of electrons. Since the gyromagnetic ratio of the core had a value twice that expected from classical electromagnetism, the core electrons should also exhibit this anomaly. The gyromagnetic ratio of the valence electrons, on the other hand, had its normal (i.e., classical) value.[28]

Pauli subsequently published this interpretation and employed it to analyze the complex structure of spectral terms. To emphasize the connection of his quantum numbers with the core model, he gave them different names ($m_r$ and $m_k$ as opposed to $\mu$ and $m_1$).[29] After he finished that work, he switched to a different line of research for a significant period.

Toward the end of 1924 Pauli resumed his work on the Zeeman effect. In a letter to Bohr on 12 December 1924, he pointed out that, because of the relativistic variation of the electron's mass, Larmor's theorem should not hold strictly.[30] Furthermore, in the formula for the gyromagnetic ratio $(e/2mc)$ one should insert the relativistic value of the electron's mass,

$$m = \frac{m_0}{\sqrt{1 - \dfrac{v^2}{c^2}}},$$

27. See Pauli to Landé, 23 May 1923, in Pauli, *Wissenschaftlicher Briefwechsel*, p. 87; and Pauli to Sommerfeld, 19 July 1923, in ibid., p. 105.

28. Cf. Heilbron, "The Origins of the Exclusion Principle," pp. 296–297.

29. See W. Pauli, "Zur Frage der Zuordnung der Komplexstrukturterme in starken und in schwachen äußeren Feldern," *Zeitschrift für Physik*, 20 (1924): 371–387 (received on 20 Oct. 1923); repr. in his *Collected Scientific Papers*, vol. 2, pp. 176–192.

30. Published in Pauli, *Wissenschaftlicher Briefwechsel*, pp. 186–189. Cf. Serwer, "Unmechanischer Zwang," p. 230; and Heilbron, "The Origins of the Exclusion Principle," p. 301.

where

$$\sqrt{1 - \frac{v^2}{c^2}}$$

is the average value of the relativistic correction factor over a period of the electron's motion.

This relativistic effect created severe problems for the core model of the atom. For an atom whose core consisted of the nucleus and the two fast-moving electrons of the K-shell, Pauli's relativistic analysis showed that the magnetic splitting of the energy levels in a strong magnetic field should be given by the formula

$$E = (2\gamma m_r + m_k) h \left( \frac{eh}{4\pi mc} \right),$$

where

$$\gamma = \sqrt{1 - \frac{v^2}{c^2}}.$$

The empirically confirmed formula, however, was

$$E = (2m_r + m_k) h \left( \frac{eh}{4\pi mc} \right).$$

Pauli's verdict was unambiguous: "I am of the opinion that the reality of the core momentum of the noble-gas shell has been greatly exaggerated. More-over, I think untenable the assumption that the anomaly of the Zeeman effect follows from the fact that the atomic core has a magnetism twice as great as it classically should. This assumption (which really originates with me, but which the others then took too seriously) contradicts my physical feeling most vigorously. For it is completely incomprehensible and *isolated* in the quantum theory. In particular there is no direct logical link between it and the fact that the alkali spectra have a doublet structure."[31] As he had pointed

---

31. Pauli to Bohr, 12 Dec. 1924, in Pauli, *Wissenschaftlicher Briefwechsel,* p. 187; the trans-lation is from Serwer, "*Unmechanischer Zwang,*" p. 231.

out sarcastically to Landé, "the assumed core momenta have not only half-integral values, but also in a way only a half reality." [32]

Pauli had a further argument against the notion that the K-shell of electrons was endowed with a nonvanishing angular momentum. In his letter to Landé on 24 November 1924 he reasoned as follows: "In reality the K-shell does not behave differently with respect to the observable properties (magnetism) than the higher-quantum shells in closed state. If one assigns a nonvanishing momentum to the K-shell, but the momentum 0 to the higher-quantum noble-gas-shells, then there is an asymmetry in the theoretical construction, with no corresponding asymmetry in experience." [33] Since the core had zero angular momentum (magnetic moment), it could not be linked with the multiplicity of spectral lines and their magnetic splittings. Rather, these phenomena would be explained by the following "working hypothesis" (Arbeitshypothese): "The doublet structure of the alkalis is essentially a property of the optical electron alone. The same classically not describable two-valuedness, which causes the doublet structure, causes also the anomaly of the Zeeman effect." [34]

32. "[D]ie angenommenen Rumpfimpulse haben nicht nur halbzahlige Werte, sondern auch gewissermaßen nur eine halbe Realität." Pauli to Landé, 10 Nov. 1924, in Pauli, *Wissenschaftlicher Briefwechsel*, p. 171. In that letter, which was very similar to the one that he later sent to Bohr, Pauli detailed the deviations from Larmor's theorem and from the standard *g*-factors, due to the relativistic mass of the electron, and wanted to know "[i]f one can definitively exclude the . . . deviations of the *g*-values . . . on the basis of the existing empirical material" (Ob man die . . . Abweichungen der *g*-Werte . . . auf Grund des vorliegenden empirischen Materials definitiv ausschließen kann; ibid.). Data from Tübingen, communicated to Pauli by Landé, confirmed the consequences of Pauli's discovery. See Pauli to Landé, 14 Nov. 1924, in ibid., p. 172; and Pauli to Landé, 24 Nov. 1924, in ibid., p. 176. Cf. Heilbron, "The Origins of the Exclusion Principle," p. 301.

33. "In Wirklichkeit verhält sich in den beobachtbaren Eigenschaften (Magnetismus) die *K*-Schale nicht anders als die höherquantigen Schalen im abgeschlossenen Zustand. Wenn man der *K*-Schale einen nicht verschwindenden Impuls zuordnet, den höherquantigen Edelgasschalen aber den Impuls 0, so liegt darin eine Unsymmetrie des theoretischen Gebäudes, dem keine Unsymmetrie in der Erfahrung entspricht." Pauli, *Wissenschaftlicher Briefwechsel*, pp. 176–177. Cf. Serwer, *"Unmechanischer Zwang,"* p. 245. It is worth pointing out that Pauli's reasoning is strikingly similar to the opening statement of Einstein's 1905 paper on the special theory of relativity: "It is known that Maxwell's electrodynamics—as usually understood at the present time—when applied to moving bodies, leads to asymmetries which do not appear to be inherent in the phenomena." A. Einstein et al., *The Principle of Relativity* (New York: Dover, 1952), p. 37.

34. "Die Dublettstruktur der Alkalien ist im wesentlichen eine Eigenschaft des Leuchtelektrons allein. Dieselbe klassisch nicht beschreibbare Zweideutigkeit, welche die Dublettstruktur verursacht, bedingt auch die Anomalie des Zeemaneffektes." Pauli to Bohr, 12 Dec. 1924, in Pauli, *Wissenschaftlicher Briefwechsel*, p. 187. Pauli had put forward the same hypothesis in his (24 Nov.) letter to Landé (ibid., p. 177). Cf. Heilbron, "The Origins of the Exclusion Principle," p. 302.

Here it is worth pausing to analyze this situation in terms of my suggestion that theoretical entities are constructions from experimental data. I have argued in previous chapters that the spectral writings attributed to the electron guided the construction of its representation and that physicists strove to correlate all (qualitative and quantitative) aspects of those writings with features of the electron's putative behavior.[35] In the above case, however, it was not originally clear that the spectral data under consideration were due directly to the electron. One could doubt that those data were authentic writings of the electron. They were produced by multielectronic atoms and could be linked with a subatomic complex structure (the "core") containing several electrons. As we saw, this possibility was undermined by Pauli's arguments, which capitalized, among other things, on the relativistic character of the electron. There could be no "direct logical link" between the properties of the core and the spectral data in question. It was more plausible that they were manifestations of the electron itself (as opposed to the core), a product of its hidden properties and behavior. Once attributed to the electron, those data indicated how its representation should be transformed. The doublet structure of the former had to obtain a counterpart in the latter; a new property (a "classically not describable two-valuedness") had to be added to the electron. The next step would be to provide a physical interpretation of that property; but more on this below. Thus, as this case shows, it is not always a straightforward issue whether an effect is the manifestation of a given entity. The authenticity of an entity's writings has to be established on the basis of theoretical and/or experimental arguments.

Pauli developed further his critique of the core model in a paper that he submitted for publication on 2 December 1924.[36] It aimed at showing that, pace the model, the K-electrons do not contribute to the total angular momentum (and magnetic moment) of the atom. First, Pauli expressed the average value of the relativistic correction factor in terms of the total energy of the atom:[37]

$$\gamma = 1 + \frac{W}{m_0 c^2}.$$

35. The writings of the electron should be distinguished from its purported behavior. The former are the observable manifestations of the latter in various experimental settings. The latter, in turn, *gives rise* to the former. It is their hidden cause, which has to be *inferred* from its effects.

36. W. Pauli, "Über den Einfluss der Geschwindigkeits-abhängigkeit der Elektronenmasse auf den Zeemaneffekt," *Zeitschrift für Physik,* 31 (1925): 373–385; repr. in his *Collected Scientific Papers,* vol. 2, pp. 201–213.

37. Pauli, *Collected Scientific Papers,* vol. 2, p. 205.

The value of that energy ($W$) had been already calculated by Sommerfeld (see chapter 6, equation [6.11]). By substituting Sommerfeld's formula in the above equation, Pauli obtained the factor in question as a function of the atomic number:

$$\gamma = \left\{ 1 + \frac{\alpha^2 Z^2}{(n - k + \sqrt{k^2 - \alpha^2 Z^2})^2} \right\}^{-1/2},$$

where $Z$ is the atomic number, $n$ and $k$ are the principal and azimuthal quantum numbers, respectively, and $\alpha$ is the fine structure constant.[38] The dependence of that factor on the atomic number implied that *"the quotient of the magnetic moment and the angular momentum of the K-shell must deviate considerably in elements with higher atomic number from its normal value."*[39] These deviations led, in turn, to "Zeeman splittings of the lines . . . , which contain a systematic dependence on the atomic number; [such splittings] are however not compatible with the observations."[40]

Pauli's analysis points to the heuristic role of the electron qua theoretical entity, which is revealed in its contribution to the demise of the core model. Here, as in some of the cases we examined in chapter 6, the heuristic function of the representation of the electron flowed from the simplifications it contained and from general theoretical constraints it was subject to, namely, the special theory of relativity. The latter determined what counted as a simplification and indicated why and how those simplified elements had to be removed in order to develop further the theory of atomic structure. The special theory of relativity, however, was brought into play only because of the particular representation of the electrons' behavior: the core electrons were supposed to move in high-speed, elliptical orbits with varying velocity. It was this specific representation that rendered relativity theory relevant to the problem situation and enabled Pauli to draw on its resources to criticize the core model, which up to that point had been empirically adequate! When one took into account the relativistic character of the electron, the core model lost its empirical adequacy.

This empirical problem could be avoided by rendering the K-shell mag-

---

38. Ibid., p. 206.

39. *"[M]uß der Quotient aus magnetischem Moment und Impulsmoment der K-Schale . . . bei Elementen mit höherer Atomnummer von seinem Normalwert beträchtlich abweichen."* Ibid., p. 207 (emphasis in original).

40. "Zeemanaufspaltungen der Linien . . . , die eine systematische Abhängigkeit von der Atomnummer enthalten, sind jedoch mit den Beobachtungen nicht vereinbar." Ibid., pp. 210–211.

netically inert and transferring to the valence electron itself the burden of the anomalous Zeeman effect: *"The closed electron configurations are not supposed to contribute to the magnetic moment and the angular momentum of the atom. Especially in the alkalis the momentum values of the atom and its energy changes in an external magnetic field will be looked on essentially as a particular effect of the optical electron, which will also be considered as the seat of the magneto-mechanical anomaly. The doublet structure of the alkali spectra, as well as the violation of Larmor's theorem, comes about according to this point of view through a peculiar, classically not describable two-valuedness of the quantum-theoretical properties of the optical electron."* [41] The occurrence of doublet spectral lines and their anomalous magnetic splittings had obtained a novel counterpart in the unobservable realm. Since the characteristics and behavior of the core could not play that role, those experimental phenomena were linked with a strange new property of the valence electron.

The consequences of the electron's "two-valuedness" for the classification of spectra were developed in Pauli's subsequent paper, where he proposed his exclusion principle.[42] To understand Pauli's proposal it is necessary to go over some prior developments in the study of x-ray and optical spectra. Two quantum numbers $(n, k)$ had proved inadequate for the classification of x-ray spectra. Since $k$ could take $n$ different values $(1, 2, \ldots, n)$, one should expect that one energy level corresponded to the K-shell $(n = 1)$, two levels to the L-shell $(n = 2)$, three levels to the M-shell $(n = 3)$, and so on. However, experiments indicated the existence of three L levels and five M levels. To remedy this problem, Bohr and Dirk Coster classified spectral terms in the x-ray region in terms of three quantum numbers $(n, k_1, k_2)$. The selection rules of $k_1$ and $k_2$ coincided with the selection rules of the quantum numbers that were used in the classification of spectral terms in the optical region $(k, j)$. The x-ray quan-

---

41. *"Die abgeshlossenen Elektronenconfigurationen sollen nichts zum magnetischen Moment und zum Impulsmoment des Atoms beitragen. Insbesondere werden bei den Alkalien die Impulswerte des Atoms und seine Energieänderungen in einem äußeren Magnetfeld im wesentlichen als eine alleinige Wirkung des Leuchtelektrons angesehen, das auch als der Sitz der magneto-mechanischen Anomalie betrachtet wird. Die Dublettstruktur der Alkalispektren, sowie die Durchbrechung des Larmortheorems kommt gemäß diesem Standpunkt durch eine eigentümliche, klassisch nicht bescreibbare Art von Zweideutigkeit der quantentheoretischen Eigenshaften des Leuchtelektrons zustande."* Ibid., p. 213 (emphasis in original).

42. W. Pauli, "Über den Zusammenhang des Abschlusses der Elektronen-gruppen im Atom mit der Komplexstruktur der Spektren," *Zeitschrift für Physik*, 31 (1925): 765–783 (received on 16 Jan. 1925); repr. in his *Collected Scientific Papers*, vol. 2, pp. 214–232. This paper has been translated in D. ter Haar (ed.), *The Old Quantum Theory* (Oxford: Pergamon Press, 1967). Although I have profited from this translation, some of the translations that follow are mine. The road to the exclusion principle has been reconstructed in detail by John Heilbron. See his "The Origins of the Exclusion Principle."

tum numbers were also similar to their optical counterparts in another respect: the relation between $k_2$ and $k_1$ ($k_2 = k_1$, $k_1 - 1$) was identical with the relation between $j$ and $k$ ($j = k$, $k - 1$). However, Bohr abstained from identifying $k_1$ with $k$ and $k_2$ with $j$.[43]

Landé pointed out the resemblance between the optical and the x-ray spectra and attributed both to a common mechanism, namely, the magnetic interaction between core and outer electrons. In his scheme $k_2$ and $j$ denoted the same *collective* property of the atom, as opposed to a property of particular electrons.[44] Following Landé, Edmund Stoner identified $k_2$ with $j$ and classified x-ray energy levels "by means of three quantum numbers—$n$ (total), $k$ (azimuthal), and $j$ (inner)."[45] In his scheme the maximum number of electrons with the same quantum numbers ($n$, $k$, $j$) was $2j$: "In the classification adopted, the remarkable feature emerges that the number of electrons in each completed level is equal to double the sum of the inner quantum numbers as assigned. . . . It is suggested that the number of electrons associated with each sub-level separately is also equal to double the inner quantum number."[46] This was also the number of atomic states, distinguished in the presence of a weak magnetic field, of an alkali atom whose inner quantum number was $j$.

It is worth pointing out that Stoner interpreted $j$ as a property of individual electrons. This was a nonstandard interpretation, since $j$ was usually attributed to the atom as a whole. There were two possibilities: either "that the atom as a whole is always the same as concerns relative orientation of core and outer electron orbit, and can take up $2j$ different orientations relative to

43. See N. Bohr and D. Coster, "Röntgenspektren und periodisches System der Elemente," *Zeitschrift für Physik,* 12 (1923): 342–374 (received on 2 Nov. 1922). Cf. Heilbron, "The Origins of the Exclusion Principle," pp. 277–278.

44. See A. Landé, "Zur Theorie der Röntgenspektren," *Zeitschrift für Physik,* 16 (1923): 391 396.

45. E. Stoner, "The Distribution of Electrons among Atomic Levels," *Philosophical Magazine,* 6th series, 48 (1924): 719–736, on p. 720. Heilbron suggests that Stoner's identification of $j$ with $k_2$ was probably "an independent discovery." His argument to that effect is that, by the time Stoner recorded his new idea in his diary (11 May 1924), he could not have known of Landé's most recent papers, in which the identification in question was put forward. Heilbron admits that Stoner "could have known the preliminary argument in Landé ["Zur Theorie der Röntgenspektren"] . . . but since he does not seem to have hunted through German journals, independent discovery is no less likely" (Heilbron, "The Origins of the Exclusion Principle," p. 282). However, in his published paper (p. 720) Stoner referred to Landé's 1923 article as the source of the classification of x-ray energy levels in terms of three quantum numbers ($n$, $k$, $j$). Thus, he must have been familiar with Landé's "preliminary argument." For biographical information on Stoner see G. Cantor, "The Making of a British Theoretical Physicist: E. C. Stoner's Early Career," *British Journal for the History of Science,* 27 (1994): 277–290.

46. Stoner, "The Distribution of Electrons," p. 722.

the (weak) field; or that the core takes up a definite orientation relative to the field, and the outer electron orbit can take up $2j$ different orientations relative to the core." [47] Stoner went for the latter possibility:

> The spectral term-values themselves, in so far as they are altered by external fields, would seem to depend primarily on the outer electron orbit itself... and remembering this, it seems reasonable to take $2j$ as the number of equally probable orbits.
>
> Without laying too much stress on any definite physical interpretation, or pressing the analogy too far, it may be suggested that for an inner sub-level, in a similar way, the number of possible orbits is equal to twice the inner quantum number, these orbits differing in their orientation relative to the atom as a whole. [48]

Even though he qualified his interpretation, it is clear that he associated $j$, in both the optical and the x-ray cases, with the orientations of the electron's orbit. It was more plausible that the physical counterpart of $j$ was a property of the electron, rather than the atom. Furthermore, the number of those orientations ($2j$) determined both the number of atomic states and the number of electrons in a sublevel. That is why those two numbers were the same. [49]

Pauli found out about Stoner's paper by reading the preface of the fourth edition of Sommerfeld's *Atombau*. [50] His reaction was very positive: "I am very enthusiastic about Stoner's paper. The more I read it the more I like it. That was an eminently clever idea to connect the number of electrons in the closed subgroups with the number of the Zeeman terms of the alkali spectra." [51] His formulation of the exclusion principle was inspired by Stoner's "eminently clever idea."

Following Bohr's and Coster's scheme for the classification of x-ray spectra, Pauli assigned three quantum numbers to the optical electron ($n, k_1, k_2$). The first was the familiar principal quantum number. The successive values

47. Ibid., pp. 725–726.

48. Ibid., p. 726.

49. Cf. Forman, "The Doublet Riddle," p. 172.

50. Pauli to Landé, 24 Nov. 1924, in Pauli, *Wissenschaftlicher Briefwechsel,* p. 180; Pauli to Sommerfeld, 6 Dec. 1924, in ibid., p. 182; and Pauli to Bohr, 12 Dec. 187, in ibid., p. 187.

51. "Von der Arbeit von Stoner bin ich sehr begeistert. Je öfter ich sie lese, desto mehr gefällt sie mir. Das war doch eine eminent gescheite Idee, die Anzahl der Elektronen in den abgeschlossenen Untergruppen mit der Anzahl der Zeeman-Terme der Alkalispektren in Beziehung zu setzen!" Pauli to Bohr, 12 Dec. 1924, in ibid., p. 188; the translation is from Heilbron, "The Origins of the Exclusion Principle," p. 303.

of the second corresponded to the various spectral terms (s, p, d, f). That number "determines the magnitude of the central force interaction of the optical electron with the rest of the atom."[52] Finally, the third took two values $(k_1, k_1 - 1)$ corresponding to the two terms of a doublet. It "determines the magnitude of the . . . relativity correction."[53] The other two quantum numbers ($j, m_1$) denoted collective properties of the atom (its total angular momentum and the component of that momentum in the direction of an external field, respectively). Note that for (monovalent) alkali atoms these numbers also denote properties of the valence electron.

In the case of strong magnetic fields, instead of $k_2$, he introduced a different quantum number ($m_2$), which denoted the component of the outer electron's magnetic moment in the direction of the field. The two values of that number ($m_1 \pm 1/2$) corresponded to the two terms of a doublet. The anomalous Zeeman splittings were linked with the fact that two quantum numbers ($m_1, m_2$) were needed to represent the angular momentum and the magnetic moment of the outer electron.[54]

An advantage of his proposal was that *"on the basis of this classification, in contrast to the usual view, we can hold on completely to the permanence of quantum numbers (building-up principle) also in the case of the complex structure of the spectra and the anomalous Zeeman effect."*[55] Thus, his scheme avoided the problem of statistical weights that plagued the core model.[56] Furthermore, Bohr's Zwang was revealed "not in a violation of the permanence of quantum numbers in the coupling of the series electron to the rest of the atom, but only in a peculiar two-valuedness of the quantum-theoretical properties of the individual electrons in the stationary states of the atom."[57] Thus, Pauli did not eliminate the notion of Zwang but only reinterpreted it in terms of the two-valuedness of the electron. That notion would be later rendered superfluous

52. "[S]ie bestimmt die Größe der Zentralkraftwechselwirkung des Leuchtelektrons mit dem Atomrest." Pauli, *Collected Scientific Papers*, vol. 2, p. 215.

53. Pauli, *Collected Scientific Papers*, vol. 2, p. 215. The translation is from ter Haar (ed.), *The Old Quantum Theory*, p. 185.

54. See Pauli, *Collected Scientific Papers*, vol. 2, pp. 215–216.

55. *"[A]uf Grund dieser Klassifikation im Gegensatz zur üblichen Auffassung an der Permanenz der Quantenzahlen (Aufbauprinzip) auch bei der Komplexstruktur der Spektren und dem anomalen Zeemaneffekt vollständig festhalten können."* Ibid., p. 216 (emphasis in original).

56. Cf. Serwer, *"Unmechanischer Zwang,"* p. 233.

57. "[N]icht in einer Durchbrechung der Permanenz der Quantenzahlen bei der Kopplung des Serienelektrons an den Atomrest, sondern nur in der eigentümlichen Zweideutigkeit der quantentheoretischen Eigenschaften der einzelnen Elektronen in den stationären Zuständen des Atoms." Pauli, *Collected Scientific Papers*, vol. 2, p. 217. Cf. Serwer, *"Unmechanischer Zwang,"* pp. 232–233.

and the electron's two-valuedness would receive a mechanical interpretation with the proposal of electron spin. Questions of interpretation aside, this novel property of individual electrons turned out to be essential for the explanation of the "anomalous" Zeeman effect, a phenomenon that had challenged the ingenuity of physicists for more than twenty-five years. After a long and arduous process, a satisfactory coordination between the observable and the unobservable realms was finally achieved through the attribution of four quantum numbers to each electron; but more on this below.

The classification of individual electrons in terms of four quantum numbers along with Stoner's proposal led to the formulation of the exclusion principle. As I mentioned above, Stoner had suggested that the total number of electrons in a sublevel was twice the corresponding inner quantum number. In Pauli's notation the number in question was, therefore, $2k_2$. Since $k_2 = k_1$, $k_1 - 1$, it followed that there were $2k_1 + 2(k_1 - 1) = 2(2k_1 - 1)$ electrons with a given quantum number $k_1$. Furthermore, the total number of electrons with a given principal quantum number $n$ was

$$\sum_{k_1=1}^{n} 2(2k_1 - 1) = 2n^2.$$

This number was also the maximum number of magnetic states of an alkali atom. Each of those states corresponded to an orbit of the valence electron and was uniquely defined by Pauli's four quantum numbers. Stoner's insight could now be reformulated as follows: "*There can never be two or more equivalent electrons in an atom for which in strong fields the values of all quantum numbers n, $k_1$, $k_2$, $m_1$ (or, equivalently, n, $k_1$, $m_1$, $m_2$) are the same. If an electron is present in the atom for which these quantum numbers (in an external field) have definite values, this state is 'occupied.'*"[58] As a result of the exclusion principle no identical electrons could coexist in the atom. According to Pauli's representation of its behavior, the electron could not tolerate its identical mates and had become "antisocial."[59]

58. Pauli, *Collected Scientific Papers*, vol. 2, p. 225 (emphasis in original). The translation is from ter Haar (ed.), *The Old Quantum Theory*, p. 196. Pauli had communicated an early formulation of the exclusion principle to Landé on 24 Nov. 1924. See Pauli, *Wissenschaftlicher Briefwechsel*, p. 180. Cf. Heilbron, "The Origins of the Exclusion Principle," p. 304.

59. "One can call particles obeying the exclusion principle the 'antisocial' particles." W. Pauli, "Remarks on the History of the Exclusion Principle," *Science*, 103 (1946): 213–215; repr. in his *Collected Scientific Papers*, vol. 2, pp. 1073–1075, on p. 1074. The expression "land of the formalism-Philistines," which I also used in the heading of this section, was coined by Heisenberg, in an ironic letter to Pauli congratulating him for his exclusion principle paper: "I rejoice that you too (et tu, Brute!) have returned with lowered head in the land of the formalism-Philistines; but don't be sad, you will be received there with open arms" (ich

Pauli gave a visualizable mechanical interpretation of the exclusion principle, which, however, was only necessary for "weak men": "For weak men, who need the crutches of the idea of unequivocally defined electron orbits and mechanical models, one could simply justify the rule by saying that: 'If more than one electron in strong fields possess the same quantum numbers, then they would have the same orbits, therefore they would collide and such cases must be excluded.'" [60] This interpretation, although possible, was not appealing to Pauli, who strongly disliked mechanically intuitive models. On the other hand, it is likely that most of his contemporaries viewed his principle through mechanical spectacles. Heisenberg, for instance, misinterpreted it as the assignment of four degrees of freedom to each individual electron.[61] Bohr also understood Pauli as implying "the use of four dimensions for a description of the electron's orbit." [62] Their interpretation of his proposal in terms of a mechanical model along with their conviction that the progress of atomic physics would be along a nonmechanical path explain why they did not try to decipher his "fourth degree of freedom." [63] It was the attempt to understand in mechanical terms Pauli's attribution of four quantum numbers to the electron itself that gave the electron its spin.[64]

---

triumphiere, daß auch Sie [et tu, Brute!] mit gesenktem Haupt ins Land der Formalismusphilister zurückgekehrt sind; aber seien Sie nicht traurig, Sie werden dort mit offenen Armen empfangen). Heisenberg to Pauli, 15 Dec. 1924, in Pauli, *Wissenschaftlicher Briefwechsel,* p. 192.

60. "Für schwache Menschen, die die Krücke der Vorstellung von eindeutig definierten Elektronenbahnen und von mechanischen Modellen brauchen, könnte man die Regel einfach so begründen, daß man sagt: 'Wenn mehr als ein Elektron in starken Feldern die gleichen Quantenzahlen besitzen, so würden sie ja dieselben Bahnen haben, sie würden daher zusammenstoßen und solche Fälle muß man ausschließen.'" Pauli to Bohr, 31 Dec. 1924, in Pauli, *Wissenschaftlicher Briefwechsel,* p. 197. Cf. Serwer, "*Unmechanischer Zwang,*" p. 235.

61. See Heisenberg to Pauli, 15 Dec. 1924, in Pauli, *Wissenschaftlicher Briefwechsel,* p. 192.

62. "[D]er Gebrauch von 4 Dimensionen für eine Beschreibung der Elektronenbahn." Bohr to Pauli (in Danish), 22 Dec. 1924, published along with a German translation in ibid., pp. 193–196; the quote is from p. 195.

63. Cf. Serwer, "*Unmechanischer Zwang,*" pp. 239–240.

64. Heilbron has argued that the widespread view that Pauli's achievement consisted in the attribution of a fourth quantum number to the electron (in addition to the three that it already had) is inaccurate. In the old quantum theory, according to Heilbron, only two quantum numbers were attributed to individual electrons, whereas the "interpretation of the third, *j*, had been the object of titanic struggle before the discovery of the exclusion principle" ("The Origins of the Exclusion Principle," p. 310). The point about the severe difficulties surrounding the interpretation of *j* is valid. One should note, however, that Pauli was not the first to attribute that quantum number to the electron itself. As we saw above, Stoner had done so before him. Cf. Forman, "The Doublet Riddle," p. 172; and MacKinnon, *Scientific Explanation and Atomic Physics,* p. 183.

## 3. A Reactionary Putsch

The proposal of spin was the culmination of the planetary analogy in the development of the old quantum theory of the atom.[65] The idea of a spinning electron was first suggested by Ralph Kronig in early 1925.[66] The twenty-year-old Kronig, who was in Tübingen at that time, found out about Pauli's recent ideas on the exclusion principle through a letter that Pauli had sent to Landé.[67] In that letter Pauli had assigned four quantum numbers $(n, k_1, m_1, m_2)$ to the valence electron of the alkali atoms (in the presence of a strong magnetic field). Empirical considerations suggested that $m_1$ was half integral ("warum, noch unerklärt") and that $m_2 = m_1 \pm 1/2$.[68] In a retrospective account of his encounter with Pauli's letter, Kronig remarks: "Pauli's letter made a great impression on me and naturally my curiosity was aroused as to the meaning of the fact that *each individual* electron of the atom was to be described in terms of quantum numbers familiar from the spectra of the alkali atoms, in particular the two angular momenta $l$ and $s = 1/2$ [these are modern terms; the terms then current were $k$ and $r$] encountered there. Evidently $s$ [$r$] could now no longer be attributed to a core, and it occurred to me immediately that it might be considered as an intrinsic angular momentum of the electron."[69]

Kronig's proposal shed new light on one of the most intriguing problems of atomic physics, the so-called doublet riddle.[70] This problem was created by the existence of two incompatible explanations of multiplet spectral lines. The first type of explanation capitalized on the fine splitting of the electron's energy levels due to the relativistic variation of its mass (see chapter 6). It was particularly successful in accounting for certain doublets, the so-called regular doublets, in the x-ray region of spectra. It led to the empirically adequate

65. Pauli applied the phrase "reactionary putsch" to the Bohr-Kramers-Slater theory of radiation, but it also captures his attitude toward spin. See Pauli to Kramers, 27 July 1925, in Pauli, *Wissenschaftlicher Briefwechsel*, p. 234. Heilbron has suggested that Pauli's dismissive phrase referred to spin. However, it is clear from the rest of that letter that Pauli's target was the BKS theory. See Heilbron, *Historical Studies in the Theory of Atomic Structure*, p. 12.

66. See R. Kronig, "The Turning Point," in M. Fierz and V. F. Weisskopf (eds.), *Theoretical Physics in the Twentieth Century: A Memorial Volume to Wolfgang Pauli* (New York: Interscience, 1960), pp. 5–39.

67. From Kronig's description it is evident that the letter in question was the one that Pauli had sent to Landé on 24 Nov. 1924.

68. Pauli, *Wissenschaftlicher Briefwechsel*, p. 178.

69. Kronig, "The Turning Point," pp. 19–20.

70. The significance of this problem for the study of atomic structure (circa 1924) has been emphasized by Forman, "The Doublet Riddle."

conclusion that the frequency difference between the components of those doublets was proportional to the factor $(Z - s)^4$, where $Z$ was the atomic number and $s$ was an empirical parameter, the so-called screening constant, corresponding to the shielding of the nucleus by the other electrons.[71] This type of explanation was also applicable to some of the multiplets observed in the optical region of spectra.

The second approach was based on the familiar core model of the atom and explained the appearance of doublets as due to the magnetic interaction between the valence electron and the core. The weakness of this approach was that it failed to account for the connection between the interval separating the doublets and $Z^4$. Thus, the doublet riddle consisted in the existence of two distinct and not wholly adequate mechanisms for the production of a phenomenon which was supposed to have a single cause.[72] It was not clear whether that phenomenon was a manifestation of the velocity dependence of the electron's mass or the outcome of the magnetic interaction between the core and the valence electron.

Kronig recast the doublet riddle by proposing a magnetic interpretation of the origin of doublets, which duplicated the successes of the relativistic explanation of that origin. By means of the hypothesis of a spinning electron he obtained the dependence of the alkali doublets' separation on the fourth power of the effective atomic number. Since he assumed that the electron, because of its intrinsic rotation, possessed a magnetic moment ($\mu$) equal to one Bohr magneton ($eh/4\pi mc$), he had to calculate its motion by taking into account the interaction between that moment and the magnetic field created by the core in the frame of reference of the electron. The magnitude of that field is equal to

$$H = \frac{E \times v}{(c^2 - v^2)^{1/2}},$$

where $E$ is the electric field of the core and $v$ the velocity of the electron. Because of this magnetic interaction the energy levels of the electron shift by $-\mu H$ if the spin vector has the same direction with the magnetic field and by $+\mu H$ if it has the opposite direction. In this way the doublet splitting of spectral terms and its dependence on the fourth power of $Z$ could be derived. This hitherto recalcitrant phenomenon had now obtained a novel counterpart in the representation of the electron.

The question whether the new hypothesis was compatible with the ex-

71. See Sommerfeld, *Atomic Structure and Spectral Lines* (1923), p. 498.
72. Cf. Serwer, "*Unmechanischer Zwang,*" pp. 202–203.

perimental data on the fine splitting of the "hydrogen-like" spectral lines then arose. The proposal of spin had to cohere with the former writings of the electron. Recall that Sommerfeld had explained those writings on the basis of the relativistic variation of its mass. Now this explanation had to be reconsidered, in light of the novel proposal. If the electron had spin, this would affect its spectral manifestations, even those that had already been explained. Kronig hoped that the combined effect of relativity and spin would somehow give the same results as those originally predicted by Sommerfeld. However, "closer examination showed that the new coupling was too large by a factor of 2." [73] Thus, the enrichment of the electron's representation resulted in a violation of the experimental constraints it had to satisfy. To put it another way, the electron qua theoretical entity resisted Kronig's attempt to transform it.

Kronig explained his proposal to Pauli, who had arrived at Tübingen one day after him. Pauli's reaction was negative. Even though he characterized spin "quite a clever idea" (ein ganz witziger Einfall), he "did not believe that the suggestion had any connection with reality." [74] As a result of Pauli's dismissive attitude, Kronig did not publish his hypothesis. [75]

It should be noted that Kronig was not the first to portray the electron as a tiny magnet. As we saw in chapter 7, several chemists (e.g., Parson, Lewis, and Langmuir) had already endowed the electron with an intrinsic magnetic disposition. Furthermore, A. H. Compton, in connection with his experiments on the scattering of x-rays by crystals, had suggested "that the elementary magnet is not the atom as a whole . . . it is something within the atom, presumably the electron, which is the ultimate magnetic particle." [76] None of those scientists, however, had employed the idea of a magnetic electron for the elucidation of the problems of spectroscopy.

Those problems led Goudsmit and Uhlenbeck to the notion of spin.

73. Kronig, "The Turning Point," p. 21. The problem of the "factor of 2" will be further discussed below.

74. Ibid.

75. The accuracy of this story is confirmed by the historical record. When Goudsmit and Uhlenbeck subsequently published a similar proposal, L. H. Thomas made the following remark about Kronig's early suggestion of a spinning electron: "I think you and Uhlenbeck have been very lucky to get your spinning electron published and talked about before Pauli heard of it. It appears that more than a year ago Kronig believed in the spinning electron and worked out something; the first person he showed it to was Pauli. Pauli ridiculed the whole thing so much that the first person became also the last and no one else heard anything of it." Thomas to Goudsmit, 25 Mar. 1926, reproduced in S. A. Goudsmit, "Guess Work: The Discovery of the Electron Spin," Delta, 15 (Summer 1972): 77–91, on p. 91.

76. A. H. Compton, "The Magnetic Electron," Journal of the Franklin Institute, 192 (1921): 145–155, on p. 149. Cf. M. Jammer, The Conceptual Development of Quantum Mechanics (New York: McGraw-Hill, 1966), p. 148.

Pauli's new scheme was unable to explain the existence of two kinds of alkaline earth spectral lines (singlets and triplets). The core model, on the other hand, handled successfully those distinct systems of lines by relating them to different values of the angular momentum of the core. In order to remedy this defect of Pauli's scheme, Goudsmit, who was an expert in the classification of spectra, reformulated it by employing quantum numbers from the core model ($m_R$ and $m_k$). By this modification, one could distinguish the alkaline earth singlets and triplets.[77] It should be noted that, in the revised scheme, the meaning of $m_R$ was different from its meaning in the context of the core model. In particular, $m_R$ referred to a property of the valence electron itself, as opposed to a property of the core. Its only possible values were $\pm 1/2$. Goudsmit did not attempt, however, to interpret this new property of the electron in terms of a physical model.

The road to spin passed through another development, the unification of the classification of the electron's states in the hydrogen atom (in general, one-electron systems) with the corresponding classification in the alkalis (many-electron systems). In the former case, three quantum numbers ($n$, $k$, $m$) were employed, whereas in the latter four were needed (as indicated by Pauli's work). The distinction was grounded on the fact that the hydrogen atom exhibited a normal Zeeman effect, which had been explained by Sommerfeld in 1916 (see chapter 6), whereas the alkalis displayed anomalous Zeeman patterns for which Sommerfeld's analysis was inadequate.[78]

This distinction was challenged by Goudsmit and Uhlenbeck.[79] As Uhlenbeck recalled in an interview with Thomas Kuhn, "I was so dissatisfied about the fact that one had to learn the theory for the alkali atoms and then one had to still keep the old theory for the hydrogen atom. I thought that was terrible. There should be at least some relation between the two. . . . [I]t was clearly very unsatisfactory that there were two theories about things which should

77. See S. A. Goudsmit, "Über die Komplexstruktur der Spektren," *Zeitschrift für Physik*, 32 (1925): 794–798 (received on 7 May 1925). Cf. N. Robotti, "Quantum Numbers and Electron Spin: The History of a Discovery," *Archives internationales d'histoire des sciences*, 40, no. 125 (1990): 305–331, on pp. 314–316.

78. In fact, there was evidence that the spectrum of hydrogen also exhibited a Paschen-Back effect, but it was overshadowed by the success of Sommerfeld's theory. See Forman, "The Doublet Riddle," esp. p. 163; and C. Jensen, "Two One-Electron Anomalies in the Old Quantum Theory," *Historical Studies in the Physical Sciences*, 15, no. 1 (1984): 81–106. Cf. also N. Robotti, "The Zeeman Effect in Hydrogen and the Old Quantum Theory," *Physis*, 29 (1992): 809–831.

79. S. A. Goudsmit, G. E. Uhlenbeck, "Opmerking over de Spectra van Waterstof en Helium," *Physica*, 5 (1925): 266–270.

be similar."[80] Both the alkali and the hydrogen spectra had to obtain a common description, since they were supposed to be manifestations of a single entity. The electron was involved in both cases, and, therefore, it was unacceptable to have two unrelated representations of its states.[81]

Goudsmit and Uhlenbeck were probably inspired by Gregor Wentzel, who had offered a unified relativistic treatment of both kinds of spectra (hydrogen and alkali).[82] Wentzel had employed three quantum numbers $(n, K, J)$ to label an atom's energy levels. The values of $J$ corresponded to the values of $k$ in Sommerfeld's representation. The quantum number $K$, on the other hand, could assume the values $J \pm 1/2$, provided that it did not exceed the value of $n$. As Goudsmit and Uhlenbeck pointed out, this new unified approach had repercussions for the fine structure of the spectral lines due to ionized helium. The splitting of one of those lines ($\lambda 4686$), observed by Paschen in 1916, could not be adequately explained by Sommerfeld's relativistic analysis. The new scheme, on the other hand, opened the road for an accommodation of the recalcitrant line. The increase of the number of possible values of $K$ resulted in a proliferation of the energy levels, which could be exploited to explain the extra fine structure components observed by Paschen. The fine structure pattern derived in that way was in agreement with the available experimental data.[83]

As Nadia Robotti has noted, the unified treatment of the hydrogen and the alkali spectra was a prerequisite for the proposal of a novel, context-independent property of the electron.[84] The story of that proposal goes as follows: In the summer of 1925 Goudsmit initiated Uhlenbeck to the intricacies of atomic spectra and informed him about Pauli's principle, reformulated in terms of the quantum numbers of the core model.[85] In Uhlenbeck's words, "[S]ince the whole argument was purely formal, it seemed like abracadabra to me. There was no picture that at least qualitatively connected Pauli's formal-

80. Goudsmit and Uhlenbeck interview, 7 Dec. 1963, Archive for the History of Quantum Physics, p. 2; quoted in Robotti, "Quantum Numbers and Electron Spin," p. 319.

81. Cf. M. Morrison, "History and Metaphysics: On the Reality of Spin," in J. Z. Buchwald and A. Warwick (eds.), *Histories of the Electron: The Birth of Microphysics* (Cambridge, Mass.: MIT Press, 2001), pp. 431, 434.

82. See G. Wentzel, "Zum Termproblem der Dublettspektren, insbesondere der Röntgenspektren," *Annalen der Physik*, 76 (1925): 803–828. Cf. Robotti, "Quantum Numbers and Electron Spin," p. 319.

83. Cf. H. Kragh, "The Fine Structure of Hydrogen and the Gross Structure of the Physics Community, 1916–26," *Historical Studies in the Physical Sciences*, 15 (1985): 67–125, pp. 114–115.

84. Robotti, "Quantum Numbers and Electron Spin," p. 327.

85. G. E. Uhlenbeck, "Personal Reminiscences," *Physics Today*, June 1976, 43–48, p. 45.

ism with the old Bohr atomic model. It was then that it occurred to me that, since (as I had learned) each quantum number corresponds to a degree of freedom of the electron, the fourth quantum number must mean that the electron had an additional degree of freedom—in other words, the electron must be rotating!"[86] Uhlenbeck suggests that the representation of the electron in the old quantum theory, which linked quantum numbers with degrees of freedom of the electron, showed him how to interpret the new quantum number proposed by Pauli. Once again, the representation of the electron played a guiding role. It indicated a plausible way to understand physically Pauli's proposal, that is, as endowing the electron with an additional degree of freedom. This interpretation required, in turn, a revision of another aspect of the representation. The notion of an electron endowed with four degrees of freedom was incompatible with its ordinary conception as a point particle with only three degrees of freedom. The only way to understand the extra degree of freedom in mechanical terms was by considering the electron "to be a small sphere that could rotate."[87]

As I mentioned above, Goudsmit had transferred the properties of the core to the electron. The electron, thus, acquired an intrinsic magnetic moment (one Bohr magneton) that was twice its magnetic moment due to its orbital motion. The question whether that property could be accommodated within the classical electromagnetic representation of the electron then arose.[88] Indeed, on Ehrenfest's suggestion, Uhlenbeck managed to explain this property, by capitalizing on Abraham's analysis of the gyromagnetic ratio of a spherical (surface) distribution of charge.[89] On the assumption that the electron was a rotating sphere whose charge was distributed on its surface, the required value of its magnetic moment followed.

However, there were two undesirable consequences of this assumption. When Goudsmit and Uhlenbeck consulted Ehrenfest about their new idea, he advised them to ask the opinion of Lorentz, an authority on classical electro-

---

86. Ibid., p. 46.

87. G. E. Uhlenbeck, *Oude en nieuwe vragen der natuurkunde,* address delivered at Leiden on 1 Apr. 1955 (Amsterdam: North-Holland, 1955). A section of this address has been translated by B. L. van der Waerden in his "Exclusion Principle and Spin," in Fierz and Weisskopf (eds.), *Theoretical Physics in the Twentieth Century,* pp. 199–244; the quote is from p. 213.

88. This question was raised by Goudsmit in a postcard to Uhlenbeck. See Uhlenbeck, "Personal Reminiscences," p. 47.

89. For Abraham's theory see M. Abraham, "Prinzipien der Dynamik des Elektrons," *Annalen der Physik,* 10 (1903): 105–179; and S. Goldberg, "The Abraham Theory of the Electron: The Symbiosis of Experiment and Theory," *Archive for History of Exact Sciences,* 7, no. 1 (1970): 7–25.

magnetic theory.[90] They followed Ehrenfest's suggestion and got in touch with Lorentz, "who was very much interested, although, I feel, somewhat sceptical too."[91] In a week Lorentz managed to derive some very disturbing results from their hypothesis. Even though his analysis was opaque to them, "it was quite clear that the picture of the rotating electron, if taken seriously, would give rise to serious difficulties. For one thing, the magnetic energy would be so large that by the equivalence of mass and energy the electron would have a larger mass than the proton, or, if one sticks to the known mass, the electron would be bigger than the whole atom! In any case, it seemed to be nonsense."[92] The magnetic energy of the spinning electron was approximately $\mu^2/\alpha^3$, where $\mu$ was its magnetic moment and $\alpha$ its radius. Since the magnetic moment was supposed to be $eh/4\pi mc$ and the magnetic energy was given by $mc^2$, it followed that the order of magnitude of $\alpha$ was $10^{-12}$ cm, a value too large to be acceptable.[93]

Furthermore, Lorentz showed that an electron with a classical radius, in order to have the required angular momentum ($h/2\pi$), should have "a surface velocity . . . about ten times the light velocity!"[94] The attempt to interpret the angular momentum of the electron in classical terms led to a violation of the special theory of relativity, which was also supposed to constrain the electron's behavior: "[O]ur enthusiasm was considerably reduced when we saw that the rotational velocity at the surface of the electron had to be many times the velocity of light!"[95]

This situation reveals, once more, the agency of the electron qua theoretical entity. There were several experimental and theoretical constraints that the electron's representation had to satisfy. On the one hand, as we saw in chapter 4, the electron had been endowed with some experimentally deter-

90. The accuracy of Uhlenbeck's recollections is confirmed by a letter from Ehrenfest to Lorentz, requesting his "judgment and advice on a *very* witty idea of Uhlenbeck about spectra." P. Ehrenfest to H. A. Lorentz, 16 Oct. 1925, Archive for the History of Quantum Physics, quoted in A. Pais, *Inward Bound* (New York: Oxford Univ. Press, 1988), pp. 277–278.

91. Uhlenbeck, *Oude en nieuwe vragen der natuurkunde*, trans. in van der Waerden, "Exclusion Principle and Spin," p. 214.

92. Ibid.

93. Cf. Pais, *Inward Bound*, p. 278. Lorentz's classical analysis of an electron with an intrinsic rotation was subsequently published. See H. A. Lorentz, "Sur la rotation d'un électron qui circule autour d'un noyau," in his *Collected Papers*, ed. P. Zeeman and A. D. Fokker, 9 vols. (The Hague: Martinus Nijhoff, 1935–1939), vol. 7, pp. 179–204.

94. Uhlenbeck, "Personal Reminiscences," p. 47. The details of this argument are discussed below.

95. Ibid.

mined properties: a certain mass, charge, and radius. On the other hand, its purported behavior was constrained by accepted scientific theories. One of those theories was the special theory of relativity, and, therefore, the electron was supposed to have an upper limit in its velocity. When a new property was added to the electron, it had to be accommodated within the theoretical framework in which the electron's behavior had been understood and, moreover, to cohere with its formerly established properties. The theoretical framework in question consisted of a hodgepodge of classical mechanics (or relativistic mechanics, depending on the context), classical electromagnetic theory, and quantum rules restricting their domain of application. So a mechanical interpretation of the new property of the electron was not excluded on theoretical grounds.[96] Goudsmit and Uhlenbeck put forward such an interpretation and represented the electron as a tiny, rotating, charged ball, with a magnetic moment that was due to its rotation. As it turned out, however, this proposal had some disturbing implications. The new property, interpreted within the context of classical mechanics and electromagnetic theory, implied unacceptable values of the electron's radius (or mass) and a surface velocity that was prohibited by the special theory of relativity. Thus, the attempt to enrich the electron's representation ended up in a negation of some of its experimentally determined properties and a violation of a theoretical principle it was supposed to obey. These unintended consequences rendered incoherent the representation of the electron, a clear symptom of the representation's resistance to manipulation.

The emergence of incoherence in the electron's representation was, to some extent, due to relativity theory. However, other features of the representation had to be in place for that theory to come into play. A new property of the electron that was the unobservable counterpart of various experimental results had to be understood in a certain way, that is, as an intrinsic angular momentum. Only given that property and its particular interpretation did the theory of relativity originally become relevant to the problem situation.[97] The incoherence in question was the result of the conjunction of all of these factors.

At any rate, some further adjustment was necessary to restore the lost coherence in the electron's representation. The novel property, as well as the special theory of relativity, was there to stay. Instead, the blame was put on its

---

96. Of course, as I already noted, there were physicists, most notably Pauli, who detested any attempt to understand the behavior of the electron in mechanical terms.

97. As it turned out later, through Dirac's work, the connection between spin and relativity was deeper and did not presuppose a mechanical interpretation of spin.

mechanical interpretation. The difficulty of incorporating the proposal of spin into the framework of classical mechanics and electromagnetic theory was one more indication of the inadequacy of that framework. The construction of an adequate representation of the electron, free of contradictions, was not possible in a classical context. It was only made possible with the new quantum mechanics, which reformulated the notion of spin without reference to any classical visualizable model.[98]

After realizing the difficulties generated by the hypothesis of spin, Goudsmit and Uhlenbeck asked Ehrenfest, who had suggested publishing their proposal in *Die Naturwissenschaften*, not to send their note, but it was too late. He had already sent it off to the journal.[99] It is evident from the title of that note that spin was proposed as a substitute for Bohr's *Zwang*.[100] It started with a brief description of the core model of the atom. Despite its utility for illuminating the structure of complicated spectra, that model, as I mentioned above, gave rise to several conceptual problems. Pauli's attribution of a two-valuedness and four quantum numbers to every single electron was a first step toward the solution of those problems. The final solution could be reached by a reformulation of his proposal in terms of the quantum numbers employed in the core model, appropriately reinterpreted. In the scheme proposed by Uhlenbeck and Goudsmit the four quantum numbers had the following meaning: $n$ and $k$ were the usual principal and azimuthal quantum numbers. "However, $R$ [which in the core model of the atom referred to the angular momentum of the core] will correspond to an intrinsic rotation of the electron."[101] The remaining quantum numbers $(m, J)$ "retain their old reference."[102] As a consequence of the shift of $R$'s reference, "the electron must now take over the still not understood property . . . , which Landé attributed

---

98. It is interesting that a very similar situation sometimes occurs in experimental practice. In Rheinberger's words, what is "intended as a mere substitution or addition within the confines of a[n experimental] system will reconfigure that very system—sometimes beyond recognition." H. Rheinberger, *Toward a History of Epistemic Things: Synthesizing Proteins in the Test Tube* (Stanford: Stanford Univ. Press, 1997), p. 4.

99. This often-told story comes from Uhlenbeck's recollections. Goudsmit, on the other hand, has not confirmed it ("I don't remember that"; Goudsmit, "Guess Work," p. 86). Besides, in the published note there is a footnote pointing out that the surface velocity of the electron should exceed that of light. Unless they added the footnote after the note was sent to the journal, they must have been aware of the relativistic difficulty before giving their draft to Ehrenfest.

100. G. E. Uhlenbeck and S. Goudsmit, "Ersetzung der Hypothese vom unmechanischen Zwang durch eine Forderung bezüglich des inneren Verhaltens jedes einzelnen Elektrons," *Die Naturwissenschaften*, 13, no. 47 (20 Nov. 1925): 953–954 (dated 17 Oct.).

101. "R aber wird man eine eigene Rotation des Elektrons zuordnen." Ibid., p. 954.

102. "[B]ehalten ihre alte Bedeutung." Ibid.

to the atomic core." [103] That is, "The ratio of the magnetic moment of the electron to its mechanical angular momentum must be for the intrinsic rotation twice as great as for the orbital motion." [104] Moreover, the occurrence of doublet spectral terms would be the observable manifestation of the different directions that $R$ could assume in relation to the orientation of the orbit.

Uhlenbeck and Goudsmit did not neglect to mention that the hypothesis of a spinning electron was incompatible with the special theory of relativity. If one assumed that the electron was a spherical rotating object whose charge was distributed on its surface, its rotational energy was, according to Abraham's theory,

$$\frac{1}{9}\frac{e^2\alpha}{c^2}\dot{\phi}^2,$$

where $\alpha$ is the radius of the electron. Therefore, its intrinsic angular momentum was

$$p_\phi = \frac{2}{9}\frac{e^2\alpha}{c^2}\dot{\phi}.$$

"However, one observes that, if this rotational motion is quantized, the peripheral velocity of the electron would exceed considerably the velocity of light." [105] We see here how the negative agency of the electron's representation unfolded. By interpreting the electron's intrinsic angular momentum in terms of a classical model, and by quantizing the value of that momentum in accordance with the empirical requirements of spectroscopy, the electron would violate the special theory of relativity, one of the theoretical constraints on its representation. The conjunction of the electron's writings, a particular model of its behavior, and part of the accepted theoretical framework turned out to be incoherent.

A few weeks after the publication of their first note, Goudsmit and Uhlenbeck, on Bohr's suggestion, employed the hypothesis of spin to develop a

103. "Das Elektron muß jetzt die noch unverstandene Eigenschaft . . . , welche LANDÉ dem Atomrest zuschrieb, übernehmen." Ibid.

104. "Das Verhältnis des magnetischen Momentes des Elektrons zum mechanischen muß für die Eigenrotation doppelt so groß sein als für die Umlaufsbewegung." Ibid. Cf. Jammer, *The Conceptual Development of Quantum Mechanics*, p. 151.

105. "Man beachte aber, daß wenn man diese Rotationsbewegung quantisiert, die periphere Geschwindigkeit des Elektrons die Lichtgeschwindigkeit weit übertreffen würde." Uhlenbeck and Goudsmit, "Ersetzung der Hypothese vom unmechanischen Zwang," p. 954.

new classification of the energy levels of the hydrogen atom.[106] This classification coincided with the one they had proposed in *Physica*, without attempting to explain it in terms of a physical model (see n. 79 above). The advantage of the new theory was that "[it] explains at once the occurrence of certain components in the fine structure of the hydrogen spectrum and of the helium spark spectrum which according to the old scheme would correspond to transitions where *K* remains unchanged. . . . [T]heir occurrence would be in disagreement with the correspondence principle, which only allows transitions in which the azimuthal quantum number changes by one unit. In the new scheme we see that, in the transitions in question, *K* will actually change by one unit and only *J* will remain unchanged."[107] Thus, the new scheme could save the phenomena without violating the correspondence principle, in contrast to the old representation.

Shortly after the publication of their note to *Nature*, in a letter to the editor of that journal Kronig elaborated on the difficulties created by the notion of a spinning electron, difficulties which Goudsmit and Uhlenbeck "fail to point out."[108] The new problem that he detected concerned the implications of spin for nuclear structure. If a magnetic moment were an intrinsic property of the electron, then nuclear electrons would also possess it.[109] It followed that the nucleus would "have a magnetic moment of the order of a Bohr magneton, unless the magnetic moments of all the nuclear electrons just happened to cancel. For such an additional moment of the nucleus there is no place in the theory of the Zeeman effect."[110] Despite his "codiscovery" of spin, Kronig's verdict was uncompromising: "The new hypothesis, therefore, appears rather to effect the removal of the family ghost from the basement to the sub-basement, instead of expelling it definitely from the house."[111]

The reception of the new hypothesis was, however, positive. Consider, for instance, the case of Bohr. Till the end of 1925 he continued to believe in the need for an *unmechanischer Zwang*.[112] In December 1925 he visited Leiden

---

106. See G. E. Uhlenbeck and S. Goudsmit, "Spinning Electrons and the Structure of Spectra," *Nature*, 117 (1926): 264–265 (dated Dec. 1925).

107. Ibid., p. 264.

108. R. Kronig, "Spinning Electrons and the Structure of Spectra," *Nature*, 117 (1926): 550.

109. At the time, physicists still believed that the nucleus contained electrons. See H. Kragh, *Quantum Generations: A History of Physics in the Twentieth Century* (Princeton: Princeton Univ. Press, 1999), pp. 174ff.

110. Kronig, "Spinning Electrons and the Structure of Spectra," p. 550.

111. Ibid. Cf. Jammer, *The Conceptual Development of Quantum Mechanics*, pp. 151–152.

112. See N. Bohr, "Atomic Theory and Mechanics," *Nature*, 116, suppl. (5 Dec. 1925): 845–852. Cf. Serwer, "*Unmechanischer Zwang*," p. 249.

for the occasion of the fiftieth anniversary of Lorentz's dissertation. There he met Ehrenfest and Einstein, who wanted to know his reaction to spin. Bohr's worry was that he could not fathom the origin of the magnetic field that interacted with the electron's spin. In the core model there was a magnetic coupling between the core and the valence electron. If, however, one abandoned the idea of a magnetic core (as Pauli had) and transferred its magnetic disposition to the electron itself (as Goudsmit and Uhlenbeck had), then one was left, so it seemed, without any magnetic interaction inside the atom. On the one hand, the core lacked magnetism, so it could not interact with the magnetic field created by the valence electron. On the other hand, there was no magnetic field to interact with the magnetic moment of the valence electron. This difficulty was resolved with the resources provided by the theory of relativity. As Einstein suggested, in the framework of the valence electron the core was in motion and, thus, generated a magnetic field which interacted with the spinning electron. Bohr's query had been answered to his satisfaction.[113] Hence, he became "convinced that it [spin] signifies an exceedingly great progress in the theory of atomic structure."[114] On his return trip he met Pauli at the Berlin train station. When Pauli found out that Bohr thought favorably of the new conjecture, he dismissed spin as "a new heresy from Copenhagen."[115] Bohr subsequently portrayed himself as "a prophet of the electron magnet gospel" and endorsed the new hypothesis in print.[116]

Pauli's resistance was, to some extent, due to another unexpected consequence of the hypothesis of a spinning electron.[117] Its double gyromagnetic ratio led to a value of the doublet splitting that was twice the corresponding empirical value. Once more, the electron's writings suggested that there was something wrong with its representation. Heisenberg pointed out this problem in a letter to Goudsmit, applauding him for his "courageous note" and

113. Cf. van der Waerden, "Exclusion Principle and Spin," p. 214.

114. "Ich bin überzeugt dass es ein überaus grosser Fortschritt in der Theorie des Atombau bedeutet." Bohr to Ehrenfest, 22 Dec. 1925, reproduced in Uhlenbeck, "Personal Reminiscences," p. 46.

115. This story was told to Pais by Bohr himself in 1946. See Pais, *Inward Bound*, pp. 278–279. There is contemporary evidence that Pauli had indeed characterized spin as an "Irrlehre." See Bohr to Pauli, 20 Feb. 1926, in Pauli, *Wissenschaftlicher Briefwechsel*, p. 294.

116. "[E]in Profet des Elektronmagnet-Evangeliums." Bohr to Ehrenfest, 22 Dec. 1925. For Bohr's public approval of spin see his note in *Nature,* 117 (1926): 265 (dated Jan. 1926).

117. For an extensive discussion of Pauli's reaction to spin and his attempt to incorporate it into a quantum-mechanical framework, see S. Richter, "Wolfgang Pauli und die Entstehung des Spin-Konzepts," *Gesnerus,* 33 (1976): 253–270, esp. pp. 261ff.

asking him "how you got rid of the factor of 2." [118] Goudsmit and Uhlenbeck managed to carry out the calculation and confirmed the difficulty detected by Heisenberg.[119]

The riddle of the infamous "factor of 2" was solved by L. H. Thomas. In a letter to *Nature*, he managed to remove the discrepancy between the theoretical and the experimental values of the doublet splitting, by drawing on the resources of relativity theory.[120] Relativity theory was not only a constraint on the electron's representation but also a heuristic element that could be exploited for the resolution of unanticipated problems. When the hypothesis of spin was proposed, certain relativistic considerations were not taken into account. In other words, the representation of the electron's behavior was, in certain respects, simplified. When anomalies appeared, a plausible move toward their elimination was the removal of those simplified elements from the representation of the electron. To put it in terms of my biographical perspective on theoretical entities, when the electron's writings turned out to be incompatible with its representation, the latter and the theoretical framework in which it was embedded suggested a way to resolve the problem.

In particular, in the calculation of the interaction between the electron's spin and the magnetic field of the nucleus the relativistic precession of the electron's orbit had been neglected. The conjunction of this relativistic effect with the spin-field interaction eliminated the unwanted factor of 2. As Thomas observed, the anomalous Zeeman effect indicated that "the spin axis of the electron precesses about an external magnetic field $H$ with angular velocity

$$\left(\frac{e}{mc}\right)H."$$

118. Heisenberg to Goudsmit, 21 Nov. 1925, Archive for the History of Quantum Physics, microfilm 60, section 5; quoted in Serwer, *"Unmechanischer Zwang,"* p. 250. Heisenberg also mentioned this problem in a postcard to Pauli, asking him "was Sie über Goudsmits Note in den Naturwissenschaften denken." Heisenberg to Pauli, 21 Nov. 1925, in Pauli, *Wissenschaftlicher Briefwechsel,* pp. 261–262. In a subsequent letter to Pauli (24 Nov. 1925) he supplied the calculation on which his argument was based. See Pauli, *Wissenschaftlicher Briefwechsel,* pp. 264ff. Cf. van der Waerden, "Exclusion Principle and Spin," p. 215.

119. Uhlenbeck and Goudsmit, "Spinning Electrons and the Structure of Spectra," p. 265. Bohr, on the other hand, did not consider this a serious difficulty and attributed it to an inadequate calculation. See Morrison, "History and Metaphysics: On the Reality of Spin," pp. 433, 435.

120. L. H. Thomas, "The Motion of the Spinning Electron," *Nature,* 117 (1926): 514.

Since the magnetic field acting on the electron was

$$H = \frac{1}{c}[E \times v],$$

the angular velocity of the precession of the electron's spin would be

$$\frac{e}{mc^2}[E \times v].$$

However, the energy shift of the electron's energy levels due to this precession "would lead to twice the observed doublet separation." [121]

This discrepancy between the electron's writings and the theoretical representation of its behavior disappeared when one took into account relativistic kinematics. The above precession was determined by assuming that the electron's velocity with respect to the nucleus was constant $(v)$, that is, by neglecting the electron's acceleration $(f)$. The effect of this acceleration was a decrease of the magnetic field experienced by the electron. The precession of the electron's spin axis[122] would also be reduced by

$$\frac{1}{2c^2}[v \times f].$$

Since $f$ was approximately $-(e/m)E$, it followed that the precession in question was

$$\frac{e}{mc^2}[E \times v] - \frac{1}{2c^2}[v \times f] = \frac{e}{mc^2}[E \times v] - \frac{e}{2mc^2}[E \times v]$$

$$= \frac{e}{2mc^2}[E \times v].$$

The magnitude of that precession was half its previously assumed value, and, therefore, the unwanted factor of 2 in the shift of the electron's energy levels vanished. Once again, the successful coordination between the electron's

121. Ibid. Thomas did not give a detailed analysis of that shift. For such an analysis see A. Sommerfeld, *Atomic Structure and Spectral Lines,* trans. from the 5th German ed. by H. L. Brose (London: Methuen, 1934), pp. 656–658. Sommerfeld's treatment was based on wave mechanics, but in a footnote he provided the results of the "older theory."

122. Thomas does not show how he obtained this result. For a detailed proof I refer the interested reader to Sommerfeld, *Atomic Structure and Spectral Lines* (1934), pp. 662–667. Cf. also J. Mehra and H. Rechenberg, *The Historical Development of Quantum Theory,* vol. 1, *The Quantum Theory of Planck, Einstein, Bohr, and Sommerfeld: Its Foundation and the Rise of Its Difficulties, 1900–1925* (New York: Springer-Verlag, 1982), pp. 708–709.

writings and its representation was achieved by bringing the theory of relativity to bear on the description of its behavior.

The result of Thomas's analysis was surprising: "[I]t seemed unbelievable that a relativistic effect could give a factor of two instead of something of order $v/c$. . . . [E]ven the cognoscenti of the relativity theory (Einstein included!) were quite surprised."[123] It was, however, crucial for the reception of spin. For instance, it neutralized Pauli, who was one of the most vigorous opponents of the new hypothesis. As he mentioned in his Nobel Lecture, it was Thomas's calculation that persuaded him of the validity of spin.[124] After an intensive correspondence with Bohr on the soundness of Thomas's reasoning, he finally surrendered: "Now nothing else is left to me than to capitulate completely!"[125]

## 4. Concluding Remarks

The proposal of spin was an important contribution to the theory of atomic structure. It explained the occurrence of half-integral quantum numbers and the two-valuedness of the electron and made possible an adequate interpretation of the complex structure of spectral lines and the anomalous Zeeman effect. These experimental phenomena had now obtained a counterpart in the representation of the electron. However, the enrichment of the representation generated conceptual and empirical problems of its own. Some of them (e.g., the factor of 2) were resolved by drawing on existing theoretical resources and the complexity of the electron's behavior. Others (e.g., the superluminal velocity of the electron's surface) could not be accommodated within the existing theoretical framework and required a more radical treatment.[126] The heuristic elements of the electron's representation and its resistance to theoretical manipulation rendered it, once more, an active participant in the development of scientific knowledge.

123. Uhlenbeck, "Personal Reminiscences," p. 48.

124. See W. Pauli, "Exclusion Principle and Quantum Mechanics," Nobel Lecture, 13 Dec. 1946, in *Nobel Lectures, Physics, 1942–1962* (Amsterdam: Elsevier, 1964), pp. 27–43, on p. 30.

125. "Jetzt bleibt mir nichts anderes übrig, als *vollständig zu kapitulieren!*" Pauli to Bohr, 12 Mar. 1926, in Pauli, *Wissenschaftlicher Briefwechsel,* p. 310. The correspondence in question took place between 10 Feb. and 12 Mar. See ibid., pp. 295–304, 308–312. Pauli also sent a postcard to Goudsmit to the same effect. Pauli to Goudsmit, 13 Mar. 1926, in ibid., p. 313.

126. For the fortunes of spin after 1925 and the continuing problems with making physical sense of this property, see Morrison, "History and Metaphysics: On the Reality of Spin."

## Chapter 9

Identifying the Electron: Meaning Variance
and the Historicity of Scientific Realism

We are speaking different languages, as always, . . . but that doesn't change the things
we talk about.

Mikhail Bulgakov, *The Master and Margarita*

The working physicist is the expert on electrons; but the historian of physics knows more
than he about the meaning of "electron."

David Lewis, "How to Define Theoretical Terms"

## 1. Introduction

From the late nineteenth century till the mid-1920s the
concept of the electron changed significantly. In the late
nineteenth century several physicists (e.g., Larmor and J. J.
Thomson) represented the electron as a structure (singularity) in the all-pervading ether, which obeyed the laws of
classical mechanics and electromagnetic theory. This representation was subsequently abandoned along with the concept of the ether, and the laws that were supposed to govern the electron's behavior were revised. The validity of
classical mechanics at the subatomic level was gradually restricted, and it turned out that the electron inside the atom
did not obey, for the most part, the laws of classical electrodynamics. Consider, for instance, how the description of the
electron's mechanism of radiation changed. Before Bohr's
1913 theory, an electron in orbital motion was supposed to
emit radiation, whose frequency coincided with the electron's frequency of revolution, whereas, according to the
research program initiated by Bohr, only the electron's

quantum "jumps" gave rise to radiation, whose frequency was now determined by a new law (Bohr's frequency condition). Furthermore, the representation of the electron's properties in the interior of the atom altered remarkably from the age of Thomson to that of Pauli and Heisenberg. New properties (quantum numbers, selection rules, spin) were proposed, and a whole new mechanics was constructed that aimed, among other things, at systematizing those properties. Other properties (such as the idea of electronic orbits within the atom) were simply abandoned.

Besides diachronic differences in the description associated with the term "electron," we came across synchronic variations of that description. As we saw in chapter 4, Larmor, Lorentz, and Thomson entertained different beliefs about the nature of the electron. For instance, Larmor, and subsequently Lorentz, considered the mass of the electron an epiphenomenon of its charge, whereas Thomson thought of it as an independent property. Furthermore, Larmor considered charge a constitutive property of electrons, whereas for Thomson it was an epiphenomenon of the interaction between ether and matter. Finally, chapter 7 revealed an incompatibility between the physicists' and the chemists' "electrons." The chemists deprived the electron of its dynamic behavior within the atom and portrayed is as a static particle. Lewis, for instance, denied that it strictly obeyed Coulomb's law and endowed it with an intrinsic magnetic disposition well before the advent of spin. All those suggestions were in conflict with how the electron was represented in the physics literature.

A philosophical question then arises: Did all those physicists and chemists refer to the same thing when they used the term "electron"? Did, say, Lorentz's "electron" and Bohr's "electron" denote the same entity? This question is crucial, not only with respect to the electron's ontological status, but also with respect to the biographical approach I have adopted, which presupposes that one can make sense of the electron's identity despite the evolution of its representation. To answer it adequately, we need to examine the philosophical problem of meaning change and its implications for scientific realism.

Since the early 1960s, when Thomas Kuhn and Paul Feyerabend, among others, stressed the importance of the history of science for the philosophy of science, two aspects of the historical development of scientific knowledge have been employed to support antirealist theses: first, the phenomenon of theory change; and second, the change of the meaning of scientific terms over time. These two aspects of the evolution of science are closely related. According to the theory of meaning favored by these philosophers, the meaning of scientific terms is determined by their place within the overall theoretical structure in which they are embedded.[1] It follows that changes in that structure result in changes in the meaning of the terms embedded in it.

The phenomenon of meaning change has direct implications for the problem of scientific realism. This problem has several dimensions. The most important for my purposes concerns the grounds that we have for believing in the reality of the unobservable entities postulated by contemporary science (atoms, electrons, photons, fields, etc.).[2] If the meanings of scientific terms that denote unobservable entities are subject to change, then to what extent are we justified in believing that the referents of these terms are unaffected by theoretical instability? To give a concrete example, what sense does it make to believe in the existence of electrons if the "electrons" of Stoney, J. J. Thomson, Larmor, Lorentz, Bohr, and Pauli differed significantly from each other?[3]

As I indicated above, the issue of meaning change has repercussions for the biographical approach that permeates this book.[4] A biographical narrative presupposes a single (albeit evolving) entity as its subject. The unity of the episodes discussed in such a narrative is guaranteed by the biographee's participation in them. Therefore, for a biographical approach to get off the ground it is imperative to show how scientists who attached different meanings to the term "electron" could, nevertheless, talk about the same entity. Otherwise, it is not clear that their experimental and theoretical investigations should be included in a single story.

In what follows I will give a critical exposition of various views on the implications of meaning change for scientific realism and discuss under what conditions a realist position is compatible with the instability of scientific concepts. I will argue that although the evolution of scientific concepts bears significantly on the issue of scientific realism, it does not, pace Kuhn and Feyerabend, necessarily lead to antirealist conclusions. At the end of the chapter,

1. Kuhn's and Feyerabend's antirealist tenets applied to all scientific terms, since they denied the distinction, central to logical positivism, between observational and theoretical terms. Furthermore, their tenets applied alike to terms denoting unobservable entities (e.g., phlogiston, ether, and electron) and terms denoting measurable properties (e.g., mass, length, and temperature).

2. Other aspects of the debate on scientific realism, for example, the aim of science (truth or empirical adequacy), are peripheral to the subject of this chapter. For a detailed analysis of this debate, from a realist point of view, see S. Psillos, *Scientific Realism: How Science Tracks Truth* (London: Routledge, 1999); and Psillos, "The Present State of the Scientific Realism Debate," *British Journal for the Philosophy of Science,* 51 (2000): 705–728.

3. This is one of the arguments employed by van Fraassen to attack the realist position. See B. C. van Fraassen, *The Scientific Image* (New York: Oxford Univ. Press, 1980), p. 214; and van Fraassen, *The Empirical Stance* (New Haven: Yale Univ. Press, 2002), p. 56.

4. See the introduction and chapter 2.

I will bring the philosophical discussion to bear on the issue of the electron's identity.

## 2. Historicizing Meaning: Kuhn's and Feyerabend's Antirealist Theses

In the early 1960s Kuhn and Feyerabend, among others, launched a novel and powerful attack against a philosophical interpretation of science which had dominated the Anglo-Saxon philosophical scene for more than thirty years, namely, logical positivism. One of the effects of that attack was a severe blow to a realist reading of the historical development of science, that is, to the view that science aims at "a permanent fixed truth, of which each stage in the development of scientific knowledge is a better exemplar."[5] Here I will focus on Feyerabend's views, since they were, I think, more radical and developed in more philosophical detail than Kuhn's early views (as expounded in *The Structure of Scientific Revolutions*).[6]

Feyerabend presented his views on meaning and scientific realism in two seminal papers (both published in the early 1960s).[7] The main purpose of his "Explanation, Reduction and Empiricism" was to show that Carl G. Hempel's theory of scientific explanation and Ernest Nagel's theory of reduction of one scientific theory to its successor do not do justice to scientific prac-

5. T. S. Kuhn, *The Structure of Scientific Revolutions,* 2nd ed. (Chicago: Univ. of Chicago Press, 1970), p. 173.

6. For Kuhn's more recent and more philosophically articulated views on scientific realism and the closely related issue of incommensurability see T. S. Kuhn, "Commensurability, Comparability, Communicability," in P. D. Asquith and T. Nickles (eds.), *PSA 1982: Proceedings of the 1982 Biennial Meeting of the Philosophy of Science Association,* 2 vols. (East Lansing, Mich.: Philosophy of Science Association, 1983), vol. 2, pp. 669–688; Kuhn, "Possible Worlds in History of Science," in Sture Allén (ed.), *Possible Worlds in Humanities, Arts and Sciences* (Berlin: Walter de Gruyter, 1989), pp. 9–32, esp. pp. 23–32; and Kuhn, "Afterwords," in P. Horwich (ed.), *World Changes: Thomas Kuhn and the Nature of Science* (Cambridge, Mass.: MIT Press, 1993), pp. 311–341. These papers are reprinted in T. S. Kuhn, *The Road since Structure,* ed. J. Conant and J. Haugeland (Chicago: Univ. of Chicago Press, 2000). I have discussed Kuhn's recent work in my "Can a Historian of Science Be a Scientific Realist?" *Philosophy of Science,* 68, suppl. (2001): S531–S541. I will refrain from discussing it here, because, as far as I can tell, the taxonomic incommensurability that he talks about cannot be found in the history of the electron's representation.

7. See P. K. Feyerabend, "Explanation, Reduction and Empiricism," in H. Feigl and G. Maxwell (eds.), *Scientific Explanation, Space and Time,* Minnesota Studies in the Philosophy of Science 3 (Minneapolis: Univ. of Minnesota Press, 1962), pp. 28–97; and Feyerabend, "Realism and Instrumentalism: Comments in the Logic of Factual Support," in M. Bunge (ed.), *The Critical Approach to Science and Philosophy* (New York: Free Press, 1964), pp. 280–308. Both papers are reprinted in his *Philosophical Papers,* vol. 1, *Realism, Rationalism and Scientific Method* (Cambridge: Cambridge Univ. Press, 1981). Others papers in that collection are equally important for my purposes. In what follows I will quote directly from this volume.

tice.[8] Hempel's and Nagel's schemes, according to Feyerabend, fail as descriptive accounts of actual scientific explanations and reductions and fail also as normative accounts of the ideal types of scientific explanation and reduction. The main reason for this failure is that the meanings of scientific terms in the explanandum are different from the meanings of the corresponding terms in the explanans (similarly with the theory to be reduced and the theory which effects the reduction). Feyerabend's denial of the assumption of meaning invariance, an assumption central to Hempel's and Nagel's theories, followed from his view of how scientific terms get their meaning. "[T]he meaning of a term is not an intrinsic property of it, but is dependent upon the way in which the term has been incorporated into a theory."[9] Since the meanings of scientific terms are dependent on the theory in which they are embedded, meaning change is a necessary consequence of theory change. It follows that "terms change their meaning with the progress of science," since "extension of knowledge leads to a decisive modification of the previous theories both as regards the quantitative assertions made and as regards the meanings of the main descriptive terms used."[10]

What are the implications of meaning variance for scientific realism? To use a concrete example, if the meanings of Lorentz's "electron," circa 1900, and Bohr's "electron," circa 1913, are not the same, then what should we conclude about the ontological status of the referents of those terms? Before laying out various possible realist responses, I should point out that this question has a Fregean presupposition, namely, that meaning determines reference. Indeed, one could avoid the antirealist implications of that example by denying this presupposition. This strategy has been adopted by Hilary Putnam and will be discussed in detail below.

Besides Putnam's maneuver there are other options available to the realist. First, he or she could accept the discrepancy between Lorentz's and Bohr's concepts and argue that only Bohr's concept of the electron has a genuine physical referent. However, the concept of the electron continued to evolve well after 1913, and our concept of the electron is very different from Bohr's. If the difference between Lorentz's and Bohr's concepts implies that the former does not have a referent, then similarly the difference between Bohr's and our concepts should imply that Bohr's "electron" is nonreferential, too. Fol-

---

8. The version of Hempel's theory of explanation criticized by Feyerabend can be found in C. G. Hempel, *Aspects of Scientific Explanation* (New York: Free Press, 1965), pp. 245–290. For Nagel's theory of reduction see E. Nagel, *The Structure of Science,* 2nd ed. (Indianapolis: Hackett, 1979; first publ. 1961), pp. 336–397.

9. Feyerabend, *Philosophical Papers,* vol. 1, p. 74.

10. Ibid., pp. 80–81.

lowing a similar reasoning, one may conclude that future theoretical developments will reveal that our "electron" also does not refer to anything.

Second, the realist could deny that Kuhn and Feyerabend have provided an accurate characterization of conceptual change in science and, thus, dispute their claim that Lorentz's "electron" does not have a referent. An argument along these lines would be that, despite the differences between Lorentz's and Bohr's concepts of the electron, there are sufficient similarities between the two concepts to enable us to maintain that both refer to the same entity.[11] This strategy for avoiding the antirealist implications of the instability of scientific concepts will be developed below (in sections 5 and 6).

Finally, another realist response would be to point out some difficulties of the theory of meaning favored by Kuhn and Feyerabend and thus to cast doubt on their conclusions. As I mentioned above, according to that theory, the meaning of scientific terms is theory dependent. One needs to know, then, what counts as part of a theory and what kinds of theory change amount to a transformation of the concepts of the theory. In this respect, as has been pointed out by Dudley Shapere, Kuhn's and Feyerabend's views are open to objection.[12] "Theory" is usually understood as a systematic and articulated body of knowledge. Feyerabend on the other hand has a more inclusive conception of "theory": "the term 'theory' will be used in a wide sense, including ordinary beliefs (e.g., the belief in the existence of material objects), myths (e.g., the myth of eternal recurrence), religious beliefs, etc. In short, any sufficiently general point of view concerning matter of fact will be termed a 'theory.'"[13] Given this notion of theory and the theory dependence of meaning it is not clear what aspects of the theoretical context are relevant to specifying the meaning of a term. Feyerabend, to the best of my knowledge, has not provided a satisfactory resolution of this difficulty. Nevertheless, this difficulty can be, at least partly, alleviated if we narrow Feyerabend's conception of theory so as to exclude ordinary beliefs, myths, religious beliefs, and so on, and to retain scientific theories only (e.g., Lorentz's theory of

---

11. Cf. Dudley Shapere, "Leplin on Essentialism," *Philosophy of Science,* 58 (1991): 655–677, on p. 656.

12. See D. Shapere, "Meaning and Scientific Change," in R. G. Colodny (ed.), *Mind and Cosmos: Essays in Contemporary Science and Philosophy* (Pittsburgh: Univ. of Pittsburgh Press, 1966), pp. 41–85; repr. in I. Hacking (ed.), *Scientific Revolutions* (Oxford: Oxford Univ. Press, 1981), pp. 28–59. The critique of Feyerabend's notion of meaning that follows is based on Shapere's paper. The attempt, however, to improve on Feyerabend's views so as to meet Shapere's criticism is mine.

13. P. K. Feyerabend, "Problems of Empiricism," in R. G. Colodny (ed.), *Beyond the Edge of Certainty* (Englewood Cliffs, N.J.: Prentice Hall, 1965), p. 219.

electrons and Bohr's theory of the atom). In such a case the meaning of a term would be the set of properties ascribed to the corresponding entity by the scientific theory in which the term is embedded.[14]

One may object that not all the properties attributed to an entity are constitutive of the meaning of the corresponding term. The attribution of some of those properties, on such a view, is a merely factual assertion about the entity and is, therefore, irrelevant to the corresponding term's meaning. A satisfactory rebuttal of that objection would require an extensive discussion of the distinction between analytic and synthetic statements, a task that is beyond the scope of this chapter. Here I can only appeal to the widely accepted view, originating from Willard Van Orman Quine, that a clear-cut distinction of this kind is not possible.[15] Furthermore, the main realist opponent of Feyerabend, Hilary Putnam, has emphasized "the impossibility of separating, in the actual use of the word, that part of the use which reflects the 'meaning' of the word and that part of the use which reflects deeply embedded collateral information."[16] I will have more to say on this topic below (in section 5).

To return to my proposed modification of Feyerabend's notion of meaning, a difficulty remains: What is the relationship between theory change and meaning change? In other words, what kinds of modifications in a theory affect the meaning of the concepts that the theory specifies? One could think of changes in a theory that would be too minor to affect the meaning of the terms embedded in it. For example, Sommerfeld's suggestion that the electrons inside the atom move in elliptical, rather than circular, orbits hardly changed the meaning of the term "electron."[17]

Feyerabend tried to alleviate this difficulty in his paper "On the 'Meaning' of Scientific Terms."[18] He proposed that we "diagnose a change of meaning either if a new theory entails that all concepts of the preceding theory have

14. A similar view is held by Dudley Shapere. The only difference is that, in Shapere's view, the meaning of a term is the set of properties attributed by the *group of scientists that are using the term* to the entity that the term denotes. See D. Shapere, "Reason, Reference, and the Quest for Knowledge," *Philosophy of Science,* 49 (1982): 1–23, on p. 21. Cf. also P. Achinstein, "On the Meaning of Scientific Terms," *Journal of Philosophy,* 61, no. 17 (1964): 497–509, on pp. 502–503.

15. For some useful reflections on this issue see D. Papineau, "Theory-Dependent Terms," *Philosophy of Science,* 63 (1996): 1–20.

16. H. Putnam, "The Analytic and the Synthetic," in his *Philosophical Papers,* vol. 2, *Mind, Language and Reality* (Cambridge: Cambridge Univ. Press, 1975), pp. 33–69, on pp. 40–41.

17. This example is also mentioned in Achinstein, "On the Meaning of Scientific Terms," on p. 505.

18. P. K. Feyerabend, "On the 'Meaning' of Scientific Terms," *Journal of Philosophy,* 12 (1965): 266–274; repr. in his *Philosophical Papers,* vol. 1, pp. 97–103.

zero extension or if it introduces rules which cannot be interpreted as attributing specific properties to objects within already existing classes, but which
change the system of classes itself." [19] The former "diagnostic" procedure is
far too strict. After all, one would want to diagnose a change of meaning if a
new theory entailed that *some* of the concepts of the preceding theory were
vacuous and refered to nothing at all.

The latter concerns the taxonomic function of scientific theories and presupposes the existence of unambiguous rules of classification, which can be
used to collect objects, events, entities, and so on, into sets. If a new theory
attributes additional properties to the entities denoted by the concepts of the
previous theory but preserves its taxonomic rules, one should conclude, according to Feyerabend, that no change of meaning has taken place. First, one
might want to dispute his claim that the properties of an entity, to the extent
that they are not related to any rules of classification, are irrelevant to the
meaning of the corresponding term. However, even if every feature associated with an entity were relevant to the specification of the meaning of the
corresponding term, it would still be the case that any change of meaning
brought about by the addition of new properties to a *fixed* set of entities
would lack any antirealist connotations. On the contrary, meaning change of
this kind would support a realist position, since it would have taken place
against a stable ontological background.

Furthermore, the existence of unambiguous taxonomic rules in science
has been denied by Shapere. Scientific classifications, in his view, are usually
based on pragmatic criteria which do not reflect any intrinsic properties of
the entities classified. For example, the question "Are mesons different 'kinds
of entities' from electrons and protons, or are they simply a different subclass
of elementary particles? . . . can be answered *either* way, depending on the
kind of information that is being requested . . . for there are differences as well
as similarities between electrons and mesons . . . [It] can be given a simple answer ('different' or 'the same') only if unwanted similarities or differences are
stipulated away as inessential." [20] Thus, an appropriate choice of taxonomic
criteria might leave the meaning of the terms in question unaffected by theory change.

Shapere's criticism is based on his emphasis on the pragmatic character
of scientific classifications, an emphasis that, I think, is overstated. Physics, for
instance, has produced an unambiguous classification of the unobservable

19. Ibid. (in *Philosophical Papers*), p. 98.
20. Shapere, "Meaning and Scientific Change," in Hacking (ed.), *Scientific Revolutions,* on
pp. 51–52.

realm into various particles (electrons, protons, neutrons, etc.). It is true that their occasional similarities enable the classification of otherwise different entities in the same category. For example, electrons, protons, neutrons, and quarks are all classified as fermions, since they obey Fermi-Dirac statistics. But these entities, despite their occasional similarities, are clearly distinguished from one another according to their intrinsic properties (e.g., their mass and the magnitude of their charge). At least some classifications are predicated on the intrinsic properties of the entities classified, and, thus, Feyerabend's notion of meaning change could escape Shapere's criticism if one limited its applicability to classifications of that kind. With these qualifications, the above difficulties in the criterion of meaning change favored by Feyerabend disappear and, thus, cannot be employed to support a realist position.

An antirealist stance is implicit in those of Feyerabend's views that I have discussed so far. In his early work he did not adopt an explicit position on the ontological status of unobservable entities, an aspect of the realism debate with which he was not directly concerned. He was occupied, rather, with the phenomenon of meaning change and its implications for the then current theories of scientific explanation and reduction. In his later writings, however, he formulated unambiguously the antirealist perspective that was implicit in his early papers. The phenomenon of meaning change at its most extreme (incommensurability) implies, in his view, that past scientific theories, which have been replaced by their incommensurable successors, were based on a nonexistent ontology. For example, *"prerelativistic* terms . . . are pretty far removed from reality (especially in view of the fact that they come from an incorrect theory implying a nonexistent ontology)."[21]

Feyerabend's conception of meaning and his skepticism toward the ontological implications of scientific theories presented a serious challenge for scientific realism. It is now time to discuss a significant attempt to evade the antirealist implications of meaning change, namely, the causal theory of reference, which was developed by Saul Kripke and Putnam.

### 3. Putnam's Theory of Meaning and Reference: A Realist Way Out?

Putnam realized that Feyerabend's view of meaning along with the idea that meaning determines reference would be a death blow to scientific realism.

---

21. P. K. Feyerabend, "Against Method: Outline of an Anarchistic Theory of Knowledge," in M. Radner and S. Winokur (eds.), *Analysis of Theories and Methods of Physics and Psychology,* Minnesota Studies in the Philosophy of Science 4 (Minneapolis: Univ. of Minnesota Press, 1970), pp. 17–130.

Thus, in the early 1970s he tried to disentangle meaning from reference by suggesting that a term's referent is an independent and essential constituent of its meaning; that is, it is a term's referent which, to a large extent, determines the term's meaning, and not vice versa. Whatever changes the other components of the meaning of a term may undergo the "referential" component will remain unaffected. Even though Putnam's more recent views on scientific realism depart considerably from the realist intuitions that motivated his earlier attempts to articulate a theory of meaning, his view on the relationship between meaning and reference has remained intact. In a recent discussion of this issue, for example, he insists that "the reference of the terms 'water,' 'leopard,' 'gold,' and so forth is partly fixed by the substances and organisms themselves . . . [while] the 'meaning' of these terms is open to indefinite future scientific discovery." [22]

Putnam was not particularly concerned with terms denoting unobservable entities. Following Kripke, whose ideas were developed in the context of modal logic, he was occupied, rather, with natural kind terms in general. The purpose of his arguments was to show that the referents of natural kind concepts have not been affected by the development of scientific knowledge. Even though "the concept [of a natural kind, e.g., fish] is not exactly correct (as a description of the corresponding natural kind) . . . that does not make it a *fiction*. . . . The concept is continually changing as a result of the impact of scientific discoveries, but that does not mean that it ceases to correspond to the same natural kind." [23] I have two comments here. First, I do not think that one can exclude the possibility that future changes in the concept of, for example, fish will affect its reference. The concept may change so that organisms that are now classified as fish will cease to be thus classified. Second, the example of "fish," or any another observable natural kind term, is not very helpful when it comes to questions of scientific realism. Even if the reference of that term changed, the existence of the individual organisms that we classify as fish would not be questioned. And that is because the existence of those organisms is established on grounds independent of our system of classification. That is not the case, however, when the natural kind term refers to unobservable entities. Besides the fact that several such terms have been abandoned, even when they have been retained the lack of independent, physical access to the entities denoted by those terms makes problematic the claim

22. H. Putnam, *Realism with a Human Face*, ed. James Conant, (Cambridge, Mass.: Harvard Univ. Press, 1990), pp. 109–110.

23. H. Putnam, "Explanation and Reference," in his *Philosophical Papers*, vol. 2, pp. 196–214. The passage quoted is on p. 197.

that, even in case of meaning change, their referents remain stable.[24] Whatever the merit of Putnam's arguments for the referential stability of observable natural kind terms, I will argue that, when it comes to terms which denote unobservable entities, the causal theory of reference does not remove the threat that meaning change poses to scientific realism.[25]

As I mentioned above, Putnam's aim was to devise a theory of meaning that would support a basic realist conviction, namely, "that there are successive scientific theories about the *same* things: about heat, about electricity, about electrons, and so forth."[26] According to such an account, "Bohr would have been referring to electrons when he used the word 'electron,' notwithstanding the fact that some of his beliefs about electrons were mistaken, and we are referring to those same particles notwithstanding the fact that some of our beliefs—even beliefs included in our scientific 'definition' of the term 'electron'—may very likely turn out to be equally mistaken."[27]

The first step toward that aim was to reject the so-called necessary and sufficient conditions view of concepts. According to that view, membership in the extension of a term is determined by a set of necessary and sufficient conditions, which constitute the concept associated with that term. It follows that even the slightest revision of those conditions affects the extension of the term that they "define." In Putnam's view, on the other hand, the referent of a term is an independent constituent of a term's meaning and is not affected by even drastic changes that the other meaning components might undergo. Putnam's rejection of the necessary and sufficient conditions view neutralizes, to some extent, the antirealist implications of meaning change. If a term could refer to the same entity or entities despite the fact that its meaning had evolved, antirealism would not be an inescapable consequence of the instability of scientific concepts. I added the qualifier "to some extent" because a belief in the existence of, say, electrons would make sense only if a core of the "electron's" meaning (i.e., some of our beliefs about electrons) had remained stable since the initial proposal of the electron hypothesis. Cata-

24. For a very long list of entities that have been abandoned see Larry Laudan, "A Confutation of Convergent Realism," in J. Leplin (ed.), *Scientific Realism* (Berkeley: Univ. of California Press, 1984), pp. 218–249.

25. Several critics of the causal theory of reference have also argued, for reasons different from the ones I will suggest, that it does not work for theoretical terms. See B. Enç, "Reference of Theoretical Terms," *Nous*, 10 (1976): 261–282; F. W. Kroon, "Theoretical Terms and the Causal View of Reference," *Australasian Journal of Philosophy*, 63, no. 2 (1985): 143–166; and R. Nola, "Fixing the Reference of Theoretical Terms," *Philosophy of Science*, 47 (1980): 505–531.

26. Putnam, *Philosophical Papers*, vol. 2, p. 197.

27. Ibid.

clysmic changes in the meanings of scientific terms, which would amount to the abandonment of every belief about the corresponding entities, would be indeed a threat to scientific realism.[28] However, barring such radical cases, Putnam's rejection of the necessary and sufficient conditions view of concepts should be an indispensable ingredient of any realist position.

The most sophisticated version of Putnam's theory of meaning is found in his "The Meaning of 'Meaning.'"[29] According to that account, the meaning of a word is a four-dimensional "vector" with the following components: "(1) the syntactic markers that apply to the word, e.g. 'noun'; (2) the semantic markers that apply to the word, e.g. 'animal,' 'period of time'; (3) a description of the additional features of the stereotype, if any; (4) a description of the extension."[30] The stereotype associated with a natural kind term consists of "a standardized description of features of the kind that are typical, or 'normal,' or at any rate stereotypical. The central features of the stereotype generally are *criteria*—features which in normal situations constitute ways of recognizing if a thing belongs to the kind."[31] Part of the stereotype associated with electrons, for example, is that they "are charged particles ('little balls' with trajectories and unit negative charge)."[32] Furthermore, the semantic markers consist of all those features of the stereotype that "attach with enormous centrality to the [corresponding] words," that "form part of a widely used and important *system of classification*," and that are "*qualitatively* harder to revise" than the rest.[33] In Putnam's view, "The centrality guarantees that

28. If this is right, then the realist has to draw a line between the beliefs about electrons that are not revisable and the beliefs that are more tentative. Historical research has an indispensable role to play in this respect, by locating the beliefs which throughout the development of the concept of the electron have remained stable. A belief in the existence of electrons would then amount to the conviction that this core of beliefs will prove immune to theory change. A variant of this approach has been adopted by Ian Hacking. His "home truths" about electrons are precisely those of our beliefs about electrons that have not been threatened with revision. Hacking, however, includes in this core only beliefs that have arisen in an experimental context. The merits and limitations of his view will be treated below.

29. H. Putnam, "The Meaning of 'Meaning,'" in his *Philosophical Papers,* vol. 2, pp. 215–271. This paper was published in 1975. In earlier papers (e.g., "How Not to Talk about Meaning" [1965], also repr. in the above volume, pp. 117–131) he held the view that the meaning of a term is identical with its referent. In such a view the meaning of a term would change only if the relevant linguistic community decided to attach that term to a different referent. See his *Philosophical Papers,* vol. 2, pp. 127–128. There is no point in criticizing Putnam's early views, since his later views allow for meaning change.

30. Putnam, *Philosophical Papers,* vol. 2, p. 269.

31. Ibid., p. 230.

32. H. Putnam, "Truth, Activation Vectors and Possession Conditions for Concepts," *Philosophy and Phenomenological Research,* 52, no. 2 (1992): 431–447, on p. 445.

33. Putnam, *Philosophical Papers,* vol. 2, p. 267.

items classified under these headings virtually never have to be *re*classified." [34]
Finally, it should be emphasized that it is the extension and not its description
that is part of the meaning of a term. The description merely helps us to identify the extension, which, in the absence of a description, could be identified
by ostension.

Putnam used the example of the term "water" to illustrate his theory. By
separating the extension of "water" ($H_2O$) from its stereotype (transparent,
tasteless, thirst-quenching, etc.) he wanted to show that the essence of water
coincides with its microscopic structure and not with the set of phenomenological properties which constitute its stereotype. His main argument to that
effect was that the discovery of a substance with the same phenomenological
properties as our familiar water but with a totally different structure (XYZ) is
conceivable. Thus, the "stereotypical" features of an entity or substance, despite their being part of the meaning of the corresponding term, do not determine the term's reference. It follows that the instability of a term's meaning due to the instability of its stereotype does not affect the stability of the
term's reference. Scientific knowledge, according to Putnam, has developed
(and will develop) on a stable ontology.

Putnam's essentialism and his distinction between contingent, superficial
properties and necessary, hidden essences (a distinction that he shared with
Saul Kripke) have been severely criticized. [35] Kuhn, in particular, has argued
that superficial and theoretical properties (Putnam's hidden essences) cannot
be distinguished on grounds of their modal status, since the former are the
observable manifestation of the latter. If the color of an element, for example, were different, then, according to contemporary atomic theory, the same
would be true of its atomic number. Thus, the contingency of a superficial
property (color) would imply the contingency of a theoretical property
(atomic number). Furthermore, superficial properties have a certain priority
over theoretical ones. The latter gain their credibility because they enable the
prediction of the former, and "superficial properties are the ones called upon
in those difficult cases of discrimination characteristically raised by new theories. Is deuterium really hydrogen, for example?" [36]

Despite that criticism, one could still argue that it can be ignored in so far
as our aim is to determine the utility of Putnam's theory of meaning in for-

---

34. Ibid., pp. 266–267.

35. See D. Shapere, "Reason, Reference"; and T. S. Kuhn, "Dubbing and Redubbing: The
Vulnerability of Rigid Designation," in C. W. Savage (ed.), *Scientific Theories,* Minnesota Studies in the Philosophy of Science 14 (Minneapolis: Univ. of Minnesota Press, 1990): 298–318.

36. Kuhn, "Dubbing," p. 313.

mulating a viable argument for the *existence*, and not the one and true essence, of unobservable entities. At first glance it seems that that theory would be a valuable tool for that purpose. One of the theory's main implications is that belief in the existence of an entity does not require that all, or even most, of our beliefs about the entity in question be true. Thus, for example, our conviction of the existence of, for example, the Empire State Building, does not presuppose the truth of all of our beliefs about that building. A discovery that we have been mistaken all along about its height or weight, to mention two of its most salient properties, would not throw doubt on its existence.

Nevertheless, Putnam's success in disentangling reference from meaning does not extend to cases where a term's referent is an unobservable entity (or a class of such entities). In order for the referent of a term to be an essential and stable constituent of its meaning, as Putnam's theory requires, it should be possible to identify the former without relying on the other constituents of the latter. In that case a change of the stereotype would not imply that the referent had also changed. That is, one should have independent, physical access to the entities denoted by a term, independent, that is, of our beliefs about them. This independent access guarantees, extreme skepticism excluded, that those entities are real.[37] However, in the case of unobservable entities no such direct epistemic access exists. All that we have is a set of indirectly confirmed beliefs about those entities, and it is not clear that we can identify them independently of those beliefs.[38] Furthermore, since the problem under examination is the ontological status of unobservable entities, it goes without saying that a theory like Putnam's, which presupposes their existence, cannot be employed without circularity to support a realist position.

My evaluation of Putnam's theory as a potential realist tool was based on the assumption that unobservable entities are beyond direct empirical access.

---

37. This assertion is not put forward as a requirement that any viable theory of meaning should fulfill. Rather, it is a condition that has to be met in order for Putnam's theory to get off the ground. Indeed, Putnam himself suggested that in his account of natural kinds "things which are given *existentially* . . . help to fix reference. Actual things, whatever their description, which have played a certain causal role in our acquisition and use of terms determine what the term refers to." H. Putnam, "Reference and Truth," in his *Philosophical Papers,* vol. 3, *Realism and Reason* (Cambridge: Cambridge Univ. Press, 1988), pp. 69–86, on p. 73. Cf. R. Harré, *Varieties of Realism* (Oxford: Basil Blackwell, 1986), p. 113.

38. The recipe that I proposed in chapter 1 (and will discuss further in section 5) for identifying the reference of terms denoting unobservable entities would allow one to include reference as a meaning component. However, it is not applicable to entities that have not been (and perhaps cannot be) subject to experimental investigation (e.g., black holes). Furthermore, even where it is applicable, the referential stability of a term requires historical investigation and should not be taken for granted prior to such an investigation.

This assumption has been forcefully challenged by Ian Hacking, whose views on scientific realism are the subject of the next section.

## 4. Hacking's Entity Realism

Hacking has advanced a realist position which is based on a close examination of experimental practice.[39] A satisfactory resolution of the problem of scientific realism would be possible, Hacking claims, only if we stopped being preoccupied with scientific theorizing and instead shifted our focus of analysis toward experimentation. Such a shift of emphasis would be enough to make us all realists with respect to some unobservable entities but would not weaken our antirealist convictions with respect to the theories that postulate those entities. This peculiar mix of realism about entities and antirealism about theories follows from two central aspects of experimental practice. On the one hand, the manipulation of unobservable entities in the laboratory provides sufficient grounds for believing in their existence. On the other hand, the fact that experimentalists use, according to the purpose at hand, a variety of sometimes incompatible theoretical models of those entities generates strong doubts that any of those models accurately represent reality. All these models, however, have some aspects in common, namely, a core of statements about the causal properties of the corresponding entities, properties which we have come to know by manipulating those entities in various experimental contexts. One can (and should) be a realist about this common core, which, however, does not deserve to be called a "theory."

Hacking's entity realism can be summarized in his aptly chosen slogan: "If you spray [e.g.] electrons then they are real."[40] Notwithstanding the charm of Hacking's slogan, it fails to impress philosophers in the empiricist tradition. Van Fraassen, for instance, when asked to evaluate Hacking's argument, responded in the following way: "If they are real then you spray them."[41] Van Fraassen does not imply, of course, that everything real can be sprayed. Rather, his point is that one can use the expression "spraying of electrons" (as the best available description of a given experimental situation) without committing oneself to believing in the existence of electrons. The coherence of his position stems from the possibility that a new theory, which would not include electrons in its ontology, could adequately account for the experimental situ-

---

39. Hacking's views on scientific realism are expounded in *Representing and Intervening* (Cambridge: Cambridge Univ. Press, 1983); and his essay "Experimentation and Scientific Realism," in Leplin (ed.), *Scientific Realism*, pp. 154–172.

40. Hacking, *Representing*, p. 24.

41. Personal communication.

ation that Hacking, following contemporary experimentalists, describes as "spraying of electrons."[42]

An example that illustrates this possibility can be found in a well-known episode from the history of chemistry, the so-called Chemical Revolution.[43] Before the establishment of the oxygen theory of combustion, late-eighteenth-century chemists explained several chemical phenomena (e.g., the formation of acids) by postulating an entity called phlogiston. Furthermore, they thought they could manipulate that entity, since they could transfer it to a substance. Georg Stahl, for instance, had discovered that "vitriolic acid [sulfuric acid] can be converted to volatile sulfurous acid by transferring to it some of the inflammable principle [phlogiston]."[44] After the Chemical Revolution and the concomitant disappearance of "phlogiston" from the chemical vocabulary, however, the process that had been previously described as "transfer of phlogiston" was redescribed in terms of a different entity, namely, oxygen. It is similarly conceivable that the process that is now described as "spraying of electrons" might be redescribed in terms of an alternative theory based on a different ontology. Thus, Hacking has not provided a conclusive argument for the existence of electrons since he has not excluded the possibility of an alternative, empirically adequate theory whose ontology would not include electrons.

Another objection has been raised against Hacking's claim that manipulability is a necessary and sufficient criterion for establishing the existence of an unobservable entity, namely, that it does not do justice to actual scientific practice.[45] In particular, his exclusive emphasis on manipulability fails to capture the variety of evaluative criteria that are employed *within* scientific practice for demonstrating the reality of an unobservable entity.[46] The most im-

42. Cf. B. C. van Fraassen, "From Vicious Circle to Infinite Regress, and Back Again," *PSA 1992: Proceedings of the Biennial Meeting of the Philosophy of Science Association* (Lansing, Mich.: Philosophy of Science Association, 1993), vol. 2, 6–29, esp. pp. 14ff.; and van Fraassen, "Michel Ghins on the Empirical versus the Theoretical," *Foundations of Physics*, 30, no. 10 (2000): 1655–1661, on p. 1658.

43. This example was suggested to me by Nancy Nersessian.

44. F. L. Holmes, *Eighteenth-Century Chemistry as an Investigative Enterprise* (Berkeley: Office for History of Science and Technology, Univ. of California, 1989), p. 100.

45. See M. Morrison, "Theory, Intervention and Realism," *Synthese*, 82 (1990): 1–22. The following passage leads me to interpret Hacking as claiming necessity for his criterion: "The sceptic like myself has a slender induction. Long-lived theoretical entities, which don't end up being manipulated, commonly turn out to have been wonderful mistakes." Hacking, *Representing*, p. 275.

46. One could deny that Hacking's criterion needs to capture the richness of scientific practice, by interpreting it as a normative as opposed to a descriptive criterion. However, Hacking himself stresses the descriptive (as well as the normative) dimensions of his criterion. For him, the main reason that *scientists themselves* believe in the reality of some unobservable

portant of those criteria seems to be the empirical adequacy of the theory that postulates the entity in question. Even though manipulability is one of the criteria that scientists often employ, it fails to carry conviction in all contexts.

An example from the history of the electron's representation will illustrate the inadequacy of manipulability as a sufficient criterion for scientific realism.[47] When J. J. Thomson was experimenting with cathode rays he was able to manipulate them in various ways. For example, he could deflect them by means of electrostatic and magnetic fields. It is well established today that cathode rays consist of electrons. Now given that Thomson manipulated cathode rays and that cathode rays are streams of electrons, it follows that Thomson manipulated electrons. However, at the time the existence of electrons was still controversial, and no choice had yet been made between a model of cathode rays which portrayed them as waves in an all-pervading ether and an alternative model which represented them as streams of particles (see chapter 4). It is clear that in that context the manipulability of cathode rays would not by itself establish the validity of the latter model and thus the reality of electrons.[48] As this example indicates, scientists need to have a fairly definite idea of what it is that they are manipulating before invoking manipulability as a demonstrative principle. In many experimental contexts we know that we are manipulating *something* without knowing what it is that we are manipulating. Manipulability, however, was supposed to provide adequate grounds for the existence of, for example, electrons and not merely for the existence of an I-know-not-what.

Thus, Hacking's manipulability criterion fails as a descriptive account of

---

entities is their ability to manipulate those entities: "The vast majority of experimental physicists are realists about some theoretical entities, namely the ones they *use*. I claim that they cannot help being so." Hacking, *Representing*, p. 262.

47. This example aims at countering an objection that could be raised against Morrison's criticism of Hacking. Her analysis rests on an episode from the history of recent physics, where the manipulation of complex entities (hadrons) did not establish the existence of their constituents (quarks). One could object that Hacking's criterion requires the manipulation of, say, electrons per se, as opposed to a more complex entity that is composed of electrons. Hacking could not have meant that the manipulation of, for example, a table, which is also (according to him) made up in part of electrons, proves the existence of electrons. The example that follows is not open to this objection, since it concerns the manipulation of cathode rays, which *are* electrons.

48. Notice, however, that Thomson's ability to manipulate cathode rays by means of electrostatic fields lent support to the hypothesis of the particulate nature of those rays. Ether waves could be deflected only by the action of magnetic fields and were not susceptible to electrostatic influences. Charged particles on the other hand could be subjected to both kinds of influence. Thus, Thomson's ability to manipulate cathode rays *by certain means* was *one* of the criteria that were employed to establish the existence of electrons.

the plurality of principles that scientists employ in the construction of "existence proofs" for unobservable entities. It also fails as a normative account, since even within actual scientific practice manipulability did not provide (and should not have provided) adequate reasons for belief in the entities that were supposedly manipulated.

A further problematic aspect of Hacking's realist position is associated with his "home truths" (low-level generalizations) about, for example, electrons that we supposedly know independently of any high-level theory. Hacking does not specify what kind of electron properties he has in mind, but one could guess that his "home truths" would include well-known causal properties of electrons, like their charge, mass, and spin, which enable us to manipulate them in order to investigate other less well known aspects of nature.[49] It is difficult to see, however, how one could isolate those properties from the background theory in which they were embedded. To use a concrete example, it is difficult to obtain an understanding of "charge" independently of any high-level theory, especially in view of the fact that "charge" meant different things for different scientists. Within Maxwellian electrodynamics, for instance, charge was not an independent substance but merely an epiphenomenon of the electromagnetic field. In Lorentz's electromagnetic theory, on the other hand, it was an independent entity that interacted with, but could not be reduced to, the electromagnetic field.

Hacking could argue that all those different and incompatible conceptions of charge had certain aspects in common, that is, the causal properties that we have associated with electric charge all along (e.g., the ability of charges to attract or repel each other). But even those properties may vary across theoretical frameworks. In classical electromagnetic theories, for instance, one of the causal properties of a charged particle was that it emitted radiation when performing accelerated motion. In the old quantum theory of the atom, on the other hand, charged particles in accelerated motion (electrons orbiting around the nucleus) did not radiate. Hacking's attempt to isolate the causal properties of electrons from any background theory was motivated by the plurality of incompatible models about electrons. However, it turns out that the causal properties themselves have been interpreted via several, incompatible theories. He could, of course, search again for a common core shared by those theories, but it is not clear that there would be an ending point to this process.

A closely associated problem is Hacking's selective realism. Let us grant

---

49. In some cases, of course, one can manipulate electrons without knowing *all* these properties. In the Thomson case, for example, the only property that was needed for the purpose of manipulation was the charge-to-mass ratio of the electron.

for the sake of argument that belief in the existence of the "well-known causal properties" of some entity does not depend on how we interpret those properties theoretically. As I already mentioned, in his view, one should maintain a realist position with respect to those properties but, nevertheless, should not commit oneself to believing that the theories involved in the interpretation of the relevant experiments are true.[50] However, as was pointed out by Duhem, experimental results falsify (or, I would add, confirm) whole chunks of knowledge consisting of high-level theory, an understanding of the instruments involved in the given experimental context, statements about initial conditions, and so on. In view of this very plausible holistic epistemology, which Hacking has not by any means discredited, his selective realism does not make much sense. To put it another way, it is far from clear why our ability to make sense of certain experimental situations confers a privileged status on the entities supposedly involved in the experiments but, nevertheless, cannot be employed as an argument for a realist stance toward the theory involved in the interpretation of those experiments. The following example would suffice to illustrate my point: The manipulation of electrons requires knowledge of their behavior in various experimental contexts (e.g., in a cloud chamber). To control that behavior one has to know, in addition to the causal properties of electrons, several background theories (electromagnetism, mechanics, etc.), which predict how an entity with those properties would behave in a given context. It is those background theories along with Hacking's "home truths" that enable intervention. Thus, if our ability to intervene is an argument for realism, that realism should be of a very broad kind, covering entities and theories alike.

Having discussed some significant positions that have been advanced with regard to the implications of meaning change for the ontological status of unobservable entities, I will now put forward my own approach to this problem.

### 5. A Historicist Approach to Meaning and Reference

Imre Lakatos has argued that the historical development of competing research programs bears significantly on their comparative appraisal. In a similar vein I want to argue that the historicity of scientific concepts (the fact that they are dynamic entities that often evolve considerably from their initial state) bears, significantly, on the issue of scientific realism. Needless to say, an argument to that effect presupposes a particular view of concepts. In what

---

50. Perhaps one should be a realist about the phenomenological theories of the instruments involved.

follows the "concept associated with a scientific term" (or, alternatively, the "meaning of a scientific term") amounts to the set of features that are ascribed, by the theory in which the term is embedded, to the corresponding entity. In the case that no such systematic theory has been developed, the "meaning of a term" would coincide with the characteristics attributed, by the group of scientists who are using the term, to the corresponding entity. Those characteristics consist of two interrelated kinds: the properties of the entity in question and the laws obeyed by it. The former are usually specified via the latter. For instance, what charge is depends on the laws obeyed by charged bodies—for example, the laws of classical electromagnetism. Conversely, the laws obeyed by an entity depend on its properties. For example, the electrons qua charged particles are supposed to obey Coulomb's law.

As I already mentioned (cf. p. 242 above), one may find objectionable the suggestion that all the characteristics ascribed by a theory to an entity are part of the meaning of the corresponding term. In particular, one may call for a distinction between the meaning-constitutive and the merely factual aspects of the information associated with a term. I think, however, that such a distinction cannot be drawn in a satisfactory way. To begin with, no distinction of this kind is inherent in the representation of the entity denoted by a scientific term. For example, electrons were represented by the old quantum theory of the atom (c. 1925) as subatomic particles, with a specific amount of charge, a specific rest mass, spin, and so on. There was nothing in that representation, however, that explicitly distinguished some of those properties as constitutive of the meaning of the term "electron."

Could one impose such a distinction and on what grounds? I can think of two alternatives, neither of which seems adequate. First, one could argue that only those properties that are relevant to the classification of an entity are part of the meaning of the corresponding term. For example, only the subset of the properties of electrons that enables us to distinguish them from other elementary particles should be considered part of the meaning of the term "electron." This recipe has the undesirable consequence that the meaning of a term may change if the domain of classification is enriched. Consider, again, the case of electrons. In the context of J. J. Thomson's model of the atom, there was no need to distinguish them from other charged particles, because they were the only particles postulated. Thomson's atom had no positive electricity. The latter was just an epiphenomenon of the arrangement of negative electrons. When the concept of the proton was introduced, however, the electron's negative charge, or the small value of its mass, became crucial for distinguishing it from a proton. Should we say that with the postulation of protons the meaning of the term "electron" changed? Would someone who learned the meaning of the term "electron" before the proposal of protons have learned something different from someone who learned about electrons after that proposal?

Second, one could argue that only those properties which are used for identifying the reference of a term belong to its meaning. These are properties, as I suggested in chapter 1, which enable the identification of an entity in an experimental situation. In the case of electrons, their charge-to-mass ratio was crucial for deciding whether various phenomena (cathode rays, the Zeeman effect, etc.) were their observable manifestations. Other properties—for example, the size of electrons—were not significant in that respect. I see two problems with this suggestion. First, the properties which are employed for identifying the reference of a term may not suffice to convey its meaning. The meaning of "electron," for instance, is not exhausted by the electron's charge-to-mass ratio. One has to know much more than that to know what an electron is. Second, whether a property is relevant to identifying the reference of a term is not settled once and for all. As we saw in chapter 8, Pauli argued, on the basis of the relativistic variation of the electron's mass, that the doublet structure of spectral lines was a manifestation of the electron and not of a more complex subatomic structure (the "core"). Thus, the relativistic character of the electron became essential for attributing an additional experimental phenomenon directly to it. Should we say that the meaning of the term "electron" changed, even though the scientists' knowledge of the electron's properties did not change?

The conception of meaning I have suggested differs both from the necessary and sufficient conditions view and from Putnam's notion of meaning. It differs from the former in that the properties associated with an entity and the laws it is supposed to obey are neither necessary nor sufficient conditions that any such entity should fulfill. The advancement of scientific knowledge may very well eliminate several of those characteristics, reveal further properties of the entity in question, or alter the laws it is supposed to obey without thereby affecting the extension of the term. For instance, after the launch of the old quantum theory of the atom the electron ceased to be considered subject to all the laws of classical electromagnetic theory (e.g., it was not supposed to radiate when performing accelerated motion). The novel constraints on its motion that were represented by quantum numbers clashed with its former representation as a mechanical particle. The last of those numbers, spin, turned out to be particularly resistant to a mechanical interpretation. These and other changes in the description of the electron did not throw doubt on its identity. That is, the physicists who struggled to make sense of its properties and behavior continued to believe that they were dealing with the same entity. This is a historical fact that any theory of meaning and reference has to come to terms with.

The notion of meaning I advocate also departs from Putnam's theory in that it does not require the referent of a term to be a stable and independent component of its meaning. Dropping this requirement has two advantages. First, no problem arises when a term denotes unobservable entities with no experimental manifestations, whose reference, therefore, cannot be identified independently of a description. Second, the stability of a term's reference is left open and, as I will argue below, becomes an issue that has to be settled through historical work (see n. 38).

Given the above view of meaning a realist position is compatible with the phenomenon of meaning change to the extent that a core of meaning has survived changes in theoretical perspective.[51] Despite my criticism of Hacking, there is a valuable element in his "experimental" realism that could (and should) be incorporated into a viable realist position. Even though his "causal properties" do not provide a theory-independent access to the corresponding entities, they do provide a core of meaning that has proved relatively immune to theory change. Any future theory about, for example, electrons will have to incorporate, but perhaps also reinterpret, these "well-known causal properties" of electrons.

My selective appropriation of Hacking's position should not obscure the fact that two main differences remain between his view and the view I advocate. First, he identifies this core of meaning with a set of properties that we have come to know by experimenting on the corresponding entities, whereas, in my view, *any stable* belief about, for example, electrons should be included in this core. Second, Hacking thinks of this stable core of meaning as necessary and sufficient for entity realism, whereas I consider this stability of meaning necessary but not sufficient for a realist position. Even if a core of beliefs about electrons has not been affected by theory change, one cannot exclude the possibility of an alternative, empirically adequate theory that would not include electrons in its ontology. One could risk the prediction, however, extrapolating from the historical development of science, that several, if not all, of those properties would be ascribed to the new entities postulated by that, radically novel, theory.[52]

51. A similar suggestion has been defended by H. Putnam. See his "Language and Reality," in his *Philosophical Papers*, vol. 2, pp. 272–290, esp. p. 275. For some other options that are open to the realist see below. Cf. also Psillos, *Scientific Realism*, p. 295.

52. The acceptance of Einstein's special theory of relativity and the concomitant abandonment of Lorentz's theory of electrons provide an actual historical example of radical theory change accompanied by the abolition of an entity (the ether) that was central to the rejected

Thus, histories of scientific concepts can play a seminal role in evaluating the tenability of a realist attitude toward the corresponding entities. On the one hand, a realist stance toward a particular entity would be discredited if the historical record revealed that the meaning of the corresponding term had undergone a radical transformation, which had affected even the most central features previously associated with that entity. On the other hand, if historical analysis showed that the core of a term's meaning had remained unaffected by changes in theoretical perspective, this would be an argument (albeit not a conclusive one) for maintaining a realist position about the entity under examination. Such a position would involve the inductively grounded belief that the core of properties attributed to the entity will survive changes in high-level theory.

One might want to challenge the view that the survival of a core of meaning enables a realist perspective to get off the ground.[53] In particular, one might ask for the specific criteria that privilege this set of beliefs (the "core") over the rest of beliefs about electrons, which proved highly unstable. I will not attempt to respond to this line of criticism, because I think that there is another way out for the aspiring realist.

This alternative amounts to maintaining that in science, descriptions are used referentially and not attributively. In the former case, a description is used to pick out a specific entity. If by the help of the description one successfully picks out the entity in question, then the description has served its purpose, even if it later turns out that it was altogether wrong. In the latter case, on the other hand, it is demanded that the entity picked out by a description satisfy completely every feature of that description.[54]

However, when it comes to unobservable entities, the claim that descriptions are used referentially cannot be easily maintained, because it is not clear in what sense a description of an unobservable entity "picks out" that entity. There are two options at this point. First, one could suggest that the description enables us to differentiate the entity in question from other unobservable entities. Thus, the description plays an essentially taxonomic function. It may adequately play that function even if it is (in some respects) wrong. For example, any description of the electron that includes the value of its charge

---

theory. However, certain of the ether's causal properties (e.g., its ability to act on charged particles) were taken over by the electromagnetic field. See N. J. Nersessian, *Faraday to Einstein: Constructing Meaning in Scientific Theories* (Dordrecht: Martinus Nijhoff, 1984).

53. Cf. Putnam, "Truth, Activation Vectors and Possession Conditions for Concepts," p. 444.

54. This distinction was introduced by Keith Donnellan. See his "Reference and Definite Descriptions," *Philosophical Review,* 75 (1966): 281–304.

and rest mass will differentiate it from other elementary particles even if it is in other respects misleading. As long as the classification of unobservable entities remains invariant, a realist attitude toward those entities remains immune to changes in their descriptions. The second option, as I suggested in chapter 1, is to identify the reference of a term denoting an unobservable entity by means of the experimental situations that are considered the observable manifestations of the entity in question.[55] Only certain aspects of the description associated with an entity enable one to pick out the experimental situations attributed to it—namely, those aspects which are inferred from those situations. For example, as we saw in chapter 4, the charge-to-mass ratio of the electron was originally derived from the magnetic splitting of spectral lines and the behavior of cathode rays under the influence of electric and magnetic fields. From then on, it was used for picking additional experimental situations as manifestations of the electron. It thus became an identification property, a way to tell whether the electron was present in an experimental situation. Again, as long as that "family" of experimental situations remains invariant (or expands in a cumulative fashion) a realist stance toward the entity involved in them is not threatened by changes in that entity's description.

Putnam has anticipated this proposal by suggesting that we can identify a magnitude (e.g., charge) "by, for example, singling it out as the magnitude which is causally responsible for certain effects."[56] However, he takes for granted that the reference of scientific terms does not change over time. In my view, on the other hand, the realist intuition that mature science has developed against a stable ontological background has to pass the test of historical scrutiny. It is an open question whether the reference of, say, "charge" has remained stable. Having provided a criterion for identifying the reference of scientific terms, the stability of the reference of a specific term over time should be investigated historically. Historical research, thus, obtains an essential role for evaluating the tenability of a realist attitude toward the ontology of science.

Furthermore, Putnam does not address a question that, I think, is important: What binds certain effects together as manifestations of a single entity?[57] As I suggested in chapter 1, different effects can be attributed to the same entity only if they have some qualitative or quantitative features in common. As-

---

55. A similar proposal has been made by Philip Kitcher. See his "Implications of Incommensurability," in Asquith and Nickles (eds.), *PSA 1982*, vol. 2, pp. 689–703, esp. p. 695.

56. Putnam, *Philosophical Papers*, vol. 2, p. ix; cf. also his *Realism with a Human Face*, p. 74.

57. Cf. T. S. Kuhn, "Metaphor in Science," in A. Ortony (ed.), *Metaphor and Thought* (Cambridge: Cambridge Univ. Press, 1979), pp. 409–419, on p. 411.

suming that a particular effect is a manifestation of a hidden entity, one can infer some of the characteristics of the latter from features of the former.[58] If inferences from different effects converge to the same results, then this is an indication that those effects are due to the same entity. For instance, the results of Zeeman's and Thomson's experiments suggested that the particles involved in both of them were negatively charged and had approximately the same charge-to-mass ratio. This convergence suggested that the same entities were "underneath" those different experimental situations.

The inference of an entity's features from its putative experimental manifestations is made possible by the conjecture of an explanatory mechanism, through which the entity in question brings about the observed phenomena. It is worth noting that those features may acquire a life of their own and outlive the postulated mechanism. The determination of the charge-to-mass ratio of the electron from the Zeeman effect illustrates this possibility. To account for that phenomenon, Lorentz conjectured a specific mechanism through which the electron produced the splitting of spectral lines. On the basis of that mechanism and the experimental results he had obtained, Zeeman inferred the charge-to-mass ratio of the electron. That ratio outlived the explanatory mechanism that was involved in its determination, which was later rejected in favor of a more complicated quantum-theoretical explanation of the Zeeman effect.[59]

To recapitulate, I have argued that historical case studies can play a dual role, either negative or positive, in disputes over scientific realism. They can either undermine realist intuitions or neutralize the antirealist implications of meaning change. However, this neutralization does not lead automatically to a realist position; the defense of scientific realism should take into account all those influential arguments that have been put forward against it, most notably by van Fraassen, and do not capitalize on conceptual change. Whether such a

58. This type of "backward" inference is sometimes called "deduction from the phenomena." I think that this expression is misleading, because these deductions always involve additional theoretical assumptions about, for instance, the entity's properties or the laws it obeys. See the very enlightening discussion by J. Worrall, "The Scope, Limits, and Distinctiveness of the Method of 'Deduction from the Phenomena': Some Lessons from Newton's 'Demonstrations' in Optics," *British Journal for the Philosophy of Science,* 51 (2000): 45–80.

59. By the way, this example shows that if "the burden of reference for . . . [a theoretical] term . . . [is partly] carried . . . by the explanatory mechanism developed in the theory," a realist account of the history of the concept of the electron is impossible. The quote is from Enç, "Reference of Theoretical Terms," p. 271. Other philosophers who have stressed the significance of explanatory mechanisms for fixing reference include Kroon, "Theoretical Terms and the Causal View of Reference," esp. pp. 154, 162–165; Nola, "Fixing the Reference of Theoretical Terms," esp. pp. 509, 524; and Psillos, *Scientific Realism,* pp. 288ff.

project is tenable is a question beyond the scope of this chapter. Moreover, the crucial philosophical issue for my purposes is the identity of theoretical entities and not the existence of their putative counterparts in nature. The viability of my biographical approach depends on the possibility of attributing a stable identity to the electron. With respect to its existence, on the other hand, this approach makes it possible to adopt an agnostic stance, since it focuses on the representation of the electron and not on the electron qua entity in nature. And this takes me to the final section, where I will bring the philosophical discussion on meaning change to bear on the question of the electron's identity.

## 6. On the Electron's Identity: What Would We Need in Order to Claim the Electron Exists?

The considerable evolution of the concept of the electron, within the time span covered in this book, raises two philosophical questions: First, was that evolution a rational process? Second, did the term "electron" refer to the same entity throughout that period? The former question can receive a straightforward answer. All the changes in the representation of the electron were the outcome of adequate solutions to specific problem situations and were, in that sense, rational.

The latter question, on the other hand, requires a more extensive discussion. Some philosophers have suggested that the rationality of the developmental process that transformed the concept of the electron implies that the reference of "electron" remained stable. Dudley Shapere, for instance, claims that "it was that continuity [in the successive uses of the term 'electron'], or, more importantly, the chain of reasons which produced that continuity, that alone justifies our speaking of 'the concept' or 'the meaning' of the term, and our speaking of the term having 'the same reference.'" [60] I do not find this claim convincing. The rationality of the development of the concept of the electron does not preclude the shifting of its reference and, thus, has little to do with the referential stability of the term "electron." Rationality and the semantics of scientific terms are two issues in the philosophy of science that should be kept apart.

---

60. D. Shapere, *Reason and the Search for Knowledge: Investigations in the Philosophy of Science,* Boston Studies in the Philosophy of Science 78 (Dordrecht: Reidel, 1984), p. xxxiv. This passage does not necessarily imply that Shapere has espoused a realist position, since the referential stability of the term "electron" does not by itself establish the existence of electrons. However, a realist perspective is manifest in other parts of that book. He suggests, for instance, that "we have found reasons to classify electrons (formerly 'theoretical') with tables and rocks." Ibid., p. 375.

As I argued in the previous section, one can maintain that the reference of "electron" has not shifted provided that a core of the meaning of that term has remained stable.[61] This was, indeed, the case in the historical developments we examined. Over the considerable evolution of the concept of the electron that took place from 1896 to 1925 a core of meaning survived changes in theoretical perspective. Throughout that period nobody doubted that the electron was a universal constituent of matter, with a certain mass and charge, and that it was the agent of radiation.

It is interesting to note that the beliefs that concerned the quantitative aspects of the electron's representation were among those that remained stable. The value of $e/m$, for instance, was measured in 1927, by means of a spectroscopic method, and found to be $(1.7606 \pm .0010) \times 10^7$ e.m.u. per gram.[62] This value agreed, in order of magnitude, with the original estimates of $e/m$. This aspect of the construction of the concept of the electron supports, albeit not conclusively, a realist attitude toward the electron. If it exists, the stability of the results of its measurement is not hard to understand. The antirealist, on the other hand, faces the challenge of developing an account of the development of the concept of the electron that comes to terms with that stability. On an antirealist view, the value of, say, the electron's charge-to-mass ratio can be interpreted as a parameter of a theory, which is specified so as to render the theory empirically adequate. However, that ratio was determined from results obtained in several different experimental situations (e.g., cathode rays and the Zeeman effect). It might be difficult to understand from an antirealist perspective why the values of $e/m$ obtained in different experimental contexts turned out to be the same. Moreover, the antirealist might have a problem explaining why the replacement of, say, Lorentz's theory of electrons by a radically different theory (the quantum theory) did not affect the parameter in question. Why did the latter theory have to incorporate the same value of $e/m$ in order to be empirically adequate?

Nevertheless, as I have already pointed out, the stability in question does not suffice to establish the belief in the existence of the electron. There are many other aspects of the realism debate that should be addressed before one

61. Cf. J. Bain and J. D. Norton, "What Should Philosophers of Science Learn from the History of the Electron?" in J. Z. Buchwald and A. Warwick (eds.), *Histories of the Electron: The Birth of Microphysics* (Cambridge, Mass.: MIT Press, 2001), pp. 451–465.

62. See W. V. Houston, "A Spectroscopic Determination of $e/m$," *Physical Review*, 30 (1927): 608–613. Other measurements, using different methods, led to similar results. See, e.g., C. T. Perry and E. L. Chaffee, "A Determination of $e/m$ for an Electron by Direct Measurement of the Velocity of Cathode Rays," *Physical Review*, 36 (1930): 904–918.

adopts a realist stance toward unobservable entities and ipso facto toward the electron. That is why I tried to tell the story of its representation from an agnostic perspective.

Two further options are open to the aspiring realist toward the electron. First, he or she can argue that the classificatory aspects of the electron's description remained stable. That is, indeed, what we find in the history of the electron's representation, provided that we take into account a turning point, namely, the first measurements of the electron's charge-to-mass ratio (1896–1897). Before those measurements electrons were not clearly differentiated from the ions of electrolysis. After those measurements, however, Lorentz's "ions," Larmor's "electrons," and Thomson's "corpuscles" were identified and their extension was clearly distinguished from the extension of "electrolytical ions." From then on, those different terms were gradually replaced by a single term—"electron"—a change that reflected the scientists' belief in the identity of the corresponding entities.

Beyond that turning point, the criteria that were used in the classification of electrons remained stable. From 1897 to 1925 the physicists' beliefs about the electron's charge, mass, and dimensions remained approximately (i.e., as to order of magnitude) stable. It was those beliefs that differentiated electrons from other unobservable entities (e.g., ions).

The second option is to establish the referential continuity of the term "electron," tracking its reference by means of the experimental situations that were putative manifestations of electrons. After 1896, those situations underwent a cumulative expansion and gradually came to cover an increasing variety of different experimental phenomena (e.g., cathode rays, the Zeeman effect, β-rays, and the photoelectric effect). Once an experimental situation was interpreted as the observable manifestation of electrons, its attribution to the agency of the electron was never challenged. Again, I want to stress that this past referential stability is necessary but not sufficient for a realist attitude toward the electron. There is the possibility that (all or some of) the experimental situations that are taken to manifest the presence of electrons will cease to be so regarded. This could happen in two ways: First, the electron may disappear altogether from the ontology of physics and chemistry. Second, new developments may throw doubt on the belief that all those experimental situations are manifestations of a single entity. This is what happened, for instance, as an aftermath of Zeeman's discovery. The realization that Lorentz's ions were distinct from the ions of electrolysis implied that the latter were not involved in the experimental phenomena attributed to the former.

In conclusion, the change of the electron's representation does not imply that it ceased to refer to the same entity. The survival of a core of meaning and the referential stability of the term "electron" from 1896 to 1925 allow a realist reading of the historical developments that I have reconstructed. Furthermore, the stability of the quantitative features in the electron's representation cannot be easily understood from an antirealist perspective. However, all these elements of continuity do not sufficiently establish the electron's existence.

# References

Abegg, R., "Die Valenz und das periodische System: Versuch einer Theorie der Molekularverbindungen," *Zeitschrift für Anorganische Chemie, 39* (1904): 330–380.

Abraham, M., "Prinzipien der Dynamik des Elektrons," *Annalen der Physik,* 10 (1903): 105–179.

Achinstein, P., "On the Meaning of Scientific Terms," *Journal of Philosophy,* 61, no. 17 (1964): 497–509.

Achinstein, P., "Discovery and Rule-Books," in Nickles (ed.), *Scientific Discovery, Logic, and Rationality,* pp. 117–132.

Achinstein, P., *Particles and Waves* (New York: Oxford Univ. Press, 1991).

Achinstein, P., "Who Really Discovered the Electron?" in Buchwald and Warwick (eds.), *Histories of the Electron,* pp. 403–424.

Agassi, J., "On the Nature of Scientific Problems and Their Roots in Metaphysics," in his *Science in Flux,* Boston Studies in the Philosophy of Science 28 (Dordrecht: Reidel, 1975), pp. 208–239.

Anderson, D. L., *The Discovery of the Electron: The Development of the Atomic Concept of Electricity* (1964; repr., New York: Arno Press, 1981).

Andrade, E. N. da C., *The Structure of the Atom,* 3rd rev. ed. (London: G. Bell and Sons, 1927; 1st ed., 1923).

Arabatzis, T., "The Discovery of the Zeeman Effect: A Case Study of the Interplay between Theory and Experiment," *Studies in History and Philosophy of Science,* 23 (1992): 365–388.

Arabatzis, T., "Rational versus Sociological Reductionism: Imre Lakatos and the Edinburgh School," in Gavroglu et al. (eds.), *Trends in the Historiography of Science,* pp. 177–192.

Arabatzis, T., "Can a Historian of Science Be a Scientific Realist?" *Philosophy of Science,* 68, suppl. (2001): S531–S541.

Arabatzis, T., "Towards a Historical Ontology?" *Studies in History and Philosophy of Science,* 34, no. 2 (2003): 431–442.

Arabatzis, T., "On the Inextricability of the Context of Discovery and the Context of Justification," in J. Schickore and F. Steinle (eds.), *Revisiting Discovery and Justification*, preprint 211 (Berlin: Max Planck Institute for the History of Science, 2002).

Arabatzis, T., "The Discovery of the Electron and the Atomism Debate" (unpublished manuscript).

Arabatzis, T., and K. Gavroglu, "The Chemists' Electron," *European Journal of Physics*, 18 (1997): 150–163.

Armstrong, H. E., presidential address, *British Association for the Advancement of Science, Transactions of Section B*, 1909, pp. 420–454.

Asquith, P. D., and T. Nickles (eds.), *PSA 1982: Proceedings of the 1982 Biennial Meeting of the Philosophy of Science Association*, 2 vols. (East Lansing, Mich.: Philosophy of Science Association, 1983).

Bain, J., and J. D. Norton, "What Should Philosophers of Science Learn from the History of the Electron?" in Buchwald and Warwick (eds.), *Histories of the Electron*, pp. 451–465.

Ball, P., *A Biography of Water: Life's Matrix* (New York: Farrar, Straus and Giroux, 2000).

Baracca, A., "Early Proposal of an Intrinsic Magnetic Moment of the Electron in Chemistry and Magnetism (1915–1921) before the Papers of Goudsmit and Uhlenbeck," *Rivista di storia della scienza*, 3, no. 3 (1986): 353–374.

Beller, M., "Matrix Theory before Schrödinger: Philosophy, Problems, Consequences," *Isis*, 74 (1983): 469–491.

Benz, U., *Arnold Sommerfeld: Lehrer und Forscher an der Schwelle zum Atomzeitalter, 1868–1951* (Stuttgart: Wissenschaftliche Verlagsgesellschaft, 1975).

Berkson, W., *Fields of Force: The Development of a World View from Faraday to Einstein* (New York: John Wiley and Sons, 1974).

Berkson, W., "Research Problems and the Understanding of Science," in Nersessian (ed.), *The Process of Science*, pp. 83–93.

Bodanis, D., $E = mc^2$: *A Biography of the World's Most Famous Equation* (New York: Walker, 2000).

Bohr, N., *Studier over metallernes elektrontheori* (first publ. 1911), in his *Collected Works*, vol. 1, pp. 167–290.

Bohr, N., *Studies on the Electron Theory of Metals*, in his *Collected Works*, vol. 1, pp. 291–395.

Bohr, N., "On the Theory of the Decrease of Velocity of Moving Electrified Particles on Passing through Matter," *Philosophical Magazine*, 6th series, 25 (1913): 10–31.

Bohr, N., "On the Constitution of Atoms and Molecules," pt. 1, *Philosophical Magazine*, 6th series, 26 (1913): 1–25; repr. in his *On the Constitution*, pp. 1–25.

Bohr, N., "On the Constitution of Atoms and Molecules," pt. 2, "Systems Containing Only a Single Nucleus," *Philosophical Magazine*, 6th series, 26 (1913): 476–502; repr. in his *On the Constitution*, pp. 28–54.

Bohr, N., "On the Constitution of Atoms and Molecules," pt. 3, "Systems Containing Several Nuclei," *Philosophical Magazine*, 6th series, 26 (1913): 857–875; repr. in his *On the Constitution*, pp. 55–73.

Bohr, N., "On the Series Spectrum of Hydrogen and the Structure of the Atom," *Philosophical Magazine*, 29 (1915): 332–335.

Bohr, N., "On the Quantum Theory of Line-Spectra," pt. 1, *Kongelige Danske Videnskabernes Selskabs Skrifter Naturvidenskabelig og mathematisk afdeling*, series 8, IV (1918), 1, 1–36; repr. in van der Waerden (ed.), *Sources of Quantum Mechanics*, pp. 95–137.

Bohr, N., "On the Series Spectra of the Elements," speech to the German Physical Society on 27 Apr. 1920; trans. in his *Theory of Spectra*, pp. 20–60.

Bohr, N., "The Structure of the Atom and the Physical and Chemical Properties of the Elements," lecture to the Physical and Chemical Societies of Copenhagen (18 Oct. 1921); trans. in his *Theory of Spectra*, pp. 61–126.

Bohr, N., "Atomic Structure," *Nature*, 107 (1921): 104–107.

Bohr, N., *The Theory of Spectra and Atomic Constitution* (Cambridge: Cambridge Univ. Press, 1922).

Bohr, N., "The Structure of the Atom," Nobel Lecture, 11 Dec. 1922, in *Nobel Lectures: Physics, 1922–1941* (Amsterdam: Elsevier, 1965), pp. 7–43.

Bohr, N., "Linienspektren und Atombau," *Annalen der Physik*, 71 (1923): 228–288.

Bohr, N., "Atomic Theory and Mechanics," *Nature*, 116, suppl. (5 Dec. 1925): 845–852.

Bohr, N., note, *Nature*, 117 (1926): 265.

Bohr, N., *On the Constitution of Atoms and Molecules* (Copenhagen: Munksgaard, 1963).

Bohr, N., *Collected Works*, vol. 1, ed. J. R. Nielsen (Amsterdam: North Holland, 1972); vol. 2, ed. U. Hoyer (Amsterdam: North Holland, 1981).

Bohr, N., and D. Coster, "Röntgenspektren und periodisches System der Elemente," *Zeitschrift für Physik*, 12 (1923): 342–374.

Brannigan, A., *The Social Basis of Scientific Discoveries* (New York: Cambridge Univ. Press, 1981).

Braudel, F., *On History* (Chicago: Univ. of Chicago Press, 1980).

Bray, W. C., and G. Branch, "Valence and Tautomerism," *Journal of the American Chemical Society*, 35 (1913): 1440–1447.

Buchwald, J. Z., *From Maxwell to Microphysics: Aspects of Electromagnetic Theory in the Last Quarter of the Nineteenth Century* (Chicago: Univ. of Chicago Press, 1985).

Buchwald, J. Z. (ed.), *Scientific Practice* (Chicago: Univ. of Chicago Press, 1995).

Buchwald, J. Z., "How the Ether Spawned the Microworld," in Daston (ed.), *Biographies of Scientific Objects*, pp. 203–225.

Buchwald, J. Z., and A. Warwick (eds.), *Histories of the Electron: The Birth of Microphysics* (Cambridge, Mass.: MIT Press, 2001).

Burian, R., "Why Philosophers Should Not Despair of Understanding Scientific Discovery," in Nickles (ed.), *Scientific Discovery, Logic, and Rationality*, pp. 317–336.

Campbell, N. R., *Modern Electrical Theory*, 2nd ed. (Cambridge: Cambridge Univ. Press, 1913).

Campbell, N. R., "Atomic Structure," *Nature*, 106 (1920): 408–409.

Caneva, K. L., *The Form and Function of Scientific Discoveries,* Dibner Library Lecture Series (Washington, D.C.: Smithsonian Institution Libraries, 2001).

Cantor, G., "The Making of a British Theoretical Physicist: E. C. Stoner's Early Career," *British Journal for the History of Science,* 27 (1994): 277–290.

Cassidy, D. C., "Heisenberg's First Core Model of the Atom: The Formation of a Professional Style," *Historical Studies in the Physical Sciences,* 10 (1979): 187–224.

Cassidy, D. C., *Uncertainty: The Life and Science of Werner Heisenberg* (New York: Freeman, 1992).

"The Centenary of the Electron," special issue, *Physics Today,* Oct. 1997.

Chayut, M , "J. J. Thomson: The Discovery of the Electron and the Chemists," *Annals of Science,* 48 (1991): 527–544.

Cole, A. D., "Recent Evidence for the Existence of the Nucleus Atom," *Science,* new series, 41, no. 1046 (15 Jan. 1915): 73–81.

Compton, A. H., "The Magnetic Electron," *Journal of the Franklin Institute,* 192 (1921): 145–155.

Conant, J. B., "The Overthrow of the Phlogiston Theory," in J. B. Conant and L. K. Nash (eds.), *Case Histories in Experimental Science* (Cambridge, Mass.: Harvard Univ. Press, 1957).

Crease, R. P., *Making Physics: A Biography of Brookhaven National Laboratory, 1946–1972* (Chicago: Univ. of Chicago Press, 1999).

Curd, M. V., "The Logic of Discovery: An Analysis of Three Approaches," in Nickles (ed.), *Scientific Discovery, Logic, and Rationality,* pp. 201–219.

Curtis, W. E., "Wave-Lengths of Hydrogen Lines and Determination of the Series Constant," *Proceedings of the Royal Society of London (A),* 90 (1914): 605–620.

Dahl, P. F., *Flash of the Cathode Rays: A History of J J Thomson's Electron* (Bristol: Institute of Physics Publishing, 1997).

Darrigol, O., *From c-Numbers to q-Numbers: The Classical Analogy in the History of Quantum Theory* (Berkeley: Univ. of California Press, 1992).

Darrigol, O., "The Electron Theories of Larmor and Lorentz: A Comparative Study," *Historical Studies in the Physical and Biological Sciences,* 24 (1994): 265–336.

Darrigol, O., "Classical Concepts in Bohr's Atomic Theory (1913–1925)," *Physis* 34 (1997): 545–567.

Darrigol, O., "Aux confins de l'électrodynamique maxwelliene: Ions et électrons vers 1897," *Revue d'Histoire des Sciences,* 51, no. 1 (1998): 5–34.

Darwin, C. G., "A Theory of the Absorption and Scattering of the α Rays," *Philosophical Magazine,* 6th series, 23 (1912): 901–920.

Daston, L. (ed.), *Biographies of Scientific Objects* (Chicago: Univ. of Chicago Press, 2000).

Debye, P., "Quantenhypothese und Zeeman-Effekt," *Nachrichten von der Gesellschaft der Wissenschaften zu Göttingen, Mathematisch-Physikalische Klasse,* 1916, pp. 142–153.

Debye, P., "Quantenhypothese und Zeeman-Effekt," *Physikalische Zeitschrift,* 17 (1916): 507–512.

Donnellan, K., "Reference and Definite Descriptions," *Philosophical Review,* 75 (1966): 281–304.

Donovan, A., L. Laudan, and R. Laudan (eds.), *Scrutinizing Science: Empirical Studies of Scientific Change* (Baltimore: Johns Hopkins Univ. Press, 1992).

Duhem, P., *The Aim and Structure of Physical Theory* (Princeton: Princeton Univ. Press, 1991).

Ehrenfest, P., "Adiabatic Invariants and the Theory of Quanta," *Philosophical Magazine,* 33 (1917): 500–513; repr. in van der Waerden (ed.), *Sources of Quantum Mechanics,* pp. 79–93.

Einstein, A., "Zur Quantentheorie der Strahlung," *Physikalische Zeitschrift,* 18 (1917): 121–128; first publ. in *Mitteilungen der Physikalischen Gesellschaft, Zürich,* 18 (1916): 47–62; trans. in van der Waerden (ed.), *Sources of Quantum Mechanics,* pp. 63–77.

Einstein, E., and A. Sommerfeld. *Briefwechsel,* ed. A. Hermann (Basel: Schwabe, 1968).

Einstein, A., et al., *The Principle of Relativity* (New York: Dover, 1952).

"The Electron: Discovery and Consequences," special issue, *European Journal of Physics,* 18, no. 3 (May 1997).

Emsley, J., *The Shocking History of Phosphorus: A Biography of the Devil's Element* (London: Macmillan, 2000).

Enç, B., "Reference of Theoretical Terms," *Nous,* 10 (1976): 261–282.

Endo, S., and S. Saito, "Zeeman Effect and the Theory of Electron of H. A. Lorentz," *Japanese Studies in History of Science,* 6 (1967): 1–18.

Epstein, P., "Zur Theorie des Starkeffektes," *Annalen der Physik,* 50 (1916): 489–520.

Falconer, I., "Corpuscles, Electrons and Cathode Rays: J. J. Thomson and the 'Discovery of the Electron,'" *British Journal for the History of Science,* 20 (1987): 241–276.

Falconer, I., "Corpuscles to Electrons," in Buchwald and Warwick (eds.), *Histories of the Electron,* pp. 77–100.

Faraday, M., *Experimental Researches in Electricity,* 3 vols. (London, 1839–1855).

Feffer, S. M., "Arthur Schuster, J. J. Thomson, and the Discovery of the Electron," *Historical Studies in the Physical Sciences,* 20, no. 1 (1989): 33–61.

Feldhay, R., "Mathematical Entities in Scientific Discourse: Paulus Guldin and His *Dissertatio De Motu Terrae,*" in Daston (ed.), *Biographies of Scientific Objects,* pp. 42–66.

Feyerabend, P. K., "Das Problem der Existenz theoretischer Entitäten," in E. Topitsch (ed.), *Probleme der Wissenschaftstheorie* (Vienna: Springer-Verlag, 1960), pp. 35–72.

Feyerabend, P. K., "Explanation, Reduction and Empiricism," in H. Feigl and G. Maxwell (eds.), *Scientific Explanation, Space and Time,* Minnesota Studies in the Philosophy of Science 3 (Minneapolis: Univ. of Minnesota Press, 1962), pp. 28–97; repr. in his *Philosophical Papers,* vol. 1, pp. 44–96.

Feyerabend, P. K., "Realism and Instrumentalism: Comments in the Logic of Factual Support," in M. Bunge (ed.), *The Critical Approach to Science and Philosophy* (New York: Free Press, 1964), pp. 280–308; repr. in his *Philosophical Papers,* vol. 1, pp. 176–202.

Feyerabend, P. K., "On the 'Meaning' of Scientific Terms," *Journal of Philosophy,* 12 (1965): 266–274; repr. in his *Philosophical Papers,* vol. 1, pp. 97–103.

Feyerabend, P. K., "Problems of Empiricism," in R. G. Colodny (ed.), *Beyond the Edge of Certainty* (Englewood Cliffs, N.J.: Prentice Hall, 1965), pp. 145–260.

Feyerabend, P. K., "Against Method: Outline of an Anarchistic Theory of Knowledge," in M. Radner and S. Winokur (eds.), *Analysis of Theories and Methods of Physics and Psychology,* Minnesota Studies in the Philosophy of Science 4 (Minneapolis: Univ. of Minnesota Press, 1970), pp. 17–130.

Feyerabend, P. K., *Philosophical Papers,* vol. 1, *Realism, Rationalism and Scientific Method* (Cambridge: Cambridge Univ. Press, 1981).

Feyerabend, P. K., "More Clothes from the Emperor's Bargain Basement: A Review of Laudan's *Progress and Its Problems,*" *British Journal for the Philosophy of Science,* 32 (1981): 57–71; repr. in his *Philosophical Papers,* vol. 2, *Problems of Empiricism* (Cambridge: Cambridge Univ. Press, 1981), pp. 231–246.

Fierz, M., and V. F. Weisskopf (eds.), *Theoretical Physics in the Twentieth Century: A Memorial Volume to Wolfgang Pauli* (New York: Interscience Publishers, 1960).

Fine, A., *The Shaky Game: Einstein, Realism, and the Quantum Theory* (Chicago: Univ. of Chicago Press, 1986).

FitzGerald, G. F., "Dissociation of Atoms," *Electrician,* 39 (1897): 103–104.

Forman, P., "The Doublet Riddle and Atomic Physics *circa* 1924," *Isis,* 59 (1968): 156–174.

Forman, P., "Alfred Landé and the Anomalous Zeeman Effect, 1919–1921," *Historical Studies in the Physical Sciences,* 2 (1970): 153–261.

Forman, P., "A Venture in Writing History," *Science,* 220 (1983): 824–827.

Forman, P., and A. Hermann, "Sommerfeld, Arnold," in Gillispie (ed.), *Dictionary of Scientific Biography,* vol. 12, pp. 525–532.

Fournier D'Albe, E. E., *The Electron Theory: A Popular Introduction to the New Theory of Electricity and Magnetism,* 2nd ed. (London: Longmans, Green, and Co., 1907).

Franck, J., and G. Hertz, "Über Zusammenstösse zwischen Gasmolekülen und langsamen Elektronen," *Verhandlungen der Deutschen Physikalischen Gesellschaft,* 15 (1913): 373–390.

Franck, J., and G. Hertz, "Über Zusammenstösse zwischen langsamen Elektronen und Gasmolekülen," pt. 2, *Verhandlungen der Deutschen Physikalischen Gesellschaft,* 15 (1913): 613–620.

Franck, J., and G. Hertz, "Über Zusammenstösse zwischen Elektronen und den Molekülen des Quecksilberdampfes und die Ionisierungspannung desselben," *Verhandlungen der Deutschen Physikalischen Gesellschaft,* 16 (1914): 457–467.

Franklin, A., "Millikan's Published and Unpublished Data on Oil Drops," *Historical Studies in the Physical Sciences,* 11 (1981): 185–201.

Friedel, R. D., and P. Israel, with B. S. Finn, *Edison's Electric Light: Biography of an Invention* (New Brunswick: Rutgers Univ. Press, 1986).

Friedman, M., "On the Sociology of Scientific Knowledge and Its Philosophical Agenda," *Studies in History and Philosophy of Science,* 29 (1998): 239–271.

Galison, P., *How Experiments End* (Chicago: Univ. of Chicago Press, 1987).

Galison, P., "Context and Constraints," in Buchwald (ed.), *Scientific Practice,* pp. 13–41.

Galison, P., *Image and Logic: A Material Culture of Microphysics* (Chicago: Univ. of Chicago Press, 1997).

Gavroglu, K., "The Physicists' Electron and Its Appropriation by the Chemists," in Buchwald and Warwick (eds.): *Histories of the Electron,* pp. 363–400.

Gavroglu, K., J. Christianidis, and E. Nicolaidis (eds.), *Trends in the Historiography of Science,* Boston Studies in the Philosophy of Science 151 (Dordrecht: Kluwer, 1994).

Gavroglu, K., and Y. Goudaroulis, *Methodological Aspects of the Development of Low Temperature Physics: Concepts out of Contexts* (Dordrecht: Kluwer, 1989).

Gavroglu, K., and A. Simoes, "The Americans, the Germans, and the Beginnings of Quantum Chemistry: The Confluence of Diverging Traditions," *Historical Studies in the Physical Sciences,* 25, no. 1 (1994): 47–110.

Gillispie, C. C. (ed.), *Dictionary of Scientific Biography,* 16 vols. (New York: Charles Scribner's Sons, 1970–1980).

Glick, T. F. (ed.), *The Comparative Reception of Relativity,* Boston Studies in the Philosophy of Science 103 (Dordrecht: Reidel, 1987).

Glitscher, K., "Spektroskopischer Vergleich zwischen den Theorien des starren und des deformierbaren Elektrons," *Annalen der Physik,* 52 (1917): 608–630.

Goldberg, S., "The Abraham Theory of the Electron: The Symbiosis of Experiment and Theory," *Archive for History of Exact Sciences,* 7, no. 1 (1970): 7–25.

Gooday, G., "The Questionable Matter of Electricity: The Reception of J. J. Thomson's 'Corpuscle' among Electrical Theorists and Technologists," in Buchwald and Warwick (eds.), *Histories of the Electron,* pp. 101–134.

Gordin, M. D., "Making Newtons: Mendeleev, Metrology, and the Chemical Ether," *Ambix,* 45, no. 2 (July 1998): 96–115.

Goudsmit, S. A., "Über die Komplexstruktur der Spektren," *Zeitschrift für Physik,* 32 (1925): 794–798.

Goudsmit, S. A., "Guess Work: The Discovery of the Electron Spin," *Delta,* 15 (Summer 1972): 77–91.

Goudsmit, S. A., and G. E. Uhlenbeck, "Opmerking over de Spectra van Waterstof en Helium," *Physica,* 5 (1925): 266–270.

Gutting, G., "Science as Discovery," *Revue Internationale de Philosophie,* 131–132 (1980): 26–48.

Hacking, I. (ed.), *Scientific Revolutions* (Oxford: Oxford Univ. Press, 1981).

Hacking, I., *Representing and Intervening* (Cambridge: Cambridge Univ. Press, 1983).

Hacking, I., "Experimentation and Scientific Realism," in Leplin (ed.), *Scientific Realism,* pp. 154–172.

Hacking, I., *The Social Construction of What?* (Cambridge, Mass.: Harvard Univ. Press, 1999).

Hacking, I., *Historical Ontology* (Cambridge, Mass.: Harvard Univ. Press, 2002).

Hanson, N. R., *Patterns of Discovery* (Cambridge: Cambridge Univ. Press, 1958).

Hanson, N. R., *The Concept of the Positron: A Philosophical Analysis* (Cambridge: Cambridge Univ. Press, 1963).

Harré, R., *Great Scientific Experiments* (New York: Oxford Univ. Press, 1983).

Harré, R., *Varieties of Realism* (Oxford: Basil Blackwell, 1986).

Heilbron, J. L., *A History of the Problem of Atomic Structure from the Discovery of the Electron to the Beginning of Quantum Mechanics* (Ph.D. dissertation, Univ. of California, Berkeley, 1964).

Heilbron, J. L., "The Kossel-Sommerfeld Theory and the Ring Atom," *Isis,* 58 (1967): 451–485; repr. in his *Historical Studies in the Theory of Atomic Structure,* pp. 261–295.

Heilbron, J. L., "The Scattering of α and β Particles and Rutherford's Atom," *Archive for History of Exact Sciences,* 4 (1968): 247–307; repr. in his *Historical Studies in the Theory of Atomic Structure,* pp. 85–145.

Heilbron, J. L., *Historical Studies in the Theory of Atomic Structure* (New York: Arno Press, 1981).

Heilbron, J. L., "The Origins of the Exclusion Principle," *Historical Studies in the Physical Sciences,* 13 (1983): 261–310.

Heilbron, J. L., and T. S. Kuhn, "The Genesis of the Bohr Atom," *Historical Studies in the Physical Sciences,* 1 (1969): 211–290.

Heisenberg, W., "Zur Quantentheorie der Linienstruktur und der anomalen Zeemaneffekte," *Zeitschrift für Physik,* 8 (1921–1922): 273–297.

Helmholtz, H. von, "On the Modern Development of Faraday's Conception of Electricity," *Journal of the Chemical Society,* 39 (1881): 277–304.

Hempel, C. G., *Aspects of Scientific Explanation* (New York: Free Press, 1965).

Hendry, J., "Weimar Culture and Quantum Causality," *History of Science,* 18 (1980): 155–180.

Hendry, J., *The Creation of Quantum Mechanics and the Bohr-Pauli Dialogue* (Dordrecht: Reidel, 1984).

Hentschel, K., "Die Entdeckung des Zeeman-Effekts als Beispiel für das komplexe Wechselspiel von wissenschaftlichen Instrumenten, Experimenten und Theorie," *Physikalische Blätter,* 52 (1996): 1232–1235.

Hexter, J. H., *More's Utopia: The Biography of an Idea* (New York: Harper and Row, 1965; 1st ed., 1952).

Hiebert, E. N., "The State of Physics at the Turn of the Century," in M. Bunge and W. R. Shea (eds.), *Rutherford and Physics at the Turn of the Century* (New York: Dawson and Science History Publications, 1979), pp. 3–22.

Hirosige, T., "Origins of Lorentz' Theory of Electrons and the Concept of the Electromagnetic Field," *Historical Studies in the Physical Sciences,* 1 (1969): 151–209.

Holmes, F. L., *Eighteenth-Century Chemistry as an Investigative Enterprise* (Berkeley: Office for History of Science and Technology, Univ. of California, 1989).

Holton, G. (ed.), *The Twentieth-Century Sciences: Studies in the Biography of Ideas* (New York: Norton, 1972).

Holton, G., "Subelectrons, Presuppositions and the Millikan-Ehrenhaft Dispute," in his *The Scientific Imagination: Case Studies* (Cambridge: Cambridge Univ. Press, 1978), pp. 25–83.

Hon, G., "Franck and Hertz versus Townsend: A Study of Two Types of Experimental Error," *Historical Studies in the Physical Sciences*, 20, no. 1 (1989): 79–106.

Hon, G., "Is the Identification of Experimental Error Contextually Dependent? The Case of Kaufmann's Experiment and Its Varied Reception," in Buchwald (ed.), *Scientific Practice*, pp. 170–223.

Horwich, P. (ed.), *World Changes: Thomas Kuhn and the Nature of Science* (Cambridge, Mass.: MIT Press, 1993).

Houston, W. V., "A Spectroscopic Determination of $e/m$," *Physical Review*, 30 (1927): 608–613.

Hoyningen-Huene, P., "Context of Discovery and Context of Justification," *Studies in History and Philosophy of Science*, 18 (1987): 501–515.

Hull, D., "Historical Entities and Historical Narratives," in C. Hookway (ed.), *Minds, Machines and Evolution: Philosophical Studies* (Cambridge: Cambridge Univ. Press, 1984), pp. 17–42.

Hunt, B. J., *The Maxwellians* (Ithaca: Cornell Univ. Press, 1991).

Jammer, M., *The Conceptual Development of Quantum Mechanics* (New York: McGraw-Hill, 1966).

Jaumann, G., "Electrostatische Ablenkung der Kathodenstrahlen," *Annalen der Physik und Chemie*, 59 (1896): 252–266.

Jensen, C., "Two One-Electron Anomalies in the Old Quantum Theory," *Historical Studies in the Physical Sciences*, 15, no. 1 (1984): 81–106.

"J. J. Thomson's Electron," special issue, *Physics Education*, 32, no. 4 (July 1997).

Jorland, G., "The Coming into Being and Passing Away of Value Theories in Economics (1776–1976)," in Daston (ed.), *Biographies of Scientific Objects*, pp. 117–131.

Jungnickel, C., and R. McCormmach, *Intellectual Mastery of Nature: Theoretical Physics from Ohm to Einstein*, 2 vols. (Chicago: Univ. of Chicago Press, 1986).

Kaiser, W., "Electron Gas Theory in Metals: Free Electrons in Bulk Matter," in Buchwald and Warwick (eds.), *Histories of the Electron*, pp. 255–303.

Kargon, R., "Mendeleev's Chemical Ether, Electrons, and the Atomic Theory," *Journal of Chemical Education*, 42, no. 7 (1965): 388–389.

Kaufmann, W., "Die magnetische Ablenkbarkeit der Kathodenstrahlen und ihre Abhängigkeit vom Entladungspotential," *Annalen der Physik und Chemie*, 61 (1897): 544–552.

Kitcher, P., "Implications of Incommensurability," in Asquith and Nickles (eds.), *PSA 1982*, vol. 2, pp. 689–703.

Kitcher, P., "The Naturalists Return," *Philosophical Review*, 101, no. 1 (1992): 53–114.

Klein, M. J., "The First Phase of the Bohr-Einstein Dialogue," *Historical Studies in the Physical Sciences*, 2 (1970): 1–39.

Klein, M. J., *Paul Ehrenfest*, vol. 1, *The Making of a Theoretical Physicist* (New York: Elsevier, 1970).

Knudsen, O., "O. W. Richardson and the Electron Theory of Matter, 1901–1916," in Buchwald and Warwick (eds.), *Histories of the Electron*, pp. 227–253.

Koertge, N., "Explaining Scientific Discovery," in Asquith and Nickles (eds.), *PSA 1982*, vol. 1, pp. 14–28.

Kohler, R. E., "The Origin of G. N. Lewis's Theory of the Shared Pair Bond," *Historical Studies in the Physical Sciences*, 3 (1971): 343–376.

Kohler, R. E., "Irving Langmuir and the 'Octet' Theory of Valence," *Historical Studies in the Physical Sciences*, 4 (1974): 39–87.

Kohler, R. E., "The Lewis-Langmuir Theory of Valence and the Chemical Community, 1920–1928," *Historical Studies in the Physical Sciences*, 6 (1975): 431–468.

Kohler, R. E., "Lewis, Gilbert Newton," in Gillispie (ed.), *Dictionary of Scientific Biography*, vol. 8, pp. 289–294.

Kordig, C. R., "Discovery and Justification," *Philosophy of Science*, 45 (1978): 110–117.

Kossel, W., and A. Sommerfeld, "Auswahlprinzip und Verschiebung bei Serienspektren," *Verhandlungen der Deutschen Physikalischen Gesellschaft*, 21 (1919): 240–259; repr. in Sommerfeld, *Gesammelte Schriften*, vol. 3, pp. 476–495.

Kox, A. J., "The Discovery of the Electron," pt. 2, "The Zeeman Effect," *European Journal of Physics*, 18 (1997): 139–144.

Kragh, H., "Niels Bohr's Second Atomic Theory," *Historical Studies in the Physical Sciences*, 10 (1979): 123–186.

Kragh, H., "Bohr's Atomic Theory and the Chemists," *Rivista di storia della scienza*, 2, no. 3 (1985): 463–486.

Kragh, H., "The Fine Structure of Hydrogen and the Gross Structure of the Physics Community, 1916–26," *Historical Studies in the Physical Sciences*, 15 (1985): 67–125.

Kragh, H., "Concept and Controversy: Jean Becquerel and the Positive Electron," *Centaurus*, 32 (1989): 203–240.

Kragh, H., *Quantum Generations: A History of Physics in the Twentieth Century* (Princeton: Princeton Univ. Press, 1999).

Kragh, H., "The Electron, the Protyle, and the Unity of Matter," in Buchwald and Warwick (eds.), *Histories of the Electron*, pp. 195–226.

Kronig, R., "Spinning Electrons and the Structure of Spectra," *Nature*, 117 (1926): 550.

Kronig, R., "The Turning Point," in Fierz and Weisskopf (eds.), *Theoretical Physics in the Twentieth Century*, pp. 5–39.

Kroon, F. W., "Theoretical Terms and the Causal View of Reference," *Australasian Journal of Philosophy*, 63, no. 2 (1985): 143–166.

Kuhn, T. S., *The Structure of Scientific Revolutions*, 2nd ed. (Chicago: Univ. of Chicago Press, 1970).

Kuhn, T. S., *The Essential Tension* (Chicago: Univ. of Chicago Press, 1977).

Kuhn, T. S., "Metaphor in Science," in A. Ortony (ed.), *Metaphor and Thought* (Cambridge: Cambridge Univ. Press, 1979), pp. 409–419.

Kuhn, T. S., "Commensurability, Comparability, Communicability," in Asquith and Nickles (eds.), *PSA 1982*, vol. 2, pp. 669–688.

Kuhn, T. S., "Revisiting Planck," *Historical Studies in the Physical Sciences*, 14, no. 2 (1984): 231–252.

Kuhn, T. S., *Black-Body Theory and the Quantum Discontinuity, 1894–1912*, 2nd ed. (Chicago: Univ. of Chicago Press, 1987).

Kuhn, T. S., "Possible Worlds in History of Science," in Sture Allén (ed.), *Possible Worlds in Humanities, Arts and Sciences* (Berlin: Walter de Gruyter, 1989), pp. 9–32.

Kuhn, T. S., "Dubbing and Redubbing: The Vulnerability of Rigid Designation," in C. W. Savage (ed.), *Scientific Theories*, Minnesota Studies in the Philosophy of Science 14 (Minneapolis: Univ. of Minnesota Press, 1990), pp. 298–318.

Kuhn, T. S., "Afterwords," in Horwich (ed.), *World Changes*, pp. 311–341.

Kuhn, T. S., *The Road since Structure*, ed. J. Conant and J. Haugeland (Chicago: Univ. of Chicago Press, 2000).

Lakatos, I., "Falsification and the Methodology of Scientific Research Programmes," in I. Lakatos and A. Musgrave (eds.), *Criticism and the Growth of Knowledge* (Cambridge: Cambridge Univ. Press, 1970), pp. 91–196.

Landé, A., "Über den anomalen Zeemaneffekt,"pt. 1, *Zeitschrift für Physik*, 5 (1921): 231–241.

Landé, A., "Termstruktur und Zeemaneffekt der Multipletts," *Zeitschrift für Physik*, 15 (1923): 189–205.

Landé, A., "Zur Theorie der Röntgenspektren," *Zeitschrift für Physik*, 16 (1923): 391–396.

Langevin, P., "Magnétisme et Théorie des Électrons," *Annales de Chimie et de Physique*, 5 (1905): 70–127.

Langley, P., H. A. Simon, G. L. Bradshaw, and J. M. Zytkow, *Scientific Discovery: Computational Explorations of the Creative Process* (Cambridge, Mass.: MIT Press, 1987).

Langmuir, I., "The Structure of Atoms and the Octet Theory of Valence," *Proceedings of the National Academy of Sciences*, vol. 5 (1919): 252–259; repr. in his *Collected Works*, vol. 6, pp. 1–8.

Langmuir, I., "The Arrangement of Electrons in Atoms and Molecules," *Journal of the American Chemical Society*, 41, no. 6 (1919): 868–934; repr. in his *Collected Works*, vol. 6, pp. 9–73.

Langmuir, I., "The Structure of Atoms and Its Bearing on Chemical Valence," *Journal of Industrial and Engineering Chemistry*, 12, no. 4 (1920): 386–389; repr. in his *Collected Works*, vol. 6, pp. 93–100.

Langmuir, I., "Theories of Atomic Structure," *Nature*, 105 (1920): 261; repr. in his *Collected Works*, vol. 6, pp. 104–105.

Langmuir, I., "The Structure of the Static Atom," *Science*, 53, no. 1369 (1921): 290–293; repr. in his *Collected Works*, vol. 6, pp. 124–127.

Langmuir, I., "Modern Concepts in Physics and Their Relation to Chemistry," *Science*, 70, no. 1817 (1929): 385–396.

Langmuir, I., *The Collected Works of Irving Langmuir*, ed. C. G. Suits, 12 vols. (Oxford: Pergamon Press, 1962).

Larmor, J., "A Dynamical Theory of the Electric and Luminiferous Medium," pt. 1, *Philosophical Transactions of the Royal Society of London A*, 185 (1894): 719–822; repr. in his *Mathematical and Physical Papers*, vol. 1, pp. 414–535.

Larmor, J., "A Dynamical Theory of the Electric and Luminiferous Medium," pt. 2, "Theory of Electrons," *Philosophical Transactions of the Royal Society of London A*, 186 (1895): 695–743; repr. in his *Mathematical and Physical Papers*, vol. 1, pp. 543–597.

Larmor, J., "A Dynamical Theory of the Electric and Luminiferous Medium," pt. 3, "Relations with Material Media," *Philosophical Transactions of the Royal Society of London A*, 190 (1897): 205–300; repr. in his *Mathematical and Physical Papers*, vol. 2, pp. 11–132.

Larmor, J., "On the Theory of the Magnetic Influence on Spectra; and on the Radiation from Moving Ions," *Philosophical Magazine*, 5th series, 44 (1897): 503–512.

Larmor, J., "The Influence of a Magnetic Field on Radiation Frequency," *Proceedings of the Royal Society of London*, 60 (1897): 514–515.

Larmor, J., *Mathematical and Physical Papers*, 2 vols. (Cambridge: Cambridge Univ. Press, 1929).

Latour, B., *Pandora's Hope: Essays on the Reality of Science Studies* (Cambridge, Mass.: Harvard Univ. Press, 1999).

Latour, B., "On the Partial Existence of Existing and Nonexisting Objects," in Daston (ed.), *Biographies of Scientific Objects*, pp. 247–269.

Laudan, L., *Progress and Its Problems: Towards a Theory of Scientific Growth* (Berkeley: Univ. of California Press, 1977).

Laudan, L., "Why Was the Logic of Discovery Abandoned?" in Nickles (ed.), *Scientific Discovery, Logic, and Rationality*, pp. 173–183.

Laudan, L., "A Problem-Solving Approach to Scientific Progress," in Hacking (ed.), *Scientific Revolutions*, pp. 144–155.

Laudan, L., "A Confutation of Convergent Realism," in Leplin (ed.), *Scientific Realism*, pp. 218–249.

Lelong, B., "Paul Villard, J. J. Thomson, and the Composition of Cathode Rays," in Buchwald and Warwick (eds.), *Histories of the Electron*, pp. 135–167.

Lenard, P., "Ueber die Absorption der Kathodenstrahlen," *Annalen der Physik und Chemie*, 56 (1895): 255–275.

Leplin, J. (ed.), *Scientific Realism* (Berkeley: Univ. of California Press, 1984).

Lewis, D., "How to Define Theoretical Terms," *Journal of Philosophy*, 67 (1970): 427–446.

Lewis, G. N., "Valence and Tautomerism," *Journal of the American Chemical Society*, 35 (1913): 1448–1455.

Lewis, G. N., "The Atom and the Molecule," *Journal of the American Chemical Society*, 38 (1916): 762–785.

Lewis, G. N., "The Static Atom," *Science*, 46 (28 Sept. 1917): 297–302.

Lewis, G. N., *Valence and the Structure of Atoms and Molecules* (New York: Chemical Catalog Co., 1923).

Lewis, G. N., "Introductory Address: Valence and the Electron," *Transactions of the Faraday Society*, 19 (1923–1924): 452–458.

Lewis, G. N., and E. Q. Adams, "Notes on Quantum Theory: A Theory of Ultimate Rational Units; Numerical Relations between Elementary Charge, Wirkungs-quantum, Constant of Stefan's Law," *Physical Review*, 3, no. 2 (1914): 92–102.

Lodge, O., "A Few Notes on Zeeman's Discovery," *Electrician*, 12 (Mar. 1897): 643–644.

Lodge, O., "The Influence of a Magnetic Field on Radiation Frequency," *Proceedings of the Royal Society of London*, 60 (1897): 513–514.

Lodge, O., "The History of Zeeman's Discovery and Its Reception in England," *Nature*, 109 (1922): 66–69.

Lodge, O., *Electrons, or the Nature and Properties of Negative Electricity* (London: George Bell and Sons, 1906).

Lorentz, H. A., "Concerning the Relation between the Velocity of Propagation of Light and the Density and Composition of Media," in his *Collected Papers*, vol. 2, pp. 1–119.

Lorentz, H. A., "La théorie électromagnétique de Maxwell et son application aux corps mouvants," in his *Collected Papers*, vol. 2, pp. 164–343.

Lorentz, H. A., "Versuch einer Theorie der electrischen und optischen Erscheinungen in bewegten Körpern," in his *Collected Papers*, vol. 5, pp. 1–137.

Lorentz, H. A., "Optical Phenomena Connected with the Charge and Mass of the Ions," pts. 1 and 2, in his *Collected Papers*, vol. 3, pp. 17–39.

Lorentz, H. A., "Théorie des phénomènes magnéto-optiques récemment découverts," *Rapports présentés au Congrés International de Physique*, vol. 3 (Paris, 1900), pp. 1–33.

Lorentz, H. A., "The Theory of Electrons and the Propagation of Light," Nobel Lecture, 11 Dec. 1902, in *Nobel Lectures: Physics, 1901–1921* (Amsterdam: Elsevier, 1967), pp. 14–29.

Lorentz, H. A., *The Theory of Electrons*, 2nd ed. (Leipzig: Teubner, 1916; 1st ed., 1909).

Lorentz, H. A., "Sur la rotation d'un électron qui circule autour d'un noyau," in his *Collected Papers*, vol. 7, pp. 179–204.

Lorentz, H. A., *Collected Papers*, ed. P. Zeeman and A. D. Fokker, 9 vols. (The Hague: Martinus Nijhoff, 1935–1939).

Mach, E., *Knowledge and Error: Sketches on the Physchology of Enquiry* (Dordrecht: Reidel, 1976; 1st German ed., 1905).

MacKinnon, E. M., *Scientific Explanation and Atomic Physics* (Chicago: Univ. of Chicago Press, 1982).

Maier, C. L., *The Role of Spectroscopy in the Acceptance of an Internally Structured Atom, 1860–1920* (Ph.D. dissertation, Univ. of Wisconsin, 1964).

Malone, M. S., *The Microprocessor: A Biography* (Santa Clara, Calif.: Telos, 1995).

Maxwell, J. C., *A Treatise on Electricity and Magnetism*, 2 vols. (Oxford: Clarendon Press, 1873).

McCormmach, R., "The Atomic Theory of John William Nicholson," *Archive for History of Exact Sciences*, 3 (1966): 160–184.

McCormmach, R., "Einstein, Lorentz, and the Electron Theory," *Historical Studies in the Physical Sciences*, 2 (1970): 41–87.

McCormmach, R., "H. A. Lorentz and the Electromagnetic View of Nature," *Isis*, 61 (1970): 459–497.

McGucken, W., *Nineteenth-Century Spectroscopy: Development of the Understanding of Spectra, 1802–1897* (Baltimore: Johns Hopkins Univ. Press, 1969).

McMullin, E., "(Panel Discussion) The Rational Explanation of Historical Discoveries," in T. Nickles (ed.), *Scientific Discovery: Case Studies*, Boston Studies in the Philosophy of Science 60 (Dordrecht: Reidel, 1980), pp. 28–33.

Mehra, J., and H. Rechenberg, *The Historical Development of Quantum Theory*, vol. 1, *The Quantum Theory of Planck, Einstein, Bohr, and Sommerfeld: Its Foundation and the Rise of Its Difficulties, 1900–1925* (New York: Springer-Verlag, 1982).

Mendeleev, D., *An Attempt towards a Chemical Conception of the Ether* (London: Longmans, Green and Co., 1904).

Miller, A. I., "Have Incommensurability and Causal Theory of Reference Anything to Do with Actual Science?—Incommensurability, No; Causal Theory, Yes," *International Studies in the Philosophy of Science*, 5, no. 2 (1991): 97–108.

Miller, A. I., *Albert Einstein's Special Theory of Relativity: Emergence (1905) and Early Interpretation (1905–1911)* (New York: Springer, 1998).

Millikan, R. A., "Radiation and Atomic Structure," *Science*, 45, no. 1162 (6 Apr. 1917): 321–330.

Millikan, R. A., *The Electron: Its Isolation and Measurement and the Determination of Some of Its Properties*, 2nd ed. (Chicago: Univ. of Chicago Press, 1924; 1st ed., 1917).

Millikan, R. A., "Atomism in Modern Physics," *Journal of the Chemical Society*, 125, pt. 2 (1924): 1405–1417.

Millikan, R. A., "The Electron and the Light-Quant from the Experimental Point of View," Nobel Lecture, 23 May 1924, in *Nobel Lectures: Physics, 1922–1941* (Amsterdam: Elsevier, 1965), pp. 54–66.

Millikan, R. A., "The Physicist's Present Conception of an Atom," *Science*, 59, no. 1535 (30 May 1924): 473–476.

Morrison, M., "Theory, Intervention and Realism," *Synthese*, 82 (1990): 1–22.

Morrison, M., "History and Metaphysics: On the Reality of Spin," in Buchwald and Warwick (eds.), *Histories of the Electron*, pp. 425–449.

Mulligan, J. F., "Emil Wiechert (1861–1928): Esteemed Seismologist, Forgotten Physicist," *American Journal of Physics*, 69, no. 3 (2001): 277–287.

Nagel, E., *The Structure of Science*, 2nd ed. (Indianapolis: Hackett, 1979; first publ. 1961).

Nersessian, N. J., *Faraday to Einstein: Constructing Meaning in Scientific Theories* (Dordrecht: Martinus Nijhoff, 1984).

Nersessian, N. J., "'Why Wasn't Lorentz Einstein?' An Examination of the Scientific Method of H. A. Lorentz," *Centaurus*, 29 (1986): 205–242.

Nersessian, N. J. (ed.), *The Process of Science* (Dordrecht: Martinus Nijhoff, 1987).

Nersessian, N. J., "Hendrik Antoon Lorentz," in *The Nobel Prize Winners: Physics* (Pasadena, Calif.: Salem Press, 1989), pp. 35–42.

Nersessian, N. J., "How Do Scientists Think? Capturing the Dynamics of Conceptual Change in Science," in R. N. Giere (ed.), *Cognitive Models of Science*, Minnesota Studies in the Philosophy of Science 15 (Minneapolis: Univ. of Minnesota Press, 1992), pp. 3–44.

Nersessian, N. J., *Opening the Black Box: Cognitive Science and History of Science*, Cognitive Science Laboratory Report 53 (Princeton: Princeton Univ., Jan. 1993).

Nersessian, N. J., "Opening the Black Box: Cognitive Science and History of Science," in Thackray (ed.), *Constructing Knowledge in the History of Science*, pp. 194–214.

Nicholson, J. W., "The Spectrum of Nebulium," *Monthly Notices of the Royal Astronomical Society*, 72 (1911): 49–64.

Nickles, T., "Scientific Problems and Constraints," in P. D. Asquith and I. Hacking (eds.), *PSA 1978: Proceedings of the 1978 Biennial Meeting of the Philosophy of Science Association* (East Lansing, Mich.: Philosophy of Science Association, 1978), vol. 1, pp. 134–148.

Nickles, T. (ed.), *Scientific Discovery, Logic, and Rationality*, Boston Studies in the Philosophy of Science 56 (Dordrecht: Reidel, 1980).

Nickles, T., "Introductory Essay: Scientific Discovery and the Future of Philosophy of Science," in Nickles (ed.), *Scientific Discovery, Logic, and Rationality*, pp. 1–59.

Nickles, T., "Can Scientific Constraints Be Violated Rationally?" in Nickles (ed.), *Scientific Discovery, Logic, and Rationality*, pp. 285–315.

Nickles, T., "Positive Science and Discoverability," in P. D. Asquith and P. Kitcher (eds.), *PSA 1984: Proceedings of the 1984 Biennial Meeting of the Philosophy of Science Association* (East Lansing, Mich.: Philosophy of Science Association, 1984), vol. 1, pp. 13–27.

Nickles, T., "Beyond Divorce: Current Status of the Discovery Debate," *Philosophy of Science*, 52 (1985): 177–206.

Nickles, T., "'Twixt Method and Madness," in Nersessian (ed.), *The Process of Science*, pp. 41–67.

Nickles, T., "Truth or Consequences? Generative versus Consequential Justification in Science," in A. Fine and J. Leplin (eds.), *PSA 1988: Proceedings of the 1988 Biennial Meeting of the Philosophy of Science Association*, 2 vols. (East Lansing, Mich.: Philosophy of Science Association, 1989), vol. 2, pp. 393–405.

Nickles, T., "Heuristic Appraisal: A Proposal," *Social Epistemology*, 3, no. 3 (1989): 175–188.

Nickles, T., "Justification and Experiment," in D. Gooding, T. Pinch, and S. Schaffer (eds.), *The Uses of Experiment: Studies in the Natural Sciences* (Cambridge: Cambridge Univ. Press, 1989), pp. 299–333.

Nickles, T., "Discovery," in R. C. Olby et al. (eds.), *Companion to the History of Modern Science* (London: Routledge, 1990), pp. 148–165.

Nickles, T., "Philosophy of Science and History of Science," in Thackray (ed.), *Constructing Knowledge in the History of Science*, pp. 139–163.

Nisio, S., "The Formation of the Sommerfeld Quantum Theory of 1916," *Japanese Studies in History of Science*, 12 (1973): 39–78.

Nola, R., "Fixing the Reference of Theoretical Terms," *Philosophy of Science*, 47 (1980): 505–531.

Norton, J., "The Logical Inconsistency of the Old Quantum Theory of Black Body Radiation," *Philosophy of Science*, 54 (1987): 327–350.

Norton, J. D., "How We Know about Electrons," in R. Nola and H. Sankey (eds.), *After Popper, Kuhn and Feyerabend: Recent Issues in Theories of Scientific Method* (Dordrecht: Kluwer, 2000), pp. 67–97.

Nye, M. J., *Molecular Reality* (New York: Elsevier, 1972).

Nye, M. J., "The Nineteenth-Century Atomic Debates and the Dilemma of an 'Indifferent Hypothesis,'" *Studies in History and Philosophy of Science*, 7, no. 3 (1976): 245–268.

Nye, M. J., "Remodeling a Classic: The Electron in Organic Chemistry, 1900–1940," in Buchwald and Warwick (eds.), *Histories of the Electron*, pp. 339–361.

O'Hara, J. G., "George Johnstone Stoney, F.R.S., and the Concept of the Electron," *Notes and Records of the Royal Society of London*, 29 (1975): 265–276.

O'Hara, J. G., "George Johnstone Stoney and the Conceptual Discovery of the Electron," in *Stoney and the Electron: Papers from a Seminar Held on November 20, 1991, to Commemorate the Centenary of the Naming of the Electron* (Dublin: Royal Dublin Society, 1993), pp. 5–28.

"100 Years of the Electron," special issue, *Physics World*, Apr. 1997.

"100 Years of Elementary Particles," *Beam Line*, 27, no. 1 (Spring 1997). Also available at http://www.slac.stanford.cdu/pubs/beamline.

Pais, A., *Inward Bound* (New York: Oxford Univ. Press, 1988).

Pais, A., *Niels Bohr's Times, in Physics, Philosophy, and Polity* (New York: Oxford Univ. Press, 1991).

Palmer, W. G., *A History of the Concept of Valency to 1930* (Cambridge: Cambridge Univ. Press, 1965).

Papineau, D., "Theory-Dependent Terms," *Philosophy of Science*, 63 (1996): 1–20.

Parson, A. L., "A Magneton Theory of the Structure of the Atom," *Smithsonian Miscellaneous Collections*, 65, no. 11 (1915):1–80.

Paschen, F., "Bohrs Heliumlinien," *Annalen der Physik*, 50 (1916): 901–940.

Paty, M., "The Scientific Reception of Relativity in France," in Glick (ed.), *The Comparative Reception of Relativity*, pp. 113–167.

Pauli, W., "Über die Gesetzmässigkeiten des anomalen Zeemaneffektes," *Zeitschrift für Physik*, 16 (1923): 155–164; repr. in his *Collected Scientific Papers*, vol. 2, pp. 151–160.

Pauli, W., "Zur Frage der Zuordnung der Komplexstrukturterme in starken und in schwachen äußeren Feldern," *Zeitschrift für Physik,* 20 (1924): 371–387; repr. in his *Collected Scientific Papers,* vol. 2, pp. 176–192.

Pauli, W., "Über den Einfluss der Geschwindigkeits-abhängigkeit der Elektronenmasse auf den Zeemaneffekt," *Zeitschrift für Physik,* 31 (1925): 373–385; repr. in his *Collected Scientific Papers,* vol. 2, pp. 201–213.

Pauli, W., "Über den Zusammenhang des Abschlusses der Elektronen-gruppen im Atom mit der Komplexstruktur der Spektren," *Zeitschrift für Physik,* 31 (1925): 765–783; repr. in his *Collected Scientific Papers,* vol. 2, pp. 214–232.

Pauli, W., "Remarks on the History of the Exclusion Principle," *Science,* 103 (1946): 213–215; repr. in his *Collected Scientific Papers,* vol. 2, pp. 1073–1075.

Pauli, W., *Collected Scientific Papers,* ed. R. Kronig and V. F. Weisskopf, 2 vols. (New York: Interscience, 1964).

Pauli, W., "Exclusion Principle and Quantum Mechanics," Nobel Lecture, 13 Dec. 1946, in *Nobel Lectures, Physics, 1942–1962* (Amsterdam: Elsevier, 1964), pp. 27–43.

Pauli, W., *Wissenschaftlicher Briefwechsel mit Bohr, Einstein, Heisenberg, u.a.,* band 1, *1919–1929,* ed. A. Hermann, K. von Meyenn, and V. F. Weisskopf (New York: Springer-Verlag, 1979).

Pauling, L., and S. Goudsmit, *The Structure of Line Spectra* (New York: McGraw-Hill, 1930).

Perry, C. T., and E. L. Chaffee, "A Determination of $e/m$ for an Electron by Direct Measurement of the Velocity of Cathode Rays," *Physical Review,* 36 (1930): 904–918.

Pickering, A., "Beyond Constraint: The Temporality of Practice and the Historicity of Knowledge," in Buchwald (ed.), *Scientific Practice,* pp. 42–55.

Pickering, A., *The Mangle of Practice: Time, Agency, and Science* (Chicago: Univ. of Chicago Press, 1995).

Pinnick, C., and G. Gale, "Philosophy of Science and History of Science: A Troubling Interaction," *Journal for General Philosophy of Science,* 31 (2000): 109–125.

Popper, K. R., *Logik der Forschung* (Vienna: Springer, 1935).

Popper, K. R., *The Poverty of Historicism* (London: Routledge and Kegan Paul, 1957).

Popper, K. R., *Conjectures and Refutations: The Growth of Scientific Knowledge* (1962; repr., New York: Harper and Row, 1968).

Popper, K. R., *Objective Knowledge: An Evolutionary Approach,* rev. ed. (Oxford: Oxford Univ. Press, 1979; first publ. 1972).

Popper, K. R., *Unended Quest: An Intellectual Autobiography* (La Salle, Ill.: Open Court, 1976).

Priestley, J., "Experiments Relating to Phlogiston, and the Seeming Conversion of Water into Air," *Philosophical Transactions of the Royal Society,* 73 (1783): 398–434.

Pringsheim, E., "Kirchhoff'sches Gesetz und die Strahlung der Gase," *Wiedmannsche Annalen der Physik,* 45 (1892): 428–459.

Przibram, K. (ed.), *Letters on Wave Mechanics: Schrödinger, Planck, Einstein, Lorentz,* trans. M. J. Klein (New York: Philosophical Library, 1967).

Psillos, S., *Scientific Realism: How Science Tracks Truth* (London: Routledge, 1999).

Psillos, S., "The Present State of the Scientific Realism Debate," *British Journal for the Philosophy of Science,* 51 (2000): 705–728.

Putnam, H., "The Analytic and the Synthetic," in his *Philosophical Papers,* vol. 2, pp. 33–69.

Putnam, H., "How Not to Talk about Meaning," in his *Philosophical Papers,* vol. 2, pp. 117–131.

Putnam, H., "Explanation and Reference," in his *Philosophical Papers,* vol. 2, pp. 196–214.

Putnam, H., "Language and Reality," in his *Philosophical Papers,* vol. 2, pp. 272–290.

Putnam, H. (1975), "The Meaning of 'Meaning,'" in his *Philosophical Papers,* vol. 2, pp. 215–271.

Putnam, H., *Philosophical Papers,* vol. 2, *Mind, Language and Reality* (Cambridge: Cambridge Univ. Press, 1975).

Putnam, H., "Reference and Truth," in his *Philosophical Papers,* vol. 3, *Realism and Reason* (Cambridge: Cambridge Univ. Press, 1983), pp. 69–86.

Putnam, H., *Representation and Reality* (Cambridge, Mass: MIT Press, 1988).

Putnam, H., *Realism with a Human Face,* ed. James Conant, (Cambridge, Mass.: Harvard Univ. Press, 1990).

Putnam, H., "The 'Corroboration' of Theories," in R. Boyd et al. (eds.), *The Philosophy of Science* (Cambridge, Mass.: MIT Press, 1991), pp. 121–137.

Putnam, H., "Truth, Activation Vectors and Possession Conditions for Concepts," *Philosophy and Phenomenological Research,* 52, no. 2 (1992): 431–447.

Pyenson, L., "The Relativity Revolution in Germany," in Glick (ed.), *The Comparative Reception of Relativity,* pp. 59–111.

Radder, H., "Philosophy and History of Science: Beyond the Kuhnian Paradigm," *Studies in History and Philosophy of Science,* 28 (1997): 633–655.

Reichenbach, H., *Experience and Prediction* (Chicago: Univ. of Chicago Press, 1938).

Reichenbach, H., *The Rise of Scientific Philosophy* (Berkeley: Univ. of California Press, 1951).

Rheinberger, H., *Toward a History of Epistemic Things: Synthesizing Proteins in the Test Tube* (Stanford: Stanford Univ. Press, 1997).

Rheinberger, H., "Cytoplasmic Particles: The Trajectory of a Scientific Object," in Daston (ed.), *Biographies of Scientific Objects,* pp. 270–294.

Richardson, O. W., *The Electron Theory of Matter* (Cambridge: Cambridge Univ. Press, 1914).

Richter, S., "Wolfgang Pauli und die Entstehung des Spin-Konzepts," *Gesnerus,* 33 (1976): 253–270.

Robotti, N., "Quantum Numbers and Electron Spin: The History of a Discovery," *Archives internationales d'histoire des sciences,* 40, no. 125 (1990): 305–331.

Robotti, N., "The Zeeman Effect in Hydrogen and the Old Quantum Theory," *Physis,* 29 (1992): 809–831.

Robotti, N., "J. J. Thomson at the Cavendish Laboratory: The History of an Electric Charge Measurement," *Annals of Science* 52 (1995): 265–284.

Robotti, N., and F. Pastorino, "Zeeman's Discovery and the Mass of the Electron," *Annals of Science,* 55 (1998): 161–183.

Romer, A., "Zeeman's Discovery of the Electron," *American Journal of Physics,* 16 (1948): 216–223.

Rosenfeld, A., "The Quintessence of Irving Langmuir," in Langmuir, *Collected Works,* vol. 12, pp. 3–229.

Rosenfeld, L., "Introduction," in Bohr, *On the Constitution,* pp. xi–liv.

Rubinowicz, A., "Bohrsche Frequenzbedingung und Erhaltung des Impulsmomentes," *Physikalische Zeitschrift,* 19 (1918): 441–445, 465–474.

Rücker, A., "Address of the President of the British Association for the Advancement of Science," *Science,* 14, no. 351 (20 Sept. 1901): 425–443.

Russell, C. A., *The History of Valency* (New York: Humanities Press, 1971).

Rutherford, E., "The Existence of Bodies Smaller than Atoms," *Transactions of the Royal Society of Canada,* 2nd series, section 3, 8 (1902): 79–86; repr. in *The Collected Papers of Lord Rutherford of Nelson,* vol. 1 (London: George Allen and Unwin, 1962), pp. 403–409.

Rutherford, E., "The Scattering of α and β Particles by Matter and the Structure of the Atom," *Philosophical Magazine,* 6th series, 21 (1911): 669–688; repr. in *The Collected Papers of Lord Rutherford of Nelson,* vol. 2 (London: George Allen and Unwin, 1963), pp. 238–254.

Ryle, G., *The Concept of Mind* (Chicago: Univ. of Chicago Press, 1949).

Schaffer, S., "Scientific Discoveries and the End of Natural Philosophy," *Social Studies of Science,* 16 (1986): 387–420.

Schaffer, S., "The Eighteenth Brumaire of Bruno Latour," *Studies in History and Philosophy of Science,* 22 (1991): 174–192.

Seife, C., *Zero: The Biography of a Dangerous Idea* (New York: Penguin, 2000).

Servos, J. W., *Physical Chemistry from Ostwald to Pauling: The Making of a Science in America* (Princeton: Princeton Univ. Press, 1990).

Serwer, D., *"Unmechanischer Zwang:* Pauli, Heisenberg, and the Rejection of the Mechanical Atom, 1923–1925," *Historical Studies in the Physical Sciences,* 8 (1977): 189–256.

Shapere, D., "Meaning and Scientific Change," in R. G. Colodny (ed.), *Mind and Cosmos: Essays in Contemporary Science and Philosophy* (Pittsburgh: Univ. of Pittsburgh Press, 1966), pp. 41–85; repr. in Hacking (ed.), *Scientific Revolutions,* pp. 28–59.

Shapere, D., "Scientific Theories and Their Domains," in F. Suppe (ed.), *The Structure of Scientific Theories,* 2nd ed. (Urbana: Univ. of Illinois Press, 1977), 518–565.

Shapere, D., "Reason, Reference, and the Quest for Knowledge," *Philosophy of Science,* 49 (1982): 1–23.

Shapere, D., *Reason and the Search for Knowledge: Investigations in the Philosophy of Science,* Boston Studies in the Philosophy of Science 78 (Dordrecht: Reidel, 1984).

Shapere, D., "Leplin on Essentialism," *Philosophy of Science,* 58 (1991): 655–677.

Shapin, S., "Discipline and Bounding: The History and Sociology of Science as Seen through the Externalism-Internalism Debate," *History of Science,* 30 (1992): 333–369.

Shortland, M., and R. Yeo, *Telling Lives in Science: Essays on Scientific Biography* (Cambridge: Cambridge Univ. Press, 1996).

Sidgwick, N. V., *The Electronic Theory of Valency* (London: Oxford Univ. Press, 1927).

Sinclair, S. B., "J. J. Thomson and the Chemical Atom: From Ether Vortex to Atomic Decay," *Ambix,* 34, pt. 2 (1987): 89–116.

Skinner, Q., "Meaning and Understanding in the History of Ideas," *History and Theory,* 8, no. 1 (1969): 3–53.

Smith, C., *The Science of Energy: A Cultural History of Energy Physics in Victorian Britain* (Chicago: Univ. of Chicago Press, 1999).

Smith, G. E., "J. J. Thomson and the Electron, 1897–1899," in Buchwald and Warwick (eds.), *Histories of the Electron,* pp. 21–76.

Smith, J. D. M., *Chemistry and Atomic Structure* (London: Ernest Benn, 1924).

Sommerfeld, A., "Zur Theorie der Balmerschen Serie," *Sitzungsberichte der Mathematisch-Physikalischen Klasse der Königlich-Bayerischen Akademie der Wissenschaften zu München,* 45 (1915): 425–458.

Sommerfeld, A., "Die Feinstruktur der Wasserstoff- und der Wasserstoff-ähnlichen Linien," *Sitzungsberichte der Mathematisch-Physikalischen Klasse der Königlich-Bayerischen Akademie der Wissenschaften zu München,* 45 1915: 459–500.

Sommerfeld, A., "Zur Quantentheorie der Spektrallinien," *Annalen der Physik,* 51 (1916): 1–94, 125–167; repr. in his *Gesammelte Schriften,* vol. 3, pp. 172–308.

Sommerfeld, A., "Zur Theorie des Zeeman-Effekts der Wasserstofflinien, mit einem Anhang über den Stark-Effekt," *Physikalische Zeitschrift,* 17 (1916): 491–507; repr. in his *Gesammelte Schriften,* vol. 3, pp. 309–325.

Sommerfeld, A., "Zur Quantentheorie der Spektrallinien: Intensitätsfragen," *Sitzungsberichte der Mathematisch-Physikalischen Klasse der Königlich-Bayerischen Akademie der Wissenschaften zu München,* 47 (1917): 83–109; repr. in his *Gesammelte Schriften,* vol. 3, pp. 432–458.

Sommerfeld, A., "Allgemeine spektroskopische Gesetze, insbesondere ein magneto-optischer Zerlegungsatz," *Annalen der Physik,* 63 (1920): 221–263; repr. in his *Gesammelte Schriften,* vol. 3, pp. 523–565.

Sommerfeld, A., "Quantentheoretische Umdeutung der Voigtschen Theorie des anomalen Zeemaneffektes vom D-Linientypus," *Zeitschrift für Physik,* 8 (1922): 257–272; repr. in his *Gesammelte Schriften,* vol. 3, pp. 609–624.

Sommerfeld, A., *Atomic Structure and Spectral Lines,* trans. from the 3rd German ed. by H. L. Brose (London: Methuen, 1923).

Sommerfeld, A., *Atombau und Spektrallinien,* 5th ed., vol. 1 (Braunschweig: F. Vieweg, 1931).

Sommerfeld, A., *Atomic Structure and Spectral Lines,* trans. from the 5th German ed. by H. L. Brose (London: Methuen, 1934).

Sommerfeld, A., *Gesammelte Schriften*, ed. F. Sauter, 4 vols. (Braunschweig: F. Vieweg, 1968).

Spencer, J. B., *An Historical Investigation of the Zeeman Effect (1896–1913)* (Ph.D. dissertation, Univ. of Wisconsin, 1964).

Spencer, J. B., "The Historical Basis for Interactions between the Bohr Theory of the Atom and Investigations of the Zeeman Effect: 1913–1925," in *XII Congrès International d'Histoire des Sciences: Actes* (Paris: Blanchard, 1971), vol. 5, pp. 95–100.

Spencer, J. B., "Zeeman, Pieter," in Gillispie (ed.), *Dictionary of Scientific Biography*, vol. 14, pp. 597–599.

Stachel, J., "Scientific Discoveries as Historical Artifacts," in Gavroglu et al. (eds.), *Trends in the Historiography of Science*, pp. 139–148.

Star, S. L., and J. R. Griesemer, "Institutional Ecology, 'Translations' and Boundary Objects: Amateurs and Professionals in Berkeley's Museum of Vertebrate Zoology, 1907–39," *Social Studies of Science*, 19, no. 3 (1989): 387–420.

Steinle, F., "Experiments in History and Philosophy of Science," *Perspectives on Science*, 10 (2002): 408–432.

Stoner, E., "The Distribution of Electrons among Atomic Levels," *Philosophical Magazine*, 6th series, 48 (1924): 719–736.

Stoney, G. J., "On the Physical Units of Nature," *Scientific Proceedings of the Royal Dublin Society*, new series, 3 (1881–1883): 51–60.

Stoney, G. J., "On the Cause of Double Lines and of Equidistant Satellites in the Spectra of Gases," *Scientific Transactions of the Royal Dublin Society*, 2nd series, 4 (1888–1892): 563–608.

Stoney, G. J., "Of the 'Electron,' or Atom of Electricity," *Philosophical Magazine*, 5th series, 38 (1894): 418–420.

Stranges, A. N., *Electrons and Valence: Development of the Theory, 1900–1925* (College Station: Texas A and M Univ. Press, 1982).

Süsskind, C., "Langmuir, Irving," in Gillispie (ed.), *Dictionary of Scientific Biography*, vol. 8, pp. 22–25.

Sutherland, W., "Cathode, Lenard and Röntgen Rays," *Philosophical Magazine*, 5th series, 47 (1899): 269–284.

ter Haar, D. (ed.), *The Old Quantum Theory* (Oxford: Pergamon Press, 1967).

Thackray, A. (ed.), *Constructing Knowledge in the History of Science*, Osiris, vol. 10 (1995).

Thomas, L. H., "The Motion of the Spinning Electron," *Nature*, 117 (1926): 514.

Thomson, J. J., "On the Electric and Magnetic Effects produced by the Motion of Electrified Bodies," *Philosophical Magazine*, 5th series, 11 (1881): 229–249.

Thomson, J. J., "Cathode Rays," *Proceedings of the Royal Institution*, 15 (1897): 419–432.

Thomson, J. J., "Cathode Rays," *Philosophical Magazine*, 5th series, 44 (1897): 293–316.

Thomson, J. J., "Note on Mr. Sutherland's Paper on the Cathode Rays," *Philosophical Magazine*, 5th series, 47 (1899): 415–416.

Thomson, J. J., "On the Masses of the Ions in Gases at Low Pressures," *Philosophical Magazine*, 48 (1899): 547–567.

Thomson, J. J., "On the Structure of the Atom: An Investigation of the Stability and Periods of Oscillation of a Number of Corpuscles Arranged at Equal Intervals around the Circumference of a Circle; with Application of the Results to the Theory of Atomic Structure," *Philosophical Magazine,* 6th series, 7 (1904): 237–265.

Thomson, J. J., "Carriers of Negative Electricity," Nobel Lecture, 11 Dec. 1906, in *Nobel Lectures: Physics, 1901–1921* (Amsterdam: Elsevier, 1967), pp. 145–153.

Thomson, J. J., "The Modern Theory of Electrical Conductivity in Metals," *Journal of the Institution of Electrical Engineers,* 38 (1906–1907): 455–465

Thomson, J. J., *The Corpuscular Theory of Matter* (London: Charles Scribner's Sons, 1907).

Thomson, J. J., "The Forces between Atoms and Chemical Affinity," *Philosophical Magazine,* 6th series, 27 (1914): 757–789.

Thomson, J. J., *The Electron in Chemistry* (Philadelphia: Franklin Institute, 1923).

Thomson, J. J., et al., "Discussion," *Journal of the Institution of Electrical Engineers,* 38 (1906–1907): 465–468.

Toulmin, S., "Do Sub-microscopic Entities Exist?" in E. D. Klemke et al. (eds.), *Introductory Readings in the Philosophy of Science,* 3rd ed. (Amherst, N.Y.: Prometheus Books, 1998), pp. 358–362.

Turpin, B. M., *The Discovery of the Electron: The Evolution of a Scientific Concept, 1800–1899* (Ph.D. dissertation, Univ. of Notre Dame, 1980).

Uhlenbeck, G. E., *Oude en nieuwe vragen der natuurkunde,* address delivered at Leiden on 1 Apr. 1955 (Amsterdam: North-Holland, 1955).

Uhlenbeck, G. E., "Personal Reminiscences," *Physics Today,* June 1976, 43–48.

Uhlenbeck, G. E., and S. Goudsmit, "Ersetzung der Hypothese vom unmechanischen Zwang durch eine Forderung bezüglich des inneren Verhaltens jedes einzelnen Elektrons," *Die Naturwissenschaften,* 13, no. 47 (20 Nov. 1925): 953–954.

Uhlenbeck, G. E., and S. Goudsmit, "Spinning Electrons and the Structure of Spectra," *Nature,* 117 (1926): 264–265.

van den Broek, A., "Die Radioelemente, das periodische System und die Konstitution der Atome," *Physikalische Zeitschrift,* 14 (1913): 32–41.

van der Waerden, B. L., "Exclusion Principle and Spin," in Fierz and Weisskopf (eds.), *Theoretical Physics in the Twentieth Century,* pp. 199–244.

van der Waerden, B. L. (ed.), *Sources of Quantum Mechanics* (New York: Dover, 1967).

van Fraassen, B. C., *The Scientific Image* (New York: Oxford Univ. Press, 1980).

van Fraassen, B. C., "From Vicious Circle to Infinite Regress, and Back Again," *PSA 1992: Proceedings of the Biennial Meeting of the Philosophy of Science Association* (East Lansing, Mich.: Philosophy of Science Association, 1993), vol. 2, pp. 6–29.

van Fraassen, B. C., "Michel Ghins on the Empirical versus the Theoretical," *Foundations of Physics,* 30, no. 10 (2000): 1655–1661.

van Fraassen, B. C., *The Empirical Stance* (New Haven: Yale Univ. Press, 2002).

Warwick, A., "On the Role of the FitzGerald-Lorentz Contraction Hypothesis in the Development of Joseph Larmor's Electronic Theory of Matter," *Archive for History of Exact Sciences,* 43, no. 1 (1991): 29–91.

Warwick, A., "Frequency, Theorem and Formula: Remembering Joseph Larmor in Electromagnetic Theory," *Notes and Records of the Royal Society of London*, 47, no. 1 (1993): 49–60.

Wentzel, G., "Zum Termproblem der Dublettspektren, insbesondere der Röntgenspektren," *Annalen der Physik*, 76 (1925): 803–828.

Wheaton, B. R., "Philipp Lenard and the Photoelectric Effect, 1889–1911," *Historical Studies in the Physical Sciences*, 9 (1978): 299–322.

Wiechert, E., "Ueber das Wesen der Elektrizität," *Schriften der Physikalisch-Ökonomischen Gesellschaft zu Königsberg*, 38 (7 Jan. 1897): 3–12.

Wiechert, E., "Experimentelles über die Kathodenstrahlen," *Schriften der Physikalisch-Ökonomischen Gesellschaft zu Königsberg*, 38 (7 Jan. 1897): 12–16.

Williams, L. P., *Michael Faraday: A Biography* (1965; repr. New York: Da Capo Press, 1987).

Wise, M. N., "German Concepts of Force, Energy, and the Electromagnetic Ether: 1845–1880," in G. N. Cantor and M. J. S. Hodge (eds.), *Conceptions of Ether: Studies in the History of Ether Theories, 1740–1900* (Cambridge: Cambridge Univ. Press, 1981), pp. 269–307.

Wise, M. N., "How Do Sums Count? On the Cultural Origins of Statistical Causality," in L. Krüger, L. J. Daston, and M. Heidelberger (eds.), *The Probabilistic Revolution*, vol. 1, *Ideas in History* (Cambridge, Mass.: MIT Press, 1987), pp. 395–425.

Wise, M. N., "Forman Reformed," unpublished manuscript.

Worrall, J., "The Scope, Limits, and Distinctiveness of the Method of 'Deduction from the Phenomena': Some Lessons from Newton's 'Demonstrations' in Optics," *British Journal for the Philosophy of Science*, 51 (2000): 45–80.

Zeeman, P., "On the Influence of Magnetism on the Nature of the Light Emitted by a Substance," pt. 1, *Communications from the Physical Laboratory at the University of Leiden*, 33 (1896): 1–8; trans. from *Verslagen van de Afdeeling Natuurkunde der Koninklijke Akademie van Wetenschappen te Amsterdam*, 31 Oct. 1896, p. 181.

Zeeman, P., "On the Influence of Magnetism on the Nature of the Light Emitted by a Substance," pt. 2, *Communications from the Physical Laboratory at the University of Leiden*, 33 (1896): 9–19; trans. from *Verslagen van de Afdeeling Natuurkunde der Koninklijke Akademie van Wetenschappen te Amsterdam*, 28 Nov. 1896, p. 242.

Zeeman, P., "Doublets and Triplets in the Spectrum Produced by External Magnetic Forces," *Philosophical Magazine*, 5th series, 44 (1897): 55–60, 255–259.

Zeeman, P., "Experimentelle Untersuchungen über Teile, welche kleiner als Atome sind," *Physikalische Zeitschrift*, 1 (1900): 562–565, 575–578.

Zeeman, P., *Researches in Magneto-Optics: With Special Reference to the Magnetic Resolution of Spectrum Lines* (London: Macmillan, 1913).

Zeeman, P., "Faraday's Researches on Magneto-Optics and Their Development," *Nature*, 128 (1931): 365–368.

# Index